countdown to mathematics

Volume 1

Lynne Graham and David Sargent
of the Open University

 **ADDISON-WESLEY PUBLISHERS LIMITED**

in association with the **OPEN UNIVERSITY PRESS** 

Wokingham, England • Reading, Massachusetts • Menlo Park, California •
New York • Don Mills, Ontario • Amsterdam • Bonn •
Sydney • Singapore • Tokyo • Madrid • San Juan •

Acknowledgements

HMSO: from *Household Survey* graphs on pages 207, 208, 211, 212, 213 (top), 221, 222 (bottom), tables on pages 192, 199, 200, 201 and chart on page 210 (bottom); from *Social Trends* 1980 graphs on pages 202 (top), 222 (top), 224 (bottom); from *Annual Abstract of Statistics* graphs on pages 173, 174, 186; from *Rent Rebates* table on page 191.

Consumer Association: from *Which?* November 1977 graphs on pages 171 (top), 175, 176; from *Which?* December 1979 chart on page 215 (centre).

The Open University: from *Maths Diagnostic and Development Booklet* charts on page 202, table on page 237; from Consumer Decisions chart on page 210 (top); from *Course D101* graphs on pages 214, 224 (top); from *Course T101* graph on page 216.

Pelican Books Ltd.: from *How to Lie with Statistics* graph on page 171 (centre).

Enterprise Holidays: from *Enterprise Summer '80 Brochure* tables on pages 196, 243.

Excess General Insurance Company Limited: from *Insurance Booklet* tables on pages 197, 198, 244.

McGraw Hill Book Co. Ltd.: from *Mathematical Handbook* Schaum Outline Series table on page 234.

Chambers Ltd.: from *Chambers Four Figure Mathematical Tables* tables on page 235.

Freeman & Sons Ltd.: from *Statistical Tables* table on page 236.

Published in 1981 by Addison-Wesley Publishers Limited.

Cartoons by David Farris
Typeset by Allset Composition, London
Printed in Great Britain by Hollen Street Press,
Berwick upon Tweed

First printed in 1981
Reprinted in 1983, 1985 (twice), 1986, 1987, 1988, 1989, 1990, 1991, 1992 (twice) and 1993.

ISBN 201 13730 5

CONTENTS

Countdown to Mathematics on Radio

A series of BBC radio programmes, designed to support the study of *Countdown to Mathematics Volume 2*, is broadcast on Radio 5 (MW 909, 693 kHz) once each year, as follows.

December	Programme 1	Introduction – *Start Countdown*
	Programme 2	Module 5 – *Countdown to Angles*
	Programme 3	Module 6 – *Countdown to Algebra*
following January	Programme 4	Module 7 – *Countdown to Graphs*
	Programme 5	Module 8 – *Countdown to Trig*
	Programme 6	Module 9 – *Countdown to Logs*

Exact broadcast dates and times may be obtained from the Information Officer, BBC OU Production Centre, Walton Hall, Milton Keynes MK7 6BH.

The series is accompanied by a booklet containing Radio Notes. All registered students on The Open University *Mathematics Foundation Course* (M101) or *An Introduction to Calculus* (MS284) will be sent this automatically before their study commences, as part of their Preparatory material. Other listeners may obtain the booklet by sending a large stamped addressed envelope, to the Information Officer at the above address.

Introduction

Countdown to Mathematics has been written primarily for intending Open University students. Although the Open University has no entrance requirements some prerequisite skills, and in particular mathematical skills, are necessary. These vary from course to course.

Volume I of *Countdown to Mathematics* concentrates on basic skills and techniques in arithmetic, algebra, graphs and statistics which are often assumed in courses in the Social Science, Science, Education and Technology disciplines. Evidence collected in the Open University over the last ten years suggests that students not only lack confidence in mathematical skills and techniques but they also lack the study skills necessary for learning in isolation. Thus *Countdown to Mathematics* has been designed both to brush up mathematical skills and to develop skills for self-study. The style is informal and the approach follows the general learning strategies adopted by many Open University courses. Although the emphasis is on preparation for the Open University we feel that many other students face the same problems and consequently hope that *Countdown to Mathematics* will appeal to a wider audience.

HOW TO USE THIS BOOK

Volume I of *Countdown to Mathematics* is broken down into four Modules covering arithmetic, algebra, graphs and statistics. Each Module is itself divided into five sections. (See Contents page)

Studying on your own requires a special discipline. We hope that breaking down the book into small sections helps you to organise a study routine which you feel you can follow. You may not need to work through all the book. To help you identify which sections you can omit and which sections you should read carefully, we have prefaced each section with some diagnostic questions: *TRY THESE QUESTIONS FIRST.* Then turn the page and check your answers. These direct you to the appropriate subsection. You may feel sufficiently confident to omit the section or you may want to read through it quickly just for extra reassurance. However, if you experience any difficulty with the diagnostic questions then we advise you to work through the section carefully.

You will find that each section contains plenty of exercises which you should work through thoroughly in order to gain most benefit. Full solutions are provided at the end of each Module. When reading mathematics at any level you should have a pencil and paper handy and expect to use them! Each section concludes with a summary and some supplementary exercises should you want some extra practice.

ASSUMED KNOWLEDGE

Inevitably we have assumed that you are already confident with some skills. We assume you

(i) can count,

(ii) can add, subtract, multiply and divide numbers between 1 and 10,

(iii) can measure length and weight,

(iv) are familiar with the money system.

We also assume that you have *seen* quite a bit of the content before but that you may well have forgotten it or never understood it properly.

CALCULATOR

We expect you to develop arithmetic skills using simple whole numbers; but with the more complicated numbers, often encountered in the real world, we expect you to use a calculator. This removes much of the drudgery from the calculation. However, as with any tool, one has to learn how to use a calculator efficiently–consequently we encourage you to investigate how *your* calculator works by frequent reference to the Maker's Handbook and by suggesting preliminary calculations based on simple numbers which you can check by hand. Sometimes we indicate the keys to press, in other cases we leave you to work out the calculation for yourself. If you find that our suggestions don't work you should always refer to the Maker's Handbook for your calculator. Several of the exercises involve complex numbers and, without a calculator, the working could be quite complicated. All arithmetic involves routine procedures; a calculator makes these quicker and easier, so it is desirable that you should use one wherever possible. Note that, if you intend to work through both volumes of *Countdown to Mathematics* then you should really have a scientific calculator.

Finally, we hope that you find the book readable. Mathematics often inspires feelings of panic. After working through *Countdown to Mathematics* we hope that you will feel that mathematical skills and techniques are just like any other skills which can be easily acquired and which you will be able to use in your future studies with confidence.

MODULE 1

1.1 Estimating and Calculating

1.2 Negative Numbers: Brackets

1.3 Decimals

1.4 Fractions

1.5 Ratios and Percentages, and More Calculator Practice

1.1 Estimating and Calculating

TRY THESE QUESTIONS FIRST

1 Round off 74,965 to (i) the nearest 10 (ii) the nearest 100 (iii) the nearest 10,000.

2 Give a rough estimate to 3,839,322 + 6,753,418.

3 Use your calculator to evaluate 3,839,322 + 6,753,418.

4 Evaluate 932 × 7481.

5 Mortgage repayments are £172 per month. Make an estimate of the cost over a year. What is the exact amount paid out over a year?

1.1(i) ROUNDING

Newspapers, adverts and television often give information in numbers. We've given a few common examples here. However, it is unlikely that the Balance of Payments Deficit, for example, would be exactly £7,000,000. It's more likely to be a number very close to £7,000,000, a number like £6,895,481 or £7,132,206. From the reader's point of view, £7,000,000 gives a good idea of the size of the deficit; it's a *good approximation*. Large numbers in particular are often approximated in this way. This process is called *rounding.*

Quick Sale
required
only £500

Balance of Payments
Deficit £7,000,000

2,000,000
unemployed

Rounding to the nearest 10

You will probably think to yourself that this coat costs about £50.

£49 is considerably closer to £50 than it is to £40, so £50 is a reasonable approximation.

In this case, 49 has been rounded up to 50.

Similarly, 42 would be rounded down to 40 because it is closer to 40 than it is to 50.

Both numbers have been *rounded to the nearest 10.*

Rounding to the nearest 10 means rounding to 10, 20, 30, 40 . . . . The number ends in zero.

CHECK YOUR ANSWERS

1 (i) 74,970 (ii) 75,000 (iii) 70,000 — *Section 1.1(i)*

2 3,839,322 rounds to 4,000,000.
6,753,418 rounds to 7,000,000.
Therefore 3,839,322 + 6,753,418 ≏ 11,000,000. — *Section 1.1(ii)*

3 10,592,740 — *Section 1.1(iii)*

4 Estimate first: 900 x 7000 = 6,300,000.
Using a calculator, 932 x 7481 = 6,972,292. — *Section 1.1(iv)*

5 172 rounds to 200.
Therefore the yearly cost is approximately 12 x £200 = £2400.
The exact amount is £2064. — *Section 1.1(v)*

The process of rounding to the nearest 10 is summarised below.

The numbers 0, 1, 2, 3, 4, 5, 6, 7, 8, 9 are called digits.

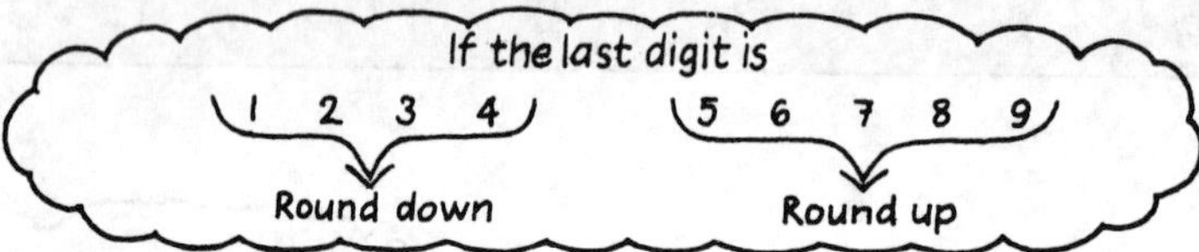

Thus, a number ending in 3 is rounded down; for example, 63 is rounded down to 60.

Of course if the number already ends in a zero (like 40) there is no need to round!

And a number ending in 7 is rounded up; for example, 67 is rounded up to 70.

In this book if a number ends in 5 we will always round it up. Some people round down; others follow rules which sometimes round up and sometimes round down. The decision to round up or down is really quite arbitrary and we shall be consistent in rounding up all the time.

Round up if a number ends in the digit 5.

TRY SOME YOURSELF

1 Round off each of the following numbers to the nearest 10:
(i) 43 (ii) 58 (iii) 204 (iv) 96 (v) 1005.

Rounding to the nearest 100

In order to round to the nearest 10 you must look at the last digit of the number and then round up or down accordingly. Similarly to round to the nearest 100 you must look at the last *two* digits. For example

267 is nearer to 300 than it is to 200

243 is nearer to 200 than it is to 300

If the last two digits are less than 50 we round down.

If the last two digits are 50 or over, then we round up.

43 is less than 50 so round down.
67 is more than 50 so round up.

TRY SOME YOURSELF

2 Round off each of the following numbers to the nearest 100:
(i) 157 (ii) 119 (iii) 1132 (iv) 4979 (v) 31,350.

But numbers are often approximated to make them easier to handle and sometimes it doesn't help very much to round to the nearest 10 or the nearest 100—particularly if the number is very large.

Balance of Payments Deficit £7,000,000

For example, suppose the Balance of Payments Deficit was actually £6,895,481. Rounded to the nearest 10, that's £6,895,480 and to the nearest 100, it's £6,895,500. But £6,895,500 is still a complicated number to deal with in your head. That's why it was rounded to £7,000,000 in the newspaper headline. In fact numbers can be rounded to any level of accuracy—to the nearest 1000, or the nearest 10,000 or even to the nearest million (1,000,000).

As in the newspaper headline.

To round to the nearest 1000 you must look at the last *three* digits. For example

6, 895, 481

Round down if the last three digits are less than 500.
Round up if the last three digits are 500 or more.

rounds down to

6,895,000.

The following table summarises the process.

Rounding to the	*Round down if the last digits are*	*Round up if the last digits are*
Nearest 10	1, 2, 3, 4	5, 6, 7, 8, 9
Nearest 100	01 to 49	50 to 99
Nearest 1000	001 to 499	500 to 999
Nearest 10,000	0001 to 4999	5000 to 9999
:	:	:
Nearest 1,000,000	000,001 to 499,999	500,000 to 999,999

Look at the ***last*** *digit.*
Look at the last ***two*** *digits.*
Look at the last ***three*** *digits.*
Look at the last ***four*** *digits.*
:
Look at the last ***six*** *digits.*

Thus, £6,895,481 rounded to the nearest 1,000,000 is £7,000,000.

Certainly 7,000,000 is an easier number to work with; it's just one digit (7) and lots of zeros.

EXAMPLE

The distance from London to Edinburgh is 373 miles. Round off the distance to the easiest number to handle (one digit and lots of zeros).

SOLUTION

We need to round to the nearest 100.

373 rounds *up* to 400

Look at the last two digits.

Therefore the distance from London to Edinburgh is approximately 400 miles.

TRY SOME YOURSELF

3(i) Round off 7,183,530 to the nearest (a) 100 (b) 1000 (c) 10,000 (d) 100,000.

(ii) The population of London in 1964 was 8,187,000. Round this number off to the easiest number to work with (one digit and lots of zeros).

(iii) The distance from London to Bedford is 57 miles. Round this off to the easiest number to handle (one digit and zeros).

(iv) The number of employees in a car factory is 29,961. Round this off to the easiest number to handle.

1.1(ii) ESTIMATING CALCULATIONS

Approximations are most useful when it comes to making rough estimates–like adding up a bill quickly to see if it's about right. To make a rough estimate the numbers must be easy, so being able to round off numbers is an extremely useful skill.

EXAMPLE

Dave was buying a round of drinks. Make a rough estimate of the total cost.

SOLUTION

We get a rough estimate by rounding to the nearest 10.

53 rounds down to 50

So 6 pints cost about 6 × 50p = £3·00.

In fact the exact cost is £3·18. So the estimate is quite close.

EXAMPLE

Dave and Sally travelled from Newcastle to London to Exeter then to Liverpool and finally home. Make a rough estimate of the total distance travelled.

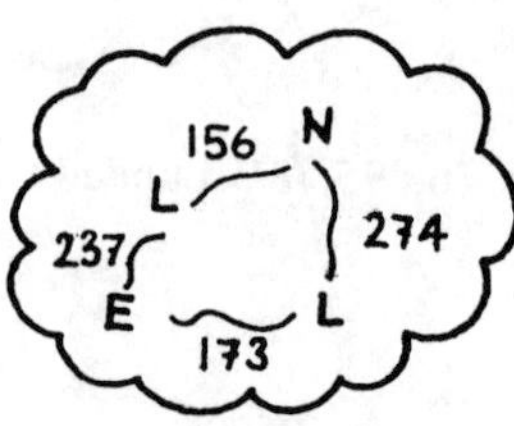

SOLUTION

We get a rough estimate by rounding to the nearest 100.

Newcastle–London	is about 300 miles
London–Exeter	is about 200 miles
Exeter–Liverpool	is about 200 miles
Liverpool–Newcastle	is about 200 miles

First make the numbers easy, then work out the estimate.

and the total distance is about 900 miles.

This can be written as

the total distance $\triangleq$ 900 miles.

(In fact the total distance is 840 miles.)

The sign $\triangleq$ means 'is approximately equal to'.

TRY SOME YOURSELF

4(i) Estimate a rough total for the supermarket bill opposite.

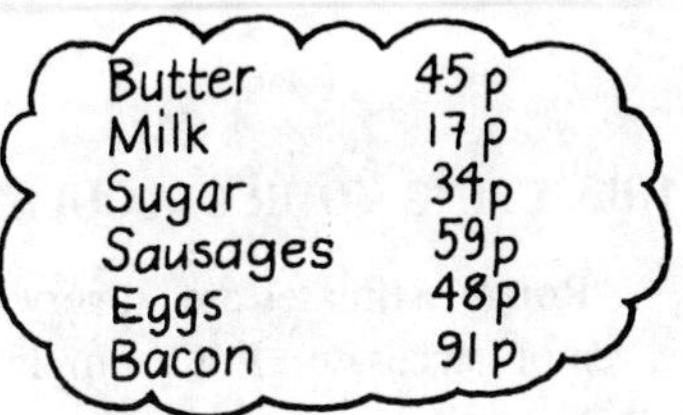

(ii) Dave earns £384 per month. Make a rough estimate of how much he earns per year.

EXAMPLE

A rectangular lawn measures 28 ft by 43 ft. Make a rough estimate of the area.

SOLUTION

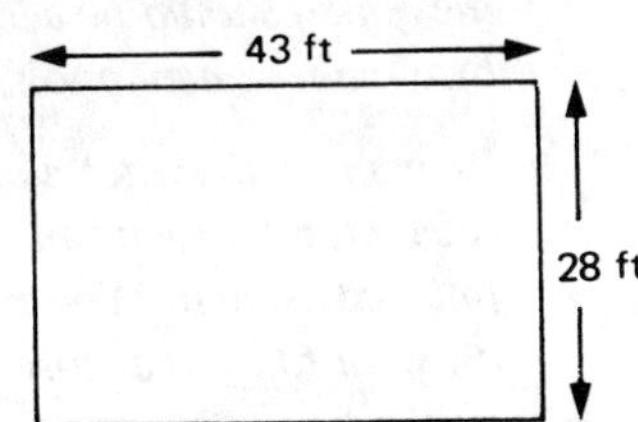

length

breadth

Area of a rectangle is length × breadth.

We get a rough estimate by rounding to the nearest 10:

43 rounds down to 40

28 rounds up to 30

So

area $\triangleq$ 40 ft × 30 ft

$\triangleq$ 1200 square feet

(In fact the exact area is 1204 square feet.)

Area is measured in ***square*** *feet or* ***square*** *yards or* ***square*** *metres etc.*

These examples illustrate that a rough estimate can give a very good idea of the result. In each case the estimate has been very close to the exact answer.

TRY SOME YOURSELF

5(i) Estimate the area of a strip of grass measuring 26 ft by 4 ft.

Watch out here. It's no good rounding 4 ft to 0 ft, since 26 x 0 = 0. It's better to leave the 4 alone.

(ii) Make a rough estimate of the amount of carpet needed for the room illustrated below.

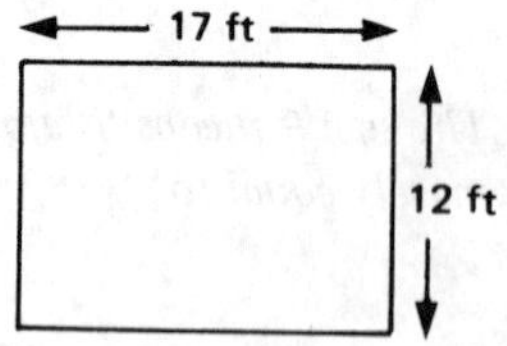

1.1(iii) USING YOUR CALCULATOR

Rough estimates are all very well but sometimes it is necessary to be more accurate. For example, if you're paying a bill you don't want to pay a penny more than is absolutely necessary!

For exact calculations it's probably a good idea to use a calculator. You'll see a symbol in the margin every time we expect you to use one in this book.

There are many different makes of calculator and many different ways of using one. The important thing is for *you* to get used to *your* calculator. If you work through our suggested activities below, with the maker's handbook for your calculator at hand, you should quickly acquire confidence. First of all, look for the following features:

Every calculator (no matter what the make) should be accompanied by a maker's handbook.

1 [ON/OFF] the on/off switch: possibly the most important feature, particularly if your calculator is battery operated.

Remember to switch your calculator off each time you've finished using it. You may find it useful to keep a store of spare batteries–just in case.

2 [C] the 'clear' key which 'clears' the display. Check how it is labelled on your calculator–it might be [C] or [CE] or something else.

3 [+] [−] [x] [÷] these are sometimes called the 'operations' keys.

4 [=] the equals key.

5 [0] [1] [2] [3] [4] [5] [6] [7] [8] [9] the digits.

Your calculator might have other keys, such as function keys or memory keys. For the moment, though, all you will need to use are the keys listed above.

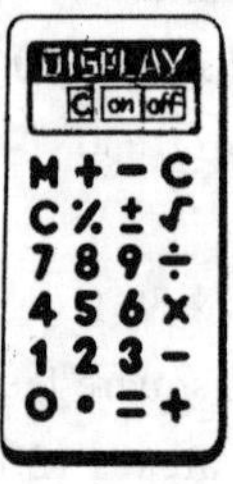

Now, get used to entering numbers into the display. Remember to clear the display each time.

Try entering 41; 52,643; 14; . . .

Sometimes we'll suggest that you perform a specific calculation on your calculator. Then we'll suggest a *key sequence* which you might want to follow.

A key sequence is just a list of keys to press in sequence.

For example, we might suggest that to calculate 1 + 3 you follow the key sequence

It's always a good idea, especially with an unfamiliar calculation, to write down your own key sequences. Then you can check that you've missed nothing out.

You probably think that 1 + 3 = 4 is an easy calculation to perform on your calculator. But it's only by trying calculations like this, which you can check in your head or on paper, that you will become confident about the *way* your calculator operates. Each time you need to use an unfamiliar key, or investigate a new process, it's a good idea to use easy numbers at first and check the answers. Having done this you will feel able to rely on your calculator for more complicated calculations.

TRY SOME YOURSELF

6 Write down some key sequences for each of the following calculations. Do the calculations on your calculator *and* check the answer in your head or on paper.
(i) 5 + 3 = (ii) 6 − 2 = (iii) 5 × 3 = (iv) 8 ÷ 4 = (v) 6 × 3 = (vi) 9 ÷ 3 =

Make up more exercises for yourself if you want more practice.

The same process (writing down key sequences, calculating and checking) can also be used for harder calculations.

EXAMPLE

Calculate 46,932 − 31,952.

SOLUTION

A key sequence is

[46932] [−] [31952] [=] [14980]

We've abbreviated [4] [6] [9] [3] [2] *to* [46932] *just to make the key sequence neater. We'll do this all the way through the book. Any number is just keyed in from left to right as it is written.*

A calculator is useful only if you can rely on the answer. When you use one you should stop and think whether the answers you get are reasonable, because it is quite easy to press the wrong key or make some other mistake. One way of checking your answer is to round off the numbers and make a rough estimate.

In the example above the answer can be checked as follows:

46,932 rounds up to 50,000
31,952 rounds down to 30,000

50,000 − 30,000 can be done in your head–it gives 20,000. This isn't very close to the exact answer, but it does indicate that the answer should be a five digit number less than 20,000.

So if we had obtained 32,980 or 78,884 or 1498 (all of which result from simple keying errors) we would have known that something was wrong.

TRY SOME YOURSELF

7(i) For each of the following, make a rough estimate first; then find the exact answer using your calculator, writing down a key sequence first:
(a) 263 + 91 (b) 66,321 − 17,923 (c) 771,325 + 806,101
(d) 10,999,674 − 9,062,136.

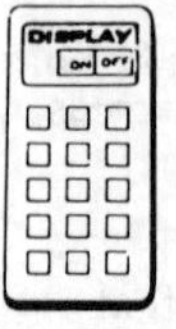

(ii) Some of the following answers are wrong, as the result of mistakes in keying the calculations into the calculator. Find which are the wrong ones, without using your calculator.
(a) 507 − 499 = 1006 (b) 1276 + 901 = 3077
(c) 1,732,644 − 973,201= 759,443
(d) 792,564 − 473,319 = 49,245.

Quite often, the estimate is not such a good approximation to the answer, particularly if the numbers are of different sizes. In these cases you should round the numbers to the same level of accuracy, as the next example illustrates.

EXAMPLE

Evaluate 63,214 − 546.

SOLUTION

Estimate first:

63,214 rounds down to 63,000
546 rounds up to 1000
63,000 − 1000 = 62,000

Round both numbers to the same level of accuracy–here it is to the nearest 1000.

A key sequence is

[63214] [−] [546] [=] [62668]

The estimate indicates that the answer should be a five digit number close to 62,000. The estimate is said to indicate the 'order of magnitude' of the answer.

*The **order of magnitude** indicates the number of digits in the answer and gives a rough idea of the size.*

There are no easy rules to follow when working out estimates in cases like this, though you should aim to get an *idea* about the order of magnitude of the answer. The accuracy of your estimation depends upon the level of accuracy selected for rounding the numbers.

TRY SOME YOURSELF

8 For each of the following, first work out the order of magnitude of the answer then calculate it exactly:
(i) 562 − 14 (ii) 1927 − 163 (iii) 14,632 + 359
(iv) 19,632 + 1534.

1.1(iv) PLACE VALUE

Before we look at some examples involving multiplication we need to digress a little to discuss place value.

We've assumed that you're familiar with the way we write down numbers, placing the digits in imaginary columns:

....	....	hundred thousand	ten thousand	thousand	hundred	ten	unit
....	....		3	4	9	2	1

34,921 means
3 (ten thousands)
4 (thousands)
9 (hundreds)
2 (tens)
1 (unit)

The column occupied by a digit is often referred to as the *place value* of the digit. Now enter 34,921 into your calculator and multiply by 10. What do you notice? Try multiplying by 10 again.

Each time a number is multiplied by 10 the digits move 1 place to the left and a zero is added to the right. This gives a quick way to multiply numbers by 10.

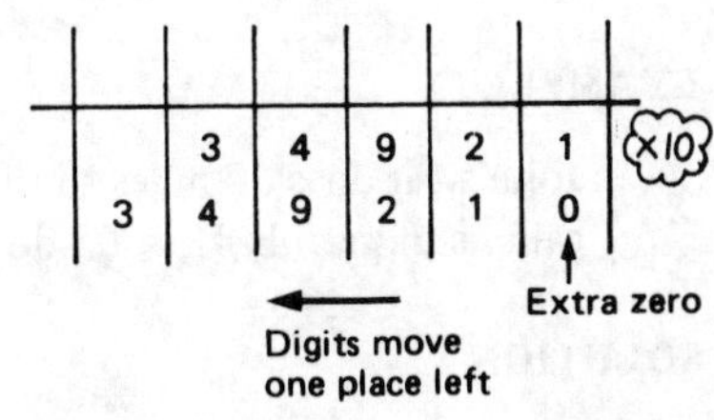

To multiply by 100, move the digits two places to the left and add two zeros. (Try this on your calculator if you're not sure.)

To multiply by 1000, move the digits three places to the left and add three extra zeros. Try this on your calculator.

Not all calculators move the digits in this way but zeroes appear as described here.

Being able to multiply by 10, 100, 1000 quickly is particularly useful when making rough estimates of calculations involving multiplication.

EXAMPLE

Evaluate 2467 × 4153.

SOLUTION

Estimate first:

Round off the numbers to make them easy to handle.

2467 rounds down to 2000

4153 rounds down to 4000

$2000 \times 4000 = 2 \times 4 \times \underbrace{1000}_{\text{gives 3 zeros}} \times \underbrace{1000}_{\text{gives 3 zeros}}$

$= 8,\underbrace{000,\ 000}_{\text{6 zeros}}$

This is because

2 x 1000 = 2000

and

4 x 1000 = 4000

A key sequence is

[2467] [x] [4153] [=] [10245451]

Again we've abbreviated [2] [4] [6] [7] to [2467] etc.

The estimate is not very good, but it does indicate the order of magnitude.

TRY SOME YOURSELF

9 For each of the following, make a rough estimate first, then find the exact answer using your calculator, writing down a key sequence first:
(i) 325 × 184 (ii) 6132 × 2591 (iii) 14,912 × 614
(iv) 632 × 149 (v) 2891 × 157.

1.1(v) USING A CALCULATOR TO SOLVE PROBLEMS

By now you should be getting used to using your calculator for straightforward calculations. But you will rarely need to do arithmetic for its own sake; there are usually good reasons for doing calculations. We now investigate some everyday situations. Here, you will need to decide *when* to use your calculator and whether your answer makes sense in the context of the problem itself.

EXAMPLE

John's car does 38 miles to the gallon. Each week he buys 14 gallons of petrol. How far does he travel each week?

SOLUTION

What do we need to calculate? He travels 38 x 14 miles each week so we need to calculate 38 x 14.

Estimate:

38 rounds up to 40

14 rounds down to 10

40 × 10 = 400

A key sequence is

[38] [x] [14] [=] [532]

This is not too far from the estimate.

TRY SOME YOURSELF

First decide what needs to be calculated.

10(i) Mortgage repayments are £131 per month and rates are £310 per annum. What is the yearly expenditure?
(ii) What is the area of a football pitch measuring 360 ft by 240 ft?
(iii) A local firm employs 367 labourers, each earning £5312 a year. What is the annual salary bill for the firm?

After you have worked through this section you should be able to

a Round off any number to the nearest 10, 100, 1000, 10,000 etc.
b Make a rough estimate of a calculation by rounding off the numbers and adding, subtracting, or multiplying
c Add, subtract and multiply whole numbers using a calculator
d Know whether the answer you obtained using a calculator is reasonable by making an estimate
e Know what a key sequence is—and be able to write one down to indicate how to add, subtract or multiply using a calculator
f Multiply by 10 by moving the digits one place to the left and adding a zero

Finally, here are some exercises if you want more practice.

TRY SOME MORE YOURSELF

11(i) Round off each of the following numbers to the nearest 10:
(a) 17 (b) 8 (c) 94 (d) 99 (e) 25 (f) 117 (g) 1019.

(ii) Round off each of the following numbers to the nearest 100:
(a) 170 (b) 345 (c) 395 (d) 1015 (e) 1050 (f) 10,050
(g) 1,001,050

(iii) Round off each of the following numbers to the nearest 1000:
(a) 1700 (b) 17 (c) 1957 (d) 19,502 (e) 17,850
(f) 1,348,652.

(iv) For each of the following statements round off the number to the easiest number you can:
(a) The population of Birmingham is 1,015,342.
(b) I spent £743 on my car last year.
(c) The distance from London to Bristol is 115 miles.
(d) There are 851 pupils at my son's school.

(v) Make rough estimates for each of the following:
(a) Sally spends £22 a week on clothes. How much is that a year?
(b) Make a rough estimate of the area of the field opposite.

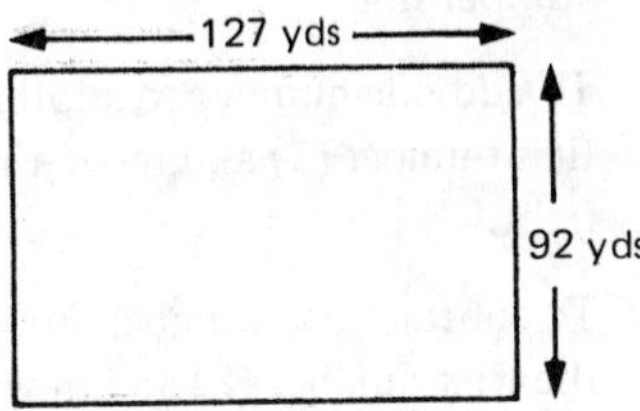

(vi) Use your calculator to evaluate each of the following (remember to check that the answer makes sense):
(a) 15 + 17 (b) 159 + 178 (c) 94,321 + 34,219
(d) 93,216 + 7417 (e) 93,216 + 17.

(vii) Use your calculator to evaluate each of the following (remember to check the answer):
(a) 94 − 37 (b) 843 − 792 (c) 4692 − 3101 (d) 4692 − 301
(e) 47,313 − 2193.

(viii) Use your calculator to evaluate each of the following:
(a) 49 x 13 (b) 473 x 21 (c) 6932 x 914 (d) 7125 x 434
(e) 43 x 3149.

(ix) Use your calculator to work out exact answers for each of the questions in part (v).

1.2 Negative Numbers: Brackets

TRY THESE QUESTIONS FIRST

1 If you had £2 in your bank account and drew out £12, how much would you have left?

2 The temperature was (−4)°C on Monday and dropped overnight by 5°C. What was the temperature on Tuesday morning?

3 Evaluate
(i) (−4) + (−11) (ii) (−5) − (−12).

4 Evaluate
(i) (−7) × 8 (ii) (−63) ÷ (−9).

5 Evaluate (7 + (−12)) × (3 − 6).

1.2(i) NEGATIVE NUMBERS

Numbers like 1, 2, 3, . . . can be represented on a diagram by drawing a line and marking off the numbers. This is called a *number line.*

To add one number to another (for example, 3 + 6) start at the first number (3) and move the appropriate number of steps (6) right.

To subtract one number from another (for example, 8 − 3) start at the first number (8) and move the appropriate number of steps left (3).

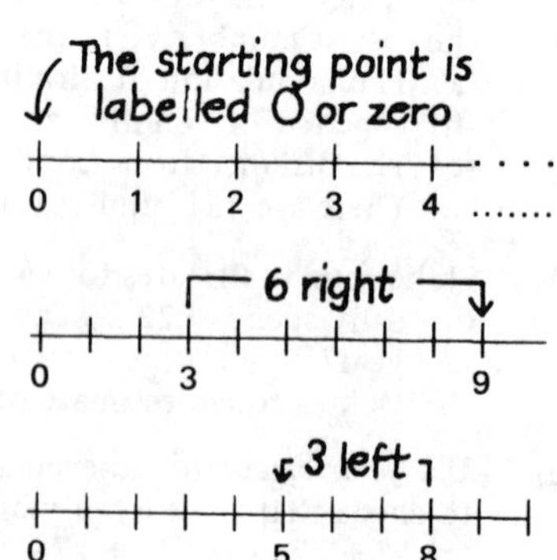

But what happens with 3 − 8?

Following the same procedure,

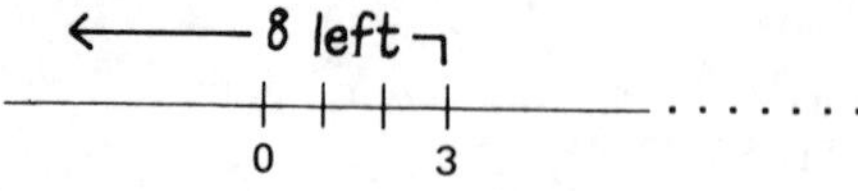

it appears that the number line needs to be extended to the left.

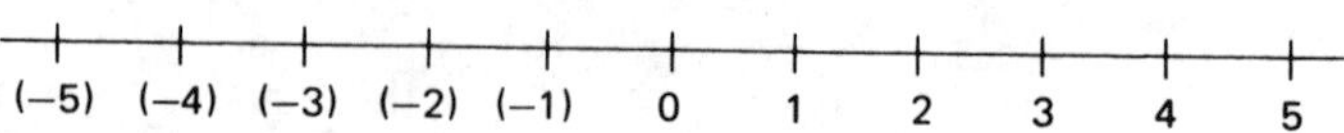

Numbers to the left of 0 are called *negative numbers.* Using the extended number line, $3 - 8 = (-5)$.

Negative numbers have a minus sign in front of them:
$(-3), \ldots, (-11), \ldots, (-14), \ldots$

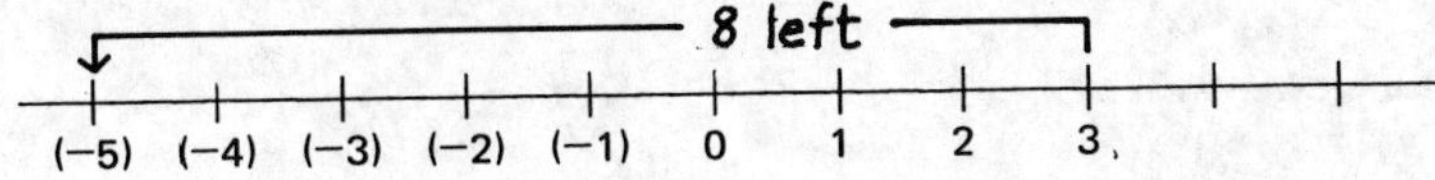

TRY SOME YOURSELF

1 Evaluate each of the following by drawing a number line:
(i) $3 - 3$ (ii) $5 - 6$ (iii) $2 - 9$.

We often use negative numbers in everyday life, perhaps without being aware of it.

You may have had a bank overdraft at some time. Bank statements now indicate an overdrawn account by placing a D or OD next to the figure in the balance column. This account indicates that on 27.4.81 Mr D. Cummings was overdrawn by £13. He owed the bank £13. In fact, so far as the bank was concerned he owned £(−13) or −£13.

Mr. D. Cummings

Date	Balance
5. 4. 81	341. 00
10. 4. 81	223. 00
21. 4. 81	110. 00
24. 4. 81	41. 00
27. 4. 81	13. 00 OD

In winter months the temperature often drops below freezing point. Nowadays temperature is measured in °C, and 0°C is freezing point. The temperature scale is just like the number line.

(−3)°C is 3 degrees below zero

3
2
1
0
(−1)
(−2)
(−3)

°C means degrees Centigrade or Celsius

Contours on a map indicate the height of land above sea level. Modern maps indicate the height in metres at 50 metre intervals. When the land is actually *below* sea level, as in some parts of Holland, the corresponding contours indicate negative heights.

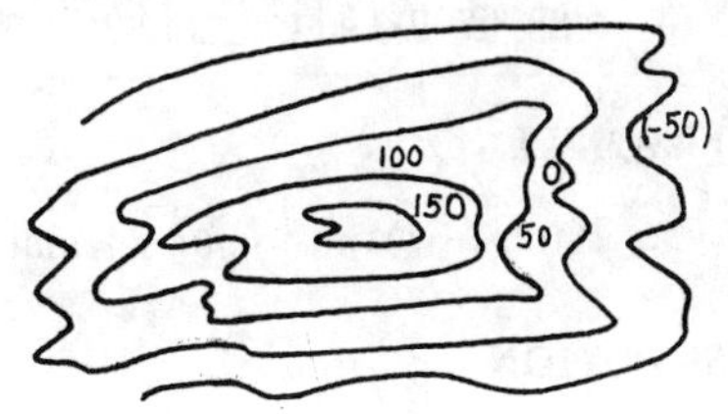

1.2(ii) MANIPULATING NEGATIVE NUMBERS: ADDING AND SUBTRACTING

Negative numbers can be manipulated just like positive numbers although it is often difficult to appreciate why we get the results we do. People often think negative numbers are a little 'magical'. Strange things can happen when they're subtracted, multiplied etc. We haven't enough space in this book to go into the complete explanation, so we're going to ask you to do some investigations with your calculator. In this way we hope to extract the rules for negative numbers. We hope that, when you have seen the patterns emerging for yourself, you will feel confident about using negative numbers later in your studies.

Positive numbers lie to the right of 0 on the number line, whereas negative numbers lie to the left. The number 0 is a special case. It is neither positive nor negative. Consequently numbers may be positive, negative or zero.

CHECK YOUR ANSWERS

1 £2 − £12 = £(−10) = − £10 *Section 1.2(i)*

2 (−4) − 5 = (−9). So the temperature was (−9)° C. *Section 1.2(ii)*

3 (i) (−4) + (−11) = −4 −11 = −15
(ii) (−5) − (−12) = −5 + 12 = 7 *Section 1.2(ii)*

4 (i) (−7) × 8 = −56
(ii) (−63) ÷ (−9) = 7 *Section 1.2(iii)*

5 Evaluate the brackets first to give *Section 1.2(iv)*

$$(7 + (-12)) = (-5)$$

and

$$(3 - 6) = (-3)$$

Now

$$(-5) \times (-3) = 15$$

First of all, it's important to learn how to enter a negative number into your calculator. Your calculator should have a change of sign key, which we're going to label as [+/−]. This key changes the sign, as you might expect.

Check with the maker's handbook.

Thus [3] [+/−] should appear on the display as [−3]. Changing the sign again, [3] [+/−] [+/−], results in [3].

*Whenever we indicate [+/−] you should press the appropriate key on **your** calculator.*

EXAMPLE

Enter (−7) onto your calculator.

SOLUTION

We need to key [7] [+/−], giving [−7].

*Notice that you should enter the number first, **then** change the sign.*

TRY SOME YOURSELF

2 Enter each of the following numbers onto your calculator:
(i) −4 (ii) −6 (iii) −1782 (iv) −2941.

Remember that it's good practice to clear the calculator after each exercise.

Adding negative numbers

The following investigation concerns what happens when you add a negative number.

Use your calculator to work out each of the following. In each case, compare the answer to part (a) with the answer to part (b).

(i) (a) $2 + (-3)$ (b) $2 - 3$
(ii) (a) $2 + (-10)$ (b) $2 - 10$
(iii) (a) $2 + (-4)$ (b) $2 - 4$
(iv) (a) $4 + (-8)$ (b) $4 - 8$

Try the key sequence
[2] [+] [3] [+/−] [=]
or check with the maker's handbook.

You should have noticed that

$2 + (-3) = 2 - 3$

$2 + (-10) = 2 - 10$

etc.

What happens if the first number is negative? Use your calculator to evaluate each of the following:

(i) (a) $(-2) + (-7)$ (b) $(-2) - 7$
(ii) (a) $(-3) + (-10)$ (b) $(-3) + 10$.

The same pattern emerges. Try some more if you're not sure. This suggests the rule:

Try
$(-1) + (-3); (-4) + (-1); \ldots$

Adding a negative number is the same as subtracting a positive number.

At the beginning of the section we introduced the idea of subtraction, using the number line. In fact we indicated how to subtract *positive numbers*—by moving the appropriate number of steps *left*. It doesn't matter if the first number is positive or negative. The next example indicates how to add negative numbers using the rule outlined above and the number line.

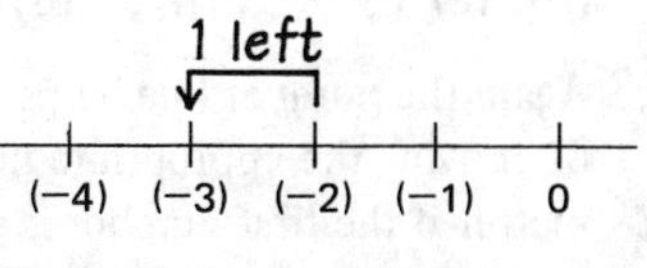

Thus $(-2) - 1 = (-3)$

EXAMPLE

Evaluate $(-2) + (-4)$.

SOLUTION

The rule for adding negative numbers tells us that

$(-2) + (-4) = (-2) - 4$

Now we can use the number line.

Start at (-2) on the number line and move 4 left.

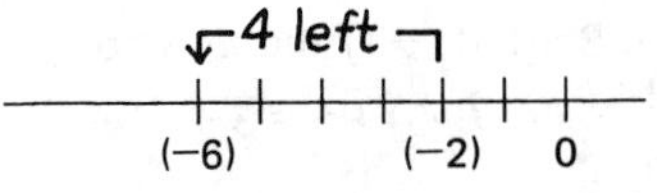

$(-2) - 4 = -6$

so

$(-2) + (-4) = -6$

TRY SOME YOURSELF

3 Without using a calculator evaluate each of the following:
(i) $(-3) - 7$ (ii) $(-2) - 9$ (iii) $2 + (-6)$
(iv) $4 + (-7)$ (v) $(-3) + (-8)$ (vi) $(-1) + (-2)$.

Use the rule above and the number line as a guide.

Subtracting negative numbers

The following investigation suggests a similar pattern for subtraction.

Try the following subtractions using your calculator. What do you notice?

What is the value of $2 + 3$? $4 + 7$? etc.

(i) $2 - (-3) =$

(ii) $4 - (-7) =$

(iii) $6 - (-2) =$

(iv) $8 - (-10) =$

(v) $1 - (-1) =$

This investigation suggests the rule:

Subtracting a negative number is the same as adding a positive number.

The rule also holds if the first number is negative. Use your calculator to evaluate each of the following. Compare the answer to part (a) with the answer to part (b).

(i) (a) $(-2) - (-3)$ (b) $(-2) + 3$

(ii) (a) $(-7) - (-4)$ (b) $(-7) + 4$

Again the number line helps to picture addition of a *positive number* by moving the appropriate number of places *right*. It doesn't matter if the first number is positive or negative. The following example indicates how to subtract negative numbers using the rule outlined above and the number line.

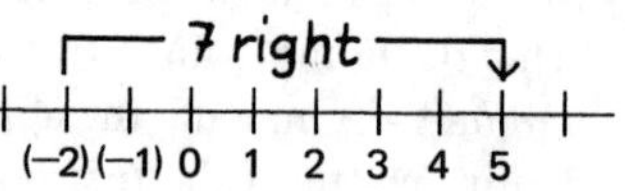

Thus $(-2) + 7 = 5$.

EXAMPLE

Evaluate $(-3) - (-6)$.

SOLUTION

The rule for subtracting negative numbers tells us that

$$(-3) - (-6) = (-3) + 6$$

Now we can use the number line.

Start at (-3) and move 6 units right.

6 right

(−3) 0 3

$$(-3) + 6 = 3$$

So

$$(-3) - (-6) = 3.$$

TRY SOME YOURSELF

4 Without using a calculator evaluate each of the following:
(i) $(-3) + 6$ (ii) $(-7) + 4$ (iii) $3 - (-2)$
(iv) $5 - (-7)$ (v) $(-2) - (-6)$ (vi) $(-1) - (-1)$.

Use the rule above and the number line as a guide.

The following exercises involve adding and subtracting negative numbers in some everyday situations.

TRY SOME YOURSELF

5(i) Sarah is £26 overdrawn. How much must she add to the bank account to be out of debt?

(ii) Jim was rock climbing in Israel. He started at 51 metres below sea level and ended up 20 metres above sea level. How far had he climbed?

(iii) The temperature was $(-2)^\circ$C on Tuesday and dropped to $(-12)^\circ$C on Wednesday. By how much had it dropped?

(iv) Dave had £12 in his bank account. He cashed a cheque for £30, then another for £16. How much did he have left after these transactions?

1.2(iii) MANIPULATING NEGATIVE NUMBERS: MULTIPLYING AND DIVIDING

The following investigation concerns what happens when a positive number is multiplied by a negative number.

Evaluate each of the following. Use a calculator for part (a); you should be able to do part (b) in your head.

(i) (a) $2 \times (-3)$ (b) 2×3

(ii) (a) $4 \times (-7)$ (b) 4×7

(iii) (a) $6 \times (-10)$ (b) 6×10

(iv) (a) $1 \times (-1)$ (b) 1×1

Is the answer positive or negative? Two positive numbers multiplied together will always give a positive number.

To evaluate $2 \times (-3)$ try the key sequence

[2] [×] [3] [+/−] [=]

Now try

(i) $(-2) \times 3$

(ii) $(-4) \times 7$

(iii) $(-6) \times 10$

(iv) $(-1) \times 1$

The digits are the same as in the examples above–but the signs are different.

This suggests that

and

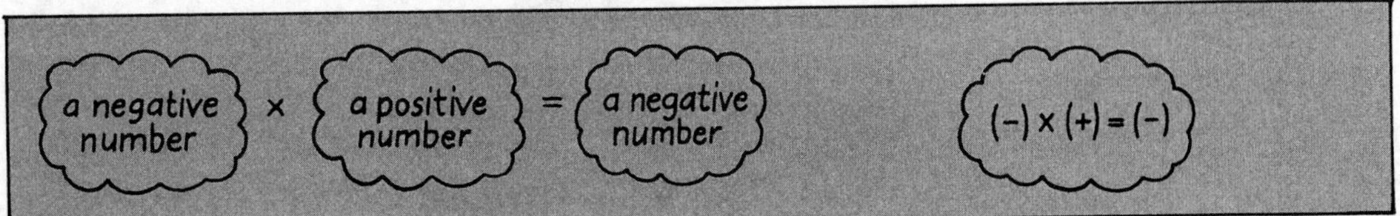

Now, what happens if both numbers are negative?

Try evaluating each of the following:

(i) $(-2) \times (-3)$
(ii) $(-4) \times (-7)$
(iii) $(-6) \times (-10)$
(iv) $(-1) \times (-1)$

Again, the digits are the same, but the signs are different.

Is the answer positive or negative?

This suggests that

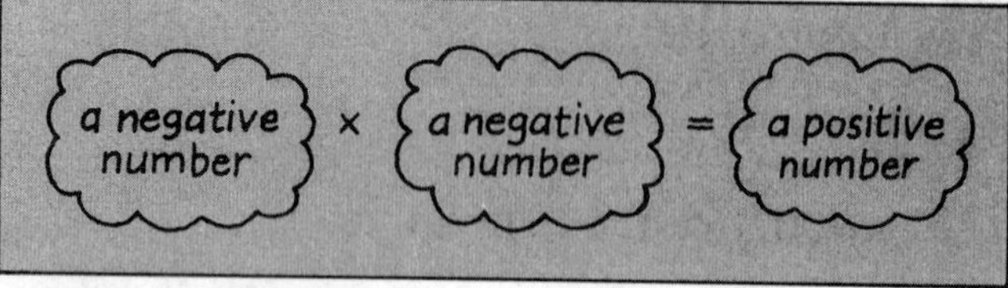

$(-) \times (-) = (+)$

This result is perhaps more mysterious than the previous ones. However at this stage it is difficult to provide an adequate intuitive explanation of *why* it works. For the moment you should just make sure that you can use the rule, even if it does appear a bit strange.

If you study mathematics at a higher level, you may be given a mathematical explanation.

The same rules apply to division.

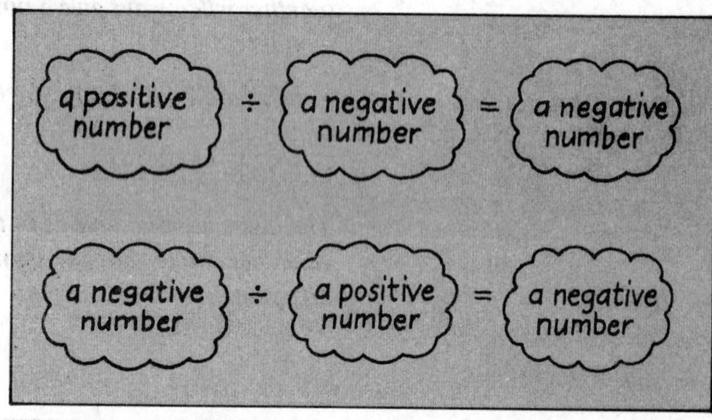

For example:
$8 \div (-2) = (-4)$
$(-8) \div 2 = (-4)$

and

a negative number ÷ a negative number = a positive number

$(-8) \div (-2) = 4$

Check for yourself using your calculator.

You should remember these rules for multiplication and division:

If the signs are the same—the answer should be positive.
If the signs are different—the answer should be negative.

Of course, if the numbers are both positive,

(+) x (+) = (+)

and

(+) ÷ (+) = (+)

The following examples show how to multiply and divide negative numbers using these rules.

EXAMPLE

Evaluate (−2) × (−7).

SOLUTION

Using the rule (−) × (−) = (+) and using the fact that 2 × 7 = 14, we get

(−2) × (−7) = 14

Work out the sign using the rules, then calculate 2 x 7.

EXAMPLE

Evaluate (−36) ÷ 9.

SOLUTION

Using the rule (−) ÷ (+) = (−) and the fact that 36 ÷ 9 = 4, we get

(−36) ÷ 9 = (−4)

Work out the sign using the rules, then calculate 36 ÷ 9.

TRY SOME YOURSELF

6 Without using a calculator, evaluate each of the following:
(i) 6 × (−2) (ii) (−3) × (−4) (iii) 2 × (−10)
(iv) 20 ÷ (−4) (v) (−36) ÷ (−12) (vi) (−42) ÷ 14.

Work out the sign first. Use your calculator to check if you're not sure.

We've used small numbers in these investigations because we've really been interested in the principles and the rules. However, the same rules apply no matter how complicated the numbers. Try some yourself on your calculator if you want to check this. If you do select some larger numbers, remember that you should always try and estimate the order of magnitude of the answer first.

Try, for example,

2473 x (−1432)

or

(−6312) + (−2134) . . .

1.2(iv) BRACKETS

We're going to change tracks a bit here and investigate calculations involving brackets. Again we're going to concentrate on the principles involved so we'll only use small numbers.

Brackets are commonly used to indicate the order in which to perform a calculation; they indicate which parts should be evaluated first.

Earlier in this section we've put brackets around negative numbers to keep the information together and to help you in the calculations.

For example, in the calculation

$$3 - (2 + 1)$$

(2 + 1) = 3

the brackets should be evaluated first to give

$$3 - 3 = 0$$

Calculations may involve two sets of brackets, such as

$$(3 + 2) \times (7 - 3)$$

In this case, both sets of brackets must be evaluated first to give

$$5 \times 4 = 20$$

*We aren't told whether to evaluate (3 + 2) first or (7 − 3) but **both** brackets must be evaluated before we multiply.*

Brackets are used in this way to avoid misunderstanding and ambiguity. For example, if you were asked to evaluate:

$$3 - 2 + 1$$

you would probably just work from left to right:

$3 - 2 = 1$ and $1 + 1 = 2$

Just working from left to right is the same as evaluating (3 − 2) + 1.

The brackets in

$$3 - (2 + 1)$$

indicate clearly that that is not what is intended.

The expression

$$3 + 2 \times 7$$

could be evaluated as

$3 + 2 = 5$ and $5 \times 7 = 35$

or

$2 \times 7 = 14$ and $3 + 14 = 17$

You can see how important it is to make it clear which of these two ways is intended. Brackets do exactly this.

Writing

$$(3 + 2) \times 7$$

indicates that the addition should be completed first; writing

$$3 + (2 \times 7)$$

indicates that the multiplication should be first.

What happens on your calculator? Try keying

[3] [+] [2] [x] [7] [=]

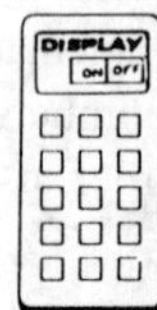

Which answer do you get? If you get 35, then your calculator works from left to right; if you get 17 then your calculator works out multiplication before addition. Some calculators work one way, some the other. It's a good idea to familiarise yourself with the way your calculator handles calculations like this so you know for the future.

Calculators which carry out the multiplication first are said to have inbuilt precedence of multiplication over addition. Check with the maker's handbook whether your calculator has inbuilt precedence.

In general, though, it is much less ambiguous to put brackets around the parts you wish to evaluate first.

TRY SOME YOURSELF

7 Evaluate
(i) $(5 + 3) - 14$ (ii) $7 \times (2 - 9)$ (iii) $4 - (2 - 6)$
(iv) $(2 - 3) \times (6 + 3)$.

Calculations look more complicated when they contain negative numbers, but they aren't really.

EXAMPLE

Evaluate $((-3) + (-7)) \times ((-1) \times 4)$.

SOLUTION

We work out the brackets first to get

$$(-3) + (-7) = (-3) - 7 = (-10)$$

and

$$(-1) \times 4 = (-4)$$

Now

$$(-10) \times (-4) = 40.$$

Use the rules for manipulating negative numbers.

TRY SOME YOURSELF

8 Evaluate
(i) $((-2) + 7) \div 5$ (ii) $((-3) - (-4)) \times 5$
(iii) $(2 + (-2)) + (6 \times (-2))$.

If your calculator has brackets, now is the time to find out how to use them. Try Exercises 7 and 8 using your calculator. You can check the answers you *should* get from the solutions to the exercises. You'll probably find that you can key the expressions by working from left to right.

Check in the maker's handbook how to use brackets (or parentheses as they are sometimes called).

For example, to evaluate

$$4 \times (3 + 5)$$

try keying

[4] [×] [(] [3] [+] [5] [)] [=]

If your calculator doesn't have brackets then you will just have to keep track of the calculation on paper, unless you are ingenious and twist the calculation round.

Brackets within brackets

Sometimes calculations involve brackets within brackets, like

$$3 + (2 - (7 + 3))$$

In this case you should work out the innermost bracket first, then the next innermost and so on.

Here $(7 + 3)$ is the innermost bracket and

$$7 + 3 = 10$$

Now

$$(2 - (7 + 3)) = (2 - 10) = (-8)$$

and

$$3 + (2 - (7 + 3)) = 3 + (-8) = -5$$

Once again, if your calculator has brackets, the calculation can probably be keyed in as it's written. Try

3 + (2 − (7 + 3)) = −5

And again, if your calculator doesn't have brackets for the moment you will either have to twist the calculation round, or write down intermediate steps on paper.

There are several alternative notations for brackets. For example, $2 - (7 + 3)$ might be written as

$$2 - [7 + 3] \text{ or } 2 - \{7 + 3\}$$

TRY SOME YOURSELF

9 Without using a calculator, evaluate each of the following:
(i) $\{(5 - 7) + 6\} - 3$
(ii) $16 + [(3 - 8) - 4]$
(iii) $(2 + 3) \times [5 - (6 + 2)]$.

These examples indicate that different notations are often used to differentiate brackets within brackets to avoid confusion.

Some calculators cannot handle brackets within brackets and *all* calculators are limited as to the number they can cope with before an error message is indicated or the calculator gives the wrong answer. Investigate your calculator by attempting to key

$$1 + (1 + (1 + (1 + (1 + (1 + 1))))) = 7$$

At what stage does your calculator give an error message? At this stage, you will probably have to resort to writing down the different parts on paper.

For example, your calculator may be able to evaluate $2 + (3 - 2)$ as it is written but not $2 + (3 - (2 + 1))$.

1.2(v) MULTIPLYING OUT BRACKETS

So far we've used brackets to keep pieces of information together and to indicate which part of the calculation should be evaluated first. But you will also need to be able to multiply out brackets. To see how to do this we'll take another look at

$$4 \times (3 + 5)$$

Think of this as the area of a rectangle, of side 4 units and 8 units (= 3 + 5).

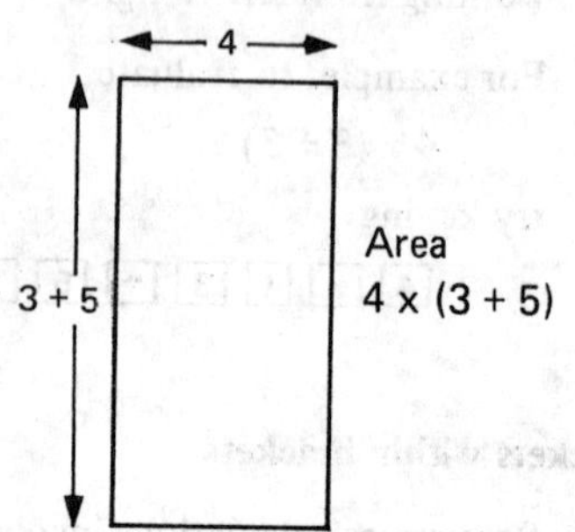

The rectangle can be split into two smaller rectangles

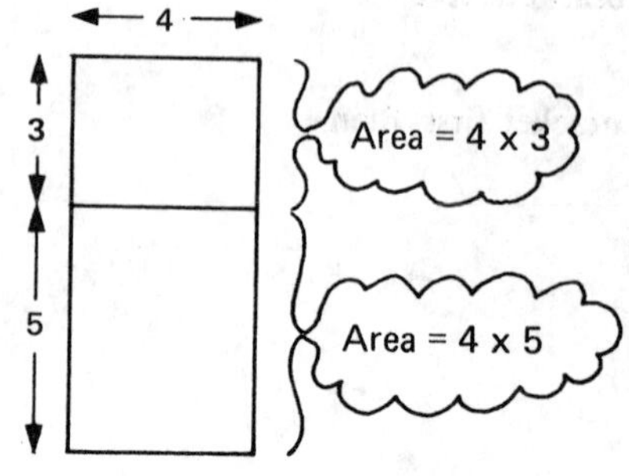

but the total area is still the same. Thus

$$4 \times (3 + 5) = (4 \times 3) + (4 \times 5)$$

(4 x 3) ↓ *4 x (3 + 5)* ↑ *4 x 5*

In the same way,

$$4 \times (3 + 5 - 4) = (4 \times 3) + (4 \times 5) + (4 \times (-4))$$
$$= 12 + 20 - 16 = 16$$

Multiply each number inside the brackets by the number outside.

Check this yourself by working out the brackets first.

TRY SOME YOURSELF

10 Evaluate each of the following by first multiplying each number inside the bracket by the number outside, then adding:
(i) $2 \times (3 + 4)$ (ii) $3 \times (2 - 6)$ (iii) $4 \times (3 - 2 + 1)$.

Check by evaluating the brackets first, then multiplying.

In fact $4 \times (3 + 5)$ can be abbreviated to

$$4(3 + 5)$$

omitting the multiplication sign.

However, although $4 \times (3 + 5)$ can be written as $4(3 + 5)$ and we understand that we should multiply by 4, this abbreviated form cannot be keyed onto your calculator. You must always insert the multiplication sign in key sequences.

Try keying

[4] [(] [3] [+] [5] [)] [=]

and

[4] [x] [(] [3] [+] [5] [)] [=]

Notice that

$$(-1) \times (3 + 5)$$

is usually written as

$$-(3 + 5)$$

Just as $(-1) \times 4$ is usually written as -4.

so that

$$-(3 + 5) = ((-1) \times 3) + ((-1) \times 5)$$
$$= (-3) + (-5)$$

Originally, each term inside the brackets was positive. Multiplying by (-1) changes the sign of each term inside the brackets.

Often the brackets around negative numbers are omitted. Thus $(-3) - 5$ is just written as $-3 - 5$.

EXAMPLE

Evaluate $-(2 + 3 - 4)$.

SOLUTION

$-(2 + 3 - 4)$ is the same as $(-1) \times (2 + 3 - 4)$, so we must change the sign of each number inside the brackets. Thus

$$-(2 + 3 - 4) = -2 - 3 + 4 = -1$$

Change all positive signs to negative signs and vice versa.

Check by working out the brackets first.

TRY SOME YOURSELF

11 Evaluate each of the following by multiplying out the brackets:
(i) $2(3 + 4)$ (ii) $3(2 - 6)$ (iii) $-(6 + 3)$ (iv) $-(2 - 4)$,
(v) $-(3 + 6 - 4)$.

Since the order in which two numbers are multiplied together doesn't matter,

For example,

$2 \times 3 = 3 \times 2 = 6$

$2 \times (3 + 4)$ is the same as $(3 + 4) \times 2$

and

$(3 + 4) \times 2 = (3 \times 2) + (4 \times 2)$

$(3 + 4) \times 2 = 2 \times (3 + 4)$

and

$2 \times (3 + 4) = (2 \times 3) + (2 \times 4)$
$= (3 \times 2) + (4 \times 2)$

However, although it is usual to write $2(3 + 4)$ it is not such good practice to write $(3 + 4)2$, since it can lead to confusion. When multiplying brackets you should always write the number before the brackets. Thus

$(3 + 5) \times 4$ is written as $4(3 + 5)$

and

$(2 + 12) \times (-2)$ is written as $-2(2 + 12)$

All the principles and rules we have introduced in this section—the rules for negative numbers and how to cope with brackets—can be extended to calculations involving larger, more complicated numbers. However, if the calculations do involve large numbers you should always make a rough estimate of the answer first, just in case you make a mistake. It's very easy to miss out a step in the key sequence, such as a change of sign or a bracket, and get a completely wrong answer.

Use the principles introduced in Section 1.1.

After you have worked through this section you should be able to

a Represent positive and negative numbers as points on a number line
b Add, subtract, multiply and divide positive and negative numbers
c Evaluate calculations involving brackets by working out the brackets first
d Evaluate calculations involving multiplication of brackets by multiplying out the brackets first

We also summarise the rules for manipulating negative numbers below, since we feel it is worth while to remember them.

Adding a negative number is the same as subtracting a positive number.
Subtracting a negative number is the same as adding a positive number.

For multiplication or division:

If the signs are the same the answer should be positive.
If the signs are different the answer should be negative.

Finally, here are some exercises if you want more practice.

TRY SOME MORE YOURSELF

12 Without using a calculator evaluate each of the following:
(i) (a) $3 + 7$ (b) $2 - 6$ (c) $5 - 11$ (d) $(-1) + 7$ (e) $(-2) - 8$
(f) $(-2) + 8$ (g) $(-3) - 2$

(ii) (a) $3 + (-2)$ (b) $4 + (-9)$ (c) $(-2) + (-9)$ (d) $-1 + (-10)$

(iii) (a) $2 - (-13)$ (b) $12 - (-2)$ (c) $(-1) - (-10)$ (d) $(-5) - (-3)$
(e) $(-16) - (-12)$

(iv) (a) $4 \times (-8)$ (b) $(-2) \times 7$ (c) $(-3) \times (-10)$ (d) $(-1) \times (-2)$
(e) $(-12) \times 9$

(v) (a) $81 \div (-9)$ (b) $(-15) \div 3$ (c) $(-27) \div (-3)$ (d) $(-39) \div (-13)$

(vi) (a) $(7 - 12) + 14$ (b) $(7 - 12) \times 3$ (c) $(4 - 9) \times (2 + 3)$

(vii) (a) $3(2 + 7)$ (b) $5(2 - (-4))$ (c) $-(7 - (-4) + 3)$
(d) $-(2 - 4 + (-7))$.

Use a calculator to evaluate each of the following:

(viii) (a) $1756 + (-3241)$ (b) $(-3497) \times (-2134)$ (c) $364 \times (342 - (-123))$
(d) $(1927 - 163) \times (12 - 149)$.

1.3 Decimals

TRY THESE QUESTIONS FIRST

1 Which of the following numbers is the bigger: 5·5; 5·05?

2 (i) Without using a calculator, evaluate $3{\cdot}52 \div 10{,}000$.
(ii) Express 349 cm in kilometres.

3 (i) Convert 0·00349 to scientific notation.
(ii) Without using a calculator, evaluate $0{\cdot}1 \div 0{\cdot}01$.

4 Round 0·14739 to (i) 2 decimal places (ii) 3 significant figures.

5 Evaluate $37{\cdot}62 \div 0{\cdot}0332$ correct to 4 decimal places.

1.3(i) WHAT IS A DECIMAL?

You may have noticed that in the last sections we mainly ignored division. That's because we were only interested in whole numbers. When one number is divided by another, more often than not it *doesn't* result in a whole number.

Whole numbers are numbers like 1, 3, 1749, (−342), . . .

Try dividing 22 by 4 on your calculator. You should get

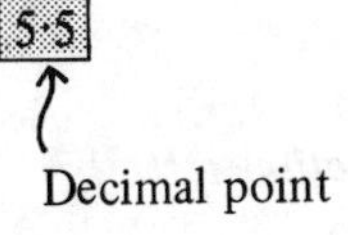

22 ÷ 4 results in a number between 5 and 6.

This is not a whole number. The point indicates that it's a whole number plus a *decimal* part.

*The word **decimal** comes from **decem**–the Latin for ten.*

Try some more: try $8 \div 3$ or $8 \div 5$ on your calculator. In each case the decimal point indicates that the answer is *not* a whole number. It's a *decimal number.*

Some examples of decimals in common use are . . .

Money

Petrol

£1·61 a gallon

Dress fabric

£4·95 a metre

CHECK YOUR ANSWERS

1 5·5 — *Section 1.3(i)*

2 (i) $3{\cdot}52 \div 10{,}000 = 0{\cdot}000352$
(ii) 349 cm = 0·00349 km — *Section 1.3(ii)*

3 (i) $0{\cdot}00349 = 3{\cdot}49 \times 10^{-3}$
(ii) $0{\cdot}1 \div 0{\cdot}01 = 10^{-1} \div 10^{-2} = 10$ — *Section 1.3 (iii)*

4 (i) 0·14739 rounds to 0·15.
(ii) $0{\cdot}14739 = 1{\cdot}4739 \times 10^{-1}$, which rounds to $1{\cdot}47 \times 10^{-1}$. — *Section 1.3 (iv)*

5 Estimate first: — *Section 1.3 (v)*

$$(4 \times 10^{1}) \div (3 \times 10^{-2}) \triangleq 1 \times 10^{3} = 1000$$

Calculate:

$$37{\cdot}62 \div 0{\cdot}0332 = 1133{\cdot}1325$$

which is quite close to the estimate.

Metric measurements

Height on passport
1·85 metres

Weight
1 kilogram

1 metre = 1·08 yards (just over 1 yard)
1 kilogram = 2·2 lbs (just over 2 lbs)

. . . and we've already mentioned that calculators work in decimals. The decimal system is just an extension of the system used for whole numbers and we can give some idea of its structure by taking another look at place value and extending the columns to the right.

See Section 1.1(iv).

We have already indicated that multiplying a number by 10 shifts all the digits one place to the left and adds a zero to the right hand side. A similar pattern emerges if you divide by 10. Use your calculator to calculate

34,921,000 ÷ 10

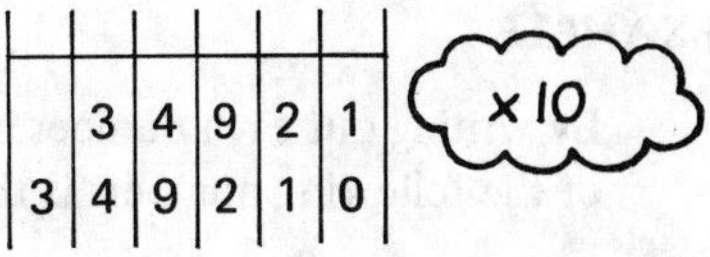

What do you notice?

Divide by 10 again . . . and again.

So dividing a number by 10 shifts all the digits one place to the right. The process can be extended. Divide 34,921 by 10. You should get a decimal number.

3	4	9	2	1	
3	4	9	2	1	0

×10

Notice that the digits still move one place to the right.

The pattern is summarised below.

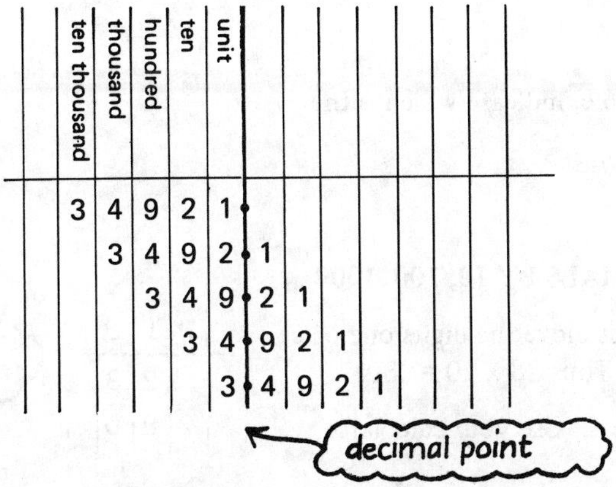

ten thousand	thousand	hundred	ten	unit					
3	4	9	2	1					
	3	4	9	2	1				
		3	4	9	2	1			
			3	4	9	2	1		
				3	4	9	2	1	

decimal point

Each time we divide by ten the digits move one place to the right.
Even if your calculator doesn't work exactly like this you'll see a similar pattern emerging.
The decimal point separates the whole number part from the decimal part.

The table above suggests that the place value system can be extended to cater for decimals. The decimal columns are labelled as follows:

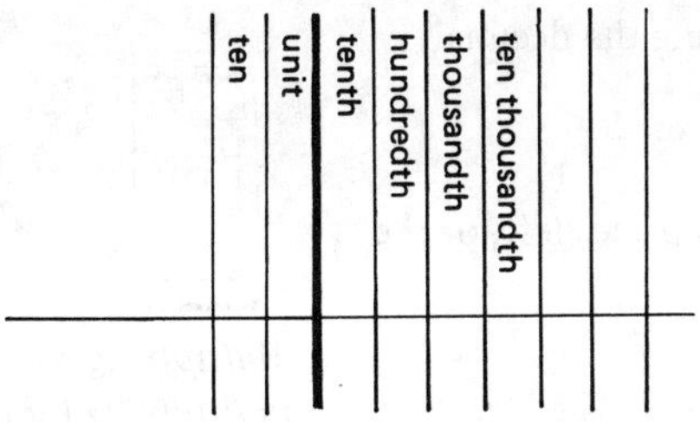

ten	unit	tenth	hundredth	thousandth	ten thousandth		

Just as the 'tens' column is obtained by multiplying the unit column by ten, the 'tenth' column is obtained by dividing the unit column by 10—and so on.

So

0·1 $= 1 \div 10$ or $\frac{1}{10}$ — pronounced one tenth

0·01 $= 1 \div 100$ or $\frac{1}{100}$ — one hundredth

0·001 $= 1 \div 1000$ or $\frac{1}{1000}$ — one thousandth

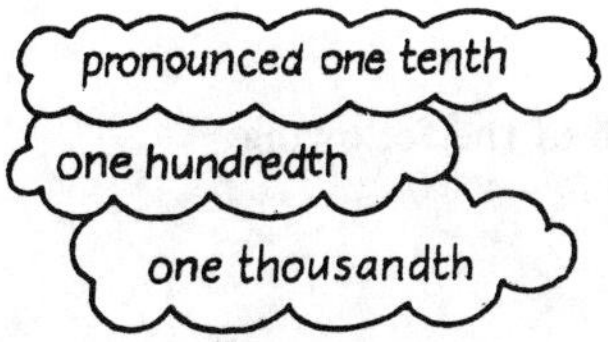

It is important to include zeros in numbers like 20 and 200, in order to indicate the place value of 2. In the same way the zeros must be included in decimal numbers like 0·2 and 0·02.

EXAMPLE

By writing out each number in a place value table, indicate which of the following numbers is the bigger: 3·5; 3·05.

SOLUTION

H	T	U	$\frac{1}{10}$	$\frac{1}{100}$	$\frac{1}{1000}$	$\frac{1}{10,000}$
		3 •	5			
		3 •	0	5		

3·5 is 3 units and 5 tenths.

3·05 is 3 units, no tenths and 5 hundredths.

So 3·5 is bigger than 3·05.

TRY SOME YOURSELF

1 By writing the numbers in a place value table, indicate which is the bigger number of each pair of numbers:
(i) 3·701, 3·071 (ii) 0·3, 0·08 (iii) 7·06, 7·6

1.3(ii) MULTIPLYING AND DIVIDING DECIMALS BY 10, 100, 1000 . . .

To multiply whole numbers by 10, you just move the digits one place left and add a zero to the right hand side. Thus 28 x 10 = 280.

H	T	U
	2	8
2	8	0

× 10

The process is the same for decimal numbers. Use your calculator to evaluate

2·83 x 10

2·83 x 10 = 28·3. Again the digits move one place left relative to the decimal point.

T	U	$\frac{1}{10}$	$\frac{1}{100}$
	2 •	8	3
2	8 •	3	

Alternatively you may like to think of this as moving the decimal point one place to the right. So

49·7812 x 10 = 497·812

Try this on your calculator.

Similarly, multiplying by 100 moves the digits two places left, or the decimal point two places right. So

49·7812 x 100 = 4978·12

Multiplying by 100 is the same as multiplying by (10 x 10) or multiplying by 10 twice.

TRY SOME YOURSELF

2 Without using a calculator, evaluate each of the following:
(i) 6·34 x 10 (ii) 14·752 x 100 (iii) 0·674 x 10
(iv) 0·00042 x 100 (v) 1·03 x 100.

Dividing by 10 moves the digits one place right, or the decimal point one place left.

Use your calculator to evaluate

48·63 ÷ 10

48·63 ÷ 10 = 4·863

What would you expect to happen when you divide by 100 or 1000? Try some examples on your calculator.

Dividing by 100 is the same as dividing by (10 x 10) or dividing by 10 twice.

TRY SOME YOURSELF

3 Without using a calculator, evaluate each of the following:
(i) 58·9 ÷ 10 (ii) 589 ÷ 10 (iii) 58·9 ÷ 1000
(iv) 0·589 ÷ 1000 (v) 0·0589 ÷ 10.

589 is the same as 589·0.

Earlier in this section we noted that the money system is based on decimals:

£1 = 100 p

£2 = 200 p

£2·63 = 263 p

To convert from pounds to pence multiply by 100—or move the decimal point two places right.

Metric systems of measurement are also based on multiples of 10.

Prefixes such as kilo-, centi- etc. have definite meanings:

kilo- means 1000

deci- means 'one tenth' or $\frac{1}{10}$

centi- means 'one hundredth' or $\frac{1}{100}$

milli- means $\frac{1}{1000}$ or 0·001

Thus

1 kilometre (km) = 1000 metres or 1000 m

1 kilogram (kg) = 1000 grams or 1000 g

1 metre = 100 centimetres or 100 cm

= 1000 millimetres or 1000 mm

metre is usually abbreviated to m, gram is abbreviated to g, etc.

We can convert from one unit to another by multiplying or dividing by 10, 100, or 1000 etc.

EXAMPLE

28·3 cm = 283 mm (Multiply by 10.)

28·3 cm = 0·283 m (Divide by 100.)

28·3 cm = 0·000283 km (Divide by 100,000.)

1 cm = 10 mm

100 cm = 1 m

100,000 cm = 1 km

TRY SOME YOURSELF

4(i) Express 589 mm in centimetres.
(ii) Express 59·5 km in metres.
(iii) Express 3·4 g in kilograms.
(iv) Express 0·3 mm in metres.

1 kg = 1000 g

1.3(iii) SCIENTIFIC NOTATION

We have suggested earlier that it is always a good idea to have some idea of the order of magnitude of the answer to serve as a check for calculations on your calculator. So before we introduce any decimal calculations, which *will* need to be worked out on a calculator, we introduce some new notation which will make the estimation easier.

You know that

$$10 \times 10 = 100$$

and

$$10 \times 10 \times 10 = 1000$$

In fact 10×10 can be abbreviated to 10^2 and $10 \times 10 \times 10$ can be abbreviated to 10^3.

10^2 is pronounced ten squared or ten to the power 2

10^3 is pronounced ten cubed or ten to the power 3

Why is this? Where do the powers come from? We explore the pattern below.

10^2 = 10×10 = 1 00

Power 2 — 10 times itself 2 times — 2 zeros

and

10^3 = $10 \times 10 \times 10$ = 1 000

Power 3 — 10 times itself 3 times — 3 zeros

The pattern can be generalised for higher powers. Thus

$$10^4 = 10 \times 10 \times 10 \times 10 = 10{,}000$$

and

$$10^5 = 10 \times 10 \times 10 \times 10 \times 10 = 100{,}000$$

In particular, 10 can be written as 10^1.

10^1 = 10 = 10

Power 1 — 1 ten only — 1 zero

Decimals, such as 0·1 or 0·01 can also be written as a power of 10; negative powers correspond to division by 10. Thus

$10^{-1} = \frac{1}{10} = 0{\cdot}1$ (1 divided by 10)
$10^{-2} = \frac{1}{100} = 0{\cdot}01$ (1 divided by 10 twice)
$10^{-3} = \frac{1}{1000} = 0{\cdot}001$ (1 divided by 10 three times)

10^{-1} is pronounced 10 to the minus 1

10^{-3} is pronounced 10 to the minus 3

The table below indicates the general pattern, but notice that there is a gap at 1.

10,000	1000	100	10	1	0·1	0·01	0·001	0·0001	0·0001
10^4	10^3	10^2	10^1		10^{-1}	10^{-2}	10^{-3}	10^{-4}	10^{-5}

This pattern suggests that

$$1 = 10^0$$

10^0 is pronounced 10 to the nought

This may seem peculiar, but remember it is just a notation, and it certainly completes the pattern.

Later on you will find that putting $10^0 = 1$ can be justified for mathematical reasons as well.

Using this notation, numbers can be expressed in a particularly neat form, as illustrated by the following examples:

$$250 = 2{\cdot}5 \times 10^2$$
$$25 = 2{\cdot}5 \times 10^1$$
$$2{\cdot}5 = 2{\cdot}5 \times 10^0$$
$$0{\cdot}25 = 2{\cdot}5 \times 10^{-1}$$

$10^2 = 100$
$10^1 = 10$
$10^0 = 1$
$10^{-1} = \frac{1}{10}$ (or divide by 10)

In general

any number = number between 1 and 10 × power of 10

A number expressed in this form is said to be in *scientific notation.* Any decimal number can be converted to scientific notation.

Scientific notation is particularly useful for handling very small numbers and very large numbers.

EXAMPLE

Convert 643·8 to scientific notation.

SOLUTION

$643{\cdot}8 = 6{\cdot}438 \times 10^?$

First write down the number between 1 and 10 and then work out the power.

Starting with 6·438 we need to move the decimal point two places right to get 643·8:

$6{\cdot}438 = 643{\cdot}8$

That is the same as multiplying by 100 or 10^2.
So $643{\cdot}8 = 6{\cdot}438 \times 10^2$.

TRY SOME YOURSELF

5 Convert each of the following numbers to scientific notation:
(i) 4942·1 (ii) 32 (iii) 734,000 (iv) 200,000.

First write down the number between 1 and 10; then work out the power.

Numbers less than 1 can be converted to scientific notation in the same way.

EXAMPLE

Convert 0·632 to scientific notation.

SOLUTION

$0{\cdot}632 = 6{\cdot}32 \times 10^?$

Starting with 6·32 we need to move the decimal point one place to the left to get 0·632.

So $0{\cdot}632 = 6{\cdot}32 \times 10^{-1}$.

Moving the decimal point one place left means dividing by 10 or multiplying by $\frac{1}{10}$ or 10^{-1}.

TRY SOME YOURSELF

6 Convert each of the following numbers to scientific notation:
(i) 0·49 (ii) 0·0062 (iii) 0·0004 (iv) 0·0000101.

First write down the number between 1 and 10; then work out the power.

It is also useful to be able to convert numbers from scientific notation to decimal.

EXAMPLE

Convert $7{\cdot}01 \times 10^{-3}$ to decimal notation.

SOLUTION

We need to multiply $7{\cdot}01 \times 10^{-3}$.

$10^{-3} = \frac{1}{1000}$

That is the same as dividing by 10 three times or moving the decimal point 3 places left:

$7{\cdot}01 \longrightarrow 0{\cdot}00701$

TRY SOME YOURSELF

7 Convert each of the following numbers to decimal notation:
(i) $3{\cdot}4 \times 10^2$ (ii) $7{\cdot}06 \times 10^{-4}$
(iii) $1{\cdot}49 \times 10^3$ (iv) $1{\cdot}3 \times 10^{-5}$

Should you move the decimal point right . . . or left? . . . and how many places?

The power notation provides a quick method for multiplying or dividing by 10, 100, 1000 etc.

For example, consider 100 x 1000 = 100,000. Writing these numbers as powers of 10, we get

$$10^2 \times 10^3 = 10^5$$

We add the powers:
$2 + 3 = 5$

Does this rule always work? Try multiplying

10 x 1000

and 1000 x 1000

and 10,000 x 1000.

$10^1 \times 10^3 = 10^4$
$10^3 \times 10^3 = 10^6$
$10^4 \times 10^3 = 10^7$

So, to multiply two powers of 10 you need only add the powers.

$10^6 \times 10^4 = 10^{(6+4)} = 10^{10}$

Similarly, to divide two powers of 10 you just subtract the second power from the first. Check this rule by working through the following:

$10^6 \div 10^4 = 10^{(6-4)} = 10^2$

10,000 ÷ 100 and $10^4 \div 10^2$

1000 ÷ 100,000 and $10^3 \div 10^5$

Check your answers using your calculator.

TRY SOME YOURSELF

8 Without using a calculator, evaluate each of the following:
(i) $10^1 \times 10^4$ (ii) $10^2 \times 10^3$ (iii) $10^0 \times 10^5$
(iv) $10^3 \div 10$ (v) $10^4 \div 10^6$ (vi) $10^2 \div 10^2$

Check the answers by writing the numbers out in full.

What answer do you expect from

$10 \div 0{\cdot}01$?

Do you expect a number smaller than 10 or a number bigger than 10? Try it on your calculator and see what happens.

It is difficult to anticipate answers when dividing by numbers less than 1. And this is one case where writing the numbers in scientific notation provides a routine method which avoids any confusion. Thus

$10 \div 0{\cdot}01$ is the same as $10^1 \div 10^{-2}$

$10 = 10^1$
$0{\cdot}01 = \frac{1}{100} = 10^{-2}$

and

$$10^1 \div 10^{-2} = 10^{(1-(-2))} = 10^{(1+2)} = 10^3$$

$10^3 = 1000$. Is this the answer you expected?

TRY SOME YOURSELF

9 Without using a calculator, evaluate each of the following:
(i) $10^3 \times 10^{-4}$ (ii) $10^2 \div 10^{-5}$ (iii) $10^{-3} \div 10^{-4}$

Think about these questions first. Do you expect a large or small answer?

TRY SOME YOURSELF

10 Use powers of ten to evaluate each of the following:
(i) 100×1000 (ii) $10 \div 1000$ (iii) $1000 \times 0{\cdot}01$
(iv) $0{\cdot}01 \times 0{\cdot}01$ (iv) $0{\cdot}01 \div 0{\cdot}01$.

First write the numbers as powers of ten. Add or subtract the powers, then rewrite the answer as a decimal.

1.3(iv) ACCURACY

Using a calculator often gives a long string of digits. For example

[1] [÷] [3] [=] [0·3333333]

But very often, for practical purposes, this level of accuracy is too precise to be useful. You try measuring 0·3333333 metres!

Decimal numbers can be rounded just like whole numbers. For example, it's easy to measure 0·3 m. Here the number has been rounded so that it contains one decimal digit; it has been *rounded to one decimal place.*

The number has been rounded to the nearest 0·1.

In practice, it's more accurate to measure to the nearest centimetre, in which case

0·3333333 m rounds to 0·33 m.

Here the number contains two decimal digits, it has been *rounded to two decimal places.*

The number has been rounded to the nearest 0·01.

EXAMPLE

Round 43·56277 to
(i) 2 decimal places (ii) 3 decimal places (iii) the nearest whole number

SOLUTION

(i) The answer must contain two decimal digits so we need to look at the third decimal digit and round up or down accordingly.

43·56277 — 2 is less than 5 so round down

43·56277 is nearer to 43·56 than it is to 43·57.

Thus 43·56277 rounds to 43·56.

(ii) Look at the fourth decimal digit. 7 is more than 5 so round up. Thus 43·56277 rounds to 43·563.

43·56277 — 7 is more than 5 so round up

(iii) To round to the nearest whole number we need to look at the first decimal digit. Thus

43·56277 rounds to 44.

The first decimal digit is 5, so round up.

TRY SOME YOURSELF

11 Round 3·0571023
(i) to 1 decimal place (ii) to 2 decimal places
(iii) to 3 decimal places (iv) to the nearest whole number
(v) to 5 decimal places.

Alternatively, answers can be rounded off to a number of *significant figures*. The best way to do this is to first write the number in scientific notation.

EXAMPLE

Round 215·32 to (i) 3 significant figures (ii) 2 significant figures.

SOLUTION

$215{\cdot}32 = 2{\cdot}1532 \times 10^2$

First write the number in scientific notation.

(i) Now, to round to 3 significant figures we count three digits from the *left* and round to that digit.

2·1532

Since the next digit is 3, round down to 2·15. The final answer contains ***three*** *digits.*

Thus 215·32 rounds to $2{\cdot}15 \times 10^2$.

(ii) To round to 2 significant figures we round to the second digit from the left.

Thus $2{\cdot}1532 \times 10^2$ rounds to $2{\cdot}2 \times 10^2$.

Since the next digit is 5, round up to 2·2. The final answer contains ***two*** *digits.*

This procedure makes it clearer how to deal with numbers less than one than it might otherwise have been.

EXAMPLE

Round 0·034921 to two significant figures.

SOLUTION

$$0{\cdot}034921 = 3{\cdot}4921 \times 10^{-2}$$

3·4921 rounds to 3·5.

Round to the second digit from the left. The answer contains two digits.

Thus 0·034921 rounds to $3{\cdot}5 \times 10^{-2}$.

The number can be rewritten as a decimal, in which case 0·034921 rounds to 0·035.

TRY SOME YOURSELF

12(i) Round 45,901 to 2 significant figures.
(ii) Round 0·049 to 1 significant figure.
(iii) Round 0·4003 to 3 significant figures.

1.3(v) DECIMAL CALCULATIONS

Now that we have outlined the principles of scientific notation (so that an estimate of the answer can be made) and rounding (so that we don't need to give all the answers correct to 7 decimal places), we can go on to look at some calculations involving decimals–and this time we *will* include division.

Addition and subtraction

Some decimal calculations can be treated in the same way as whole numbers. The next examples illustrate how useful scientific notation can be when estimating the order of magnitude.

EXAMPLE

Evaluate 38·5 + 49·6 − 76·2.

SOLUTION

Estimate first. Writing the numbers in scientific notation we get

$$38{\cdot}5 = 3{\cdot}85 \times 10 \simeq 4 \times 10$$
$$49{\cdot}6 = 4{\cdot}96 \times 10 \simeq 5 \times 10$$
$$76{\cdot}2 = 7{\cdot}62 \times 10 \simeq 8 \times 10$$

Now $(4 \times 10) + (5 \times 10) - (8 \times 10) = 10$.

In each case we have rounded the number between 1 and 10 to the nearest whole number:

$$(4 \times 10) + (5 \times 10) - (8 \times 10)$$
$$= (4 + 5 - 8) \times 10 = 1 \times 10$$

A key sequence is

[38·5] [+] [49·6] [−] [76·2] [=] [11·9]

The answer is reasonably close to our estimate (10), so we aren't likely to have made a mistake.

TRY SOME YOURSELF

13 Evaluate each of the following:
(i) 3·8624 + 2·6793 − 1·2134
(ii) 493·2 + 597·63 − 212·7
(iii) 0·64 − 0·21 + 0·43.

Use your calculator but remember to estimate the answer first—just in case you make a mistake.

It's all very well when each number involves the same power of 10; but what happens if the estimate gives something like

$$(5 \times 10^4) + (7 \times 10^3) - (2 \times 10^4)?$$

Generally, since we just want to estimate the order of magnitude, we need only consider the highest powers of 10. So here we need only consider

$$(5 \times 10^4) - (2 \times 10^4) = 3 \times 10^4$$

Although, if the calculation is very long you may need to be more cautious. This method works reasonably well if you have to add or subtract 3 or 4 numbers.
10^4 is a higher power than 10^3, so we can ignore 7×10^3.

EXAMPLE

Evaluate 0·003102 − 0·00147 + 0·00004.

SOLUTION

Estimate first:

$$0{\cdot}00312 = 3{\cdot}102 \times 10^{-3} \simeq 3 \times 10^{-3}$$
$$0{\cdot}00147 = 1{\cdot}47 \times 10^{-3} \simeq 1 \times 10^{-3}$$
$$0{\cdot}00004 = 4 \times 10^{-5}$$
$$(3 \times 10^{-3}) - (1 \times 10^{-3}) + (4 \times 10^{-5}) \simeq (2 \times 10^{-3}) = 0{\cdot}002$$

Ignoring (4×10^{-5}), we need only consider $(3 \times 10^{-3}) - (1 \times 10^{-3})$.

A key sequence is

[0·003102] [−] [0·00147] [+] [0·00004] [=] [0·001672]

Again, the estimate gives a reasonable approximation and we are unlikely to have made any mistakes.

TRY SOME YOURSELF

14 Evaluate each of the following. In each case, round your answer to two significant figures.
(i) 0·032 + 21·4 (ii) 0·0043 − 0·00043
(iii) 149·7 − 18·6 + 192·8

Use your calculator but estimate the order of magnitude of the answer first.

Your calculator may have a key which converts numbers into scientific notation. It might be labelled [EE] or [Exp] since scientific notation is sometimes called *exponential notation*.

Check in the maker's handbook how to convert into scientific notation using your calculator.

Your calculator may also be able to evaluate calculations in scientific notation. Now is a good time to investigate this facility on your calculator.

Multiplication and division

The next example indicates how to estimate the order of magnitude in a calculation involving division.

EXAMPLE

Evaluate 546·3 ÷ 0·031.

SOLUTION

Estimate first:

$546{\cdot}3 = 5{\cdot}463 \times 10^2 \simeq 5 \times 10^2$

$0{\cdot}031 = \quad 3{\cdot}1 \times 10^{-2} \simeq 3 \times 10^{-2}$

Now we can rearrange the expression

$(5 \times 10^2) \div (3 \times 10^{-2})$

to put the digit part, $(5 \div 3)$, first and the powers of 10, $(10^2 \div 10^{-2})$, second. Thus

$(5 \times 10^2) \div (3 \times 10^{-2}) = (5 \div 3) \times (10^2 \div 10^{-2})$

$5 \div 3 \simeq 2$ and $10^2 \div 10^{-2} = 10^{2-(-2)} = 10^4$

Since

$(5 \times 10^2) \div (3 \times 10^{-2})$

$= \frac{5 \times 10^2}{3 \times 10^{-2}} = (\frac{5}{3}) \times (\frac{10^2}{10^{-2}})$

$= (5 \div 3) \times (10^2 \div 10^{-2})$

So that

$(5 \div 3) \times (10^2 \div 10^{-2}) \simeq 2 \times 10^4 = 20{,}000$

A key sequence is

[546·3] [÷] [0·031] [=] [17622·581]

Notice that the answer is larger than the first number because we are dividing by a number less than one.

Again, the answer is quite close to the estimate, indicating that we're unlikely to have made a mistake.

You may think that the rearrangement of the expression

$(5 \times 10^2) \div (3 \times 10^{-2})$

was rather peculiar. For the moment we leave this without further explanation but the reasons should be made more clear after you have worked through Section 1.4, which discusses fractions.

For the moment, just follow the method outlined in the example.

TRY SOME YOURSELF

15 Use the method outlined in the example above to evaluate each of the following. In each case, round your answer to three significant figures.
(i) 593 ÷ 27 (ii) 5·367 × 819·3 (iii) 1392 ÷ 34
(iv) 0·014 × 0·0092

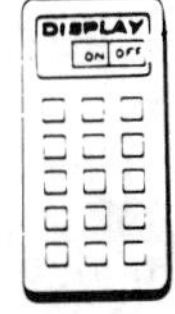

Make a rough estimate first.

Decimal calculations often involve brackets. As with whole numbers you should always evaluate the brackets first.

EXAMPLE

Evaluate 472·9 × (6·321 + 40·62).

SOLUTION

Estimate first:

$6{\cdot}321 + 40{\cdot}62 \simeq (6 \times 10^0) + (4 \times 10^1)$

$\simeq 4 \times 10^1$

Evaluate the brackets first.

$472{\cdot}9 \simeq 5 \times 10^2$

$(5 \times 10^2) \times (4 \times 10^1) = (5 \times 4) \times (10^2 \times 10^1)$

$= 20 \times 10^3$

$= 2 \times 10^4$

$= 20{,}000$

Complete the estimation.

A key sequence is

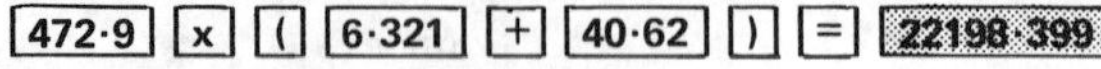

22,198·399 ≏ 20,000, so the answer is probably right.

TRY SOME YOURSELF

16 Evaluate each of the following. In each case, round your answer to three decimal places.
(i) 670 + (6·32 x 4·6) (ii) 0·012 ÷ (0·013 + 0·004)
(iii) (14·6 + 15·21) x (17·02 − 206)

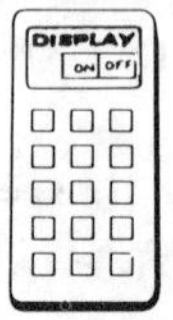

Make a rough estimate first.

After you have worked through this section you should be able to

a Understand the meaning of place value
b Multiply or divide any decimal number by 10, 100, 1000 etc. without using a calculator
c Convert metric measurements from one unit to another by multiplying or dividing by 10, 100, 1000 etc.
d Convert any decimal number to scientific notation and vice versa
e Multiply or divide powers of 10 by adding or subtracting the powers
f Round any decimal number to a given number of decimal places
g Round any decimal number to a given number of significant figures
h Add, subtract, multiply or divide decimal numbers using a calculator, estimating the answer first

Finally, here are some exercises if you want more practice.

TRY SOME MORE YOURSELF

17(i) For each of the following pairs of numbers indicate the larger number:
(a) 0·032; 0·2 (b) 36·721; 36·076 (c) 0·0045; 0·0005.

(ii) Evaluate each of the following without using a calculator:
(a) 541·632 x 100 (b) 7·29 x 10 (c) 0·0031 x 100
(d) 0·00001 x 1000 (e) 541·632 ÷ 100 (f) 7·29 ÷ 10
(g) 0·0031 ÷ 100 (h) 0·342 ÷ 1000.

(iii) Rewrite
(a) 549 g in kilograms (b) 67 cm in metres
(c) 14·5 mm in metres (d) 3·71 kg in grams.

(iv) Write down each of the following as a power of 10:
(a) 0·1 (b) 1000 (c) 1 (d) 10 (e) 0·00001 (f) 0·00100.

(v) Convert each of the following numbers to scientific notation:
(a) 14·952 (b) 178 (c) 0·034 (d) 0·000100 (e) 0·013.

(vi) Convert each of the following numbers to decimal notation:
(a) 3×10^4 (b) $2{\cdot}1 \times 10^{-2}$ (c) $1{\cdot}97 \times 10^3$ (d) $1{\cdot}01 \times 10^{-5}$.

(vii) Evaluate each of the following without using a calculator:
(a) $10^2 \times 10^3$ (b) $10^{-2} \times 10^2$ (c) $10^{-4} \times 10^{-2}$
(d) $10^2 \times 10^4$ (e) $10^3 \div 10$ (f) $10^1 \div 10^{-1}$ (g) $10^3 \div 10^{-1}$
(h) $10^{-2} \div 10^{-3}$.

(viii) Calculate and in each case give an estimate for
(a) $14{\cdot}3 + 17{\cdot}9$ (b) $149{\cdot}341 - 21{\cdot}75$ (c) $149{\cdot}72 + 1{\cdot}32$
(d) $0{\cdot}0032 + 0{\cdot}01 - 0{\cdot}002$.

(ix) Calculate and in each case give an estimate for
(a) $17{\cdot}9 \times 123{\cdot}81$ (b) $149 \div 163{\cdot}342$ (c) $1927 \div 149$
(d) $0{\cdot}013 \times 0{\cdot}132$ (e) $0{\cdot}00123 \div (0{\cdot}013 + 0{\cdot}627)$.

(x) Round each of the answers to part (viii) to 1 decimal place.

(xi) Round each of the answers to part (ix) to 2 significant figures.

1.4 Fractions

TRY THESE QUESTIONS FIRST

1 A farmer ploughs 10 acres of a 30 acre field. What fraction remains unploughed?

2 Determine whether or not $\frac{64}{81}$ and $\frac{8}{9}$ are equivalent fractions.

3 Evaluate $\frac{11}{15} + \frac{2}{9} - \frac{2}{5}$.

4 Evaluate $\frac{8}{11} \times \frac{33}{48} \div \frac{3}{4}$.

5 Evaluate $6\frac{1}{3} \times 2\frac{2}{19} \div \frac{5}{9}$.

6 Convert $1\frac{5}{8}$ to a decimal.

1.4(i) WHAT IS A FRACTION?

You may think that in the age of the calculator and computer, fractions are no longer needed, but there are still times when it is necessary to use fractions.

For example, fractions are needed in the manipulation of algebraic expressions. See Module 2.

A fraction is part of a whole. A tape measure might be marked off in inches but it is still possible to measure lengths like $2\frac{1}{2}$ inches—halfway between 2 and 3.

Here are some more examples of fractions in everyday use:

$\frac{1}{2}$ lb butter

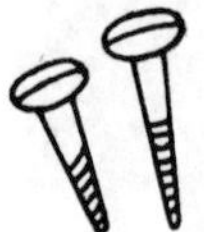

$1\frac{7}{8}''$ screws

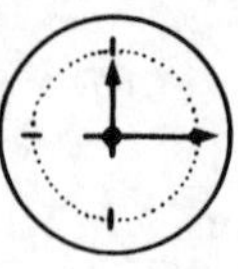

a quarter past twelve

Imperial units, such as inches, feet, pounds etc. are still measured using fractions.

You may find it useful to think of fractions using the following illustrations:

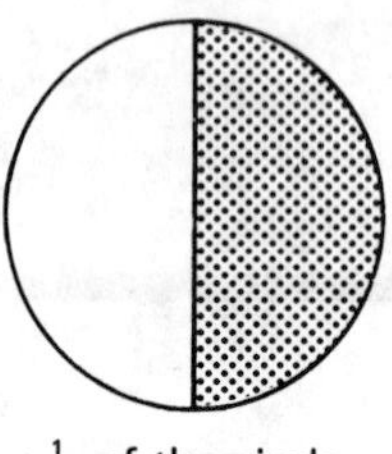

$\frac{1}{2}$ of the circle

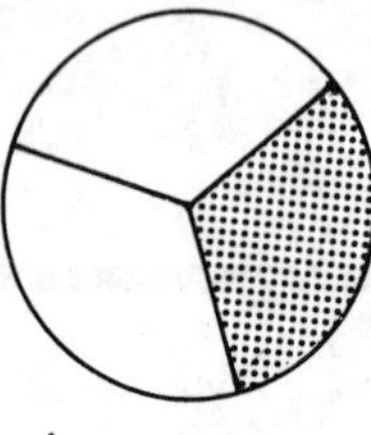

$\frac{1}{3}$ of the circle

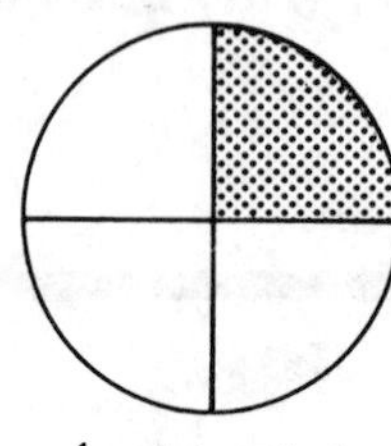

$\frac{1}{4}$ of the circle

Measurements in metric units are usually read as decimals. For example

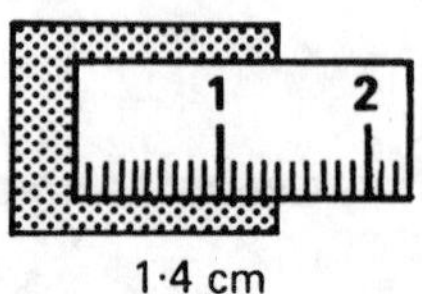

1·4 cm

Each of the circles above is divided into a number of equal pieces: the first into 2 equal pieces, the second into 3 equal pieces and the third into 4.

That is, the pieces have equal area.

The shaded part in the first circle is then $\frac{1}{2}$ of the circle;

the shaded part in the second circle is $\frac{1}{3}$ of the circle;

the shaded part in the third circle is $\frac{1}{4}$ of the circle.

Its area is $\frac{1}{2}$ of the area of the whole circle.

CHECK YOUR ANSWERS

1 20 acres remains unploughed. That is, $\frac{20}{30}$ or $\frac{2}{3}$ of the field. *Section 1.4(i)*

2 Equivalent fractions for $\frac{8}{9}$ are *Section 1.4(ii)*

$$\frac{2 \times 8}{2 \times 9} = \frac{16}{18}; \frac{3 \times 8}{3 \times 9} = \frac{24}{27}; \dots; \frac{9 \times 8}{9 \times 9} = \frac{72}{81}$$

So $\frac{64}{81}$ and $\frac{8}{9}$ are not equivalent.

3 $\frac{11}{15} + \frac{2}{9} - \frac{2}{5} = \frac{33 + 10 - 18}{45} = \frac{25}{45} = \frac{5}{9}$ *Section 1.4(iii)*

4 $\frac{8}{11} \times \frac{33}{48} \div \frac{3}{4} = \frac{8}{11} \times \frac{33}{48} \times \frac{4}{3} = \frac{2}{3}$ *Section 1.4(iv)*

5 $6\frac{1}{3} \times 2\frac{2}{19} \div \frac{5}{9} = \frac{19}{3} \times \frac{40}{19} \times \frac{9}{5} = 24$ *Section 1.4(v)*

6 $1\frac{5}{8} = 1{\cdot}625$ *Section 1.4(vi)*

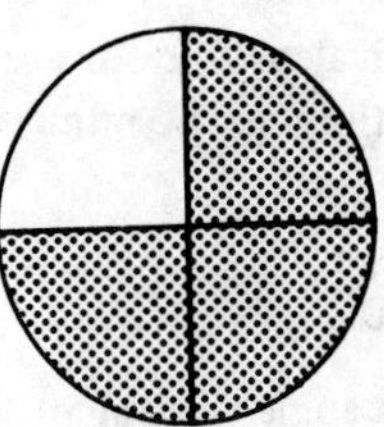

This time, we have shaded several of the equal pieces into which the circle is divided. The shaded area is $\frac{3}{4}$ of the whole circle.

Of course, it isn't necessary to use a circle; here are some sub-divisions of a rectangle:

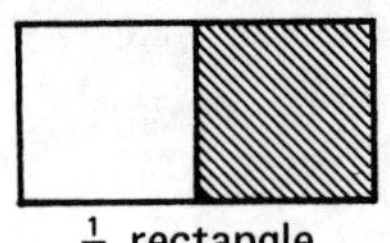

$\frac{1}{2}$ rectangle

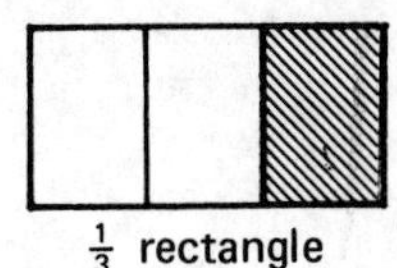

$\frac{1}{3}$ rectangle

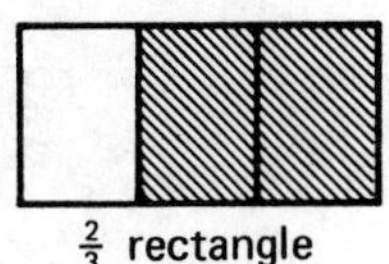

$\frac{2}{3}$ rectangle

The rectangle on the right has been divided into 3 equal parts and 2 parts are shaded.

$$\frac{2}{3} = \frac{\text{Number of shaded parts}}{\text{Total number of parts}} = \frac{\textit{Numerator}}{\textit{Denominator}}$$

The number on the top is called the ***numerator****.*
The number on the bottom is called the ***denominator****.*

TRY SOME YOURSELF

1(i) Write down each of the following measurements:

(a)

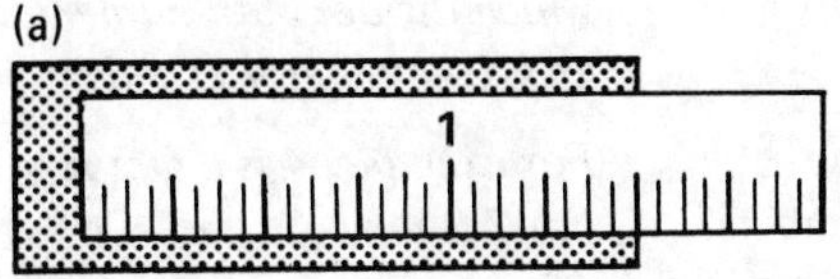

(b)

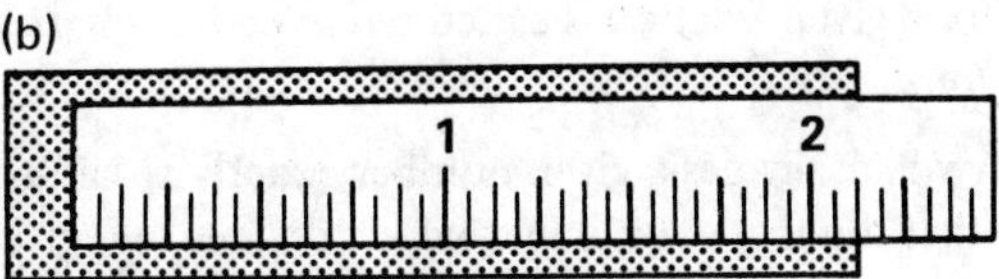

(ii) For each of the following diagrams indicate the fraction of the shape represented by the shaded parts:

(a)

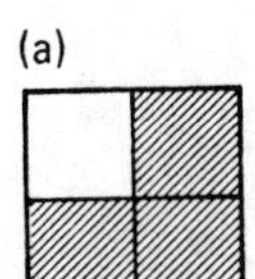

(b)

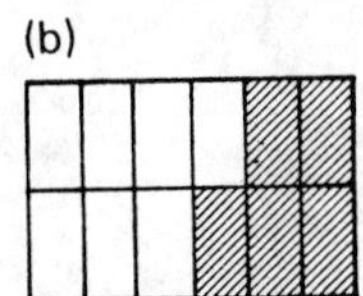

(c)

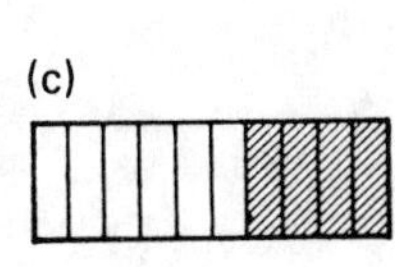

(d)

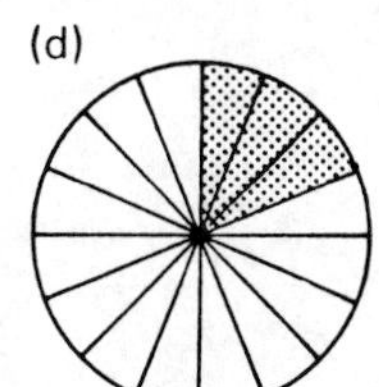

1.4(ii) EQUIVALENT FRACTIONS

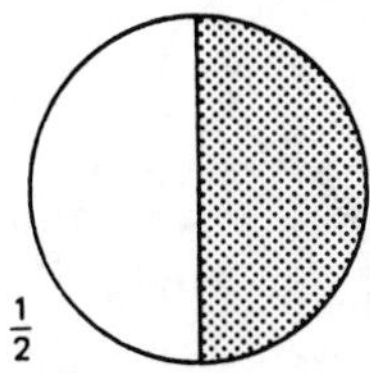

$\frac{1}{2}$

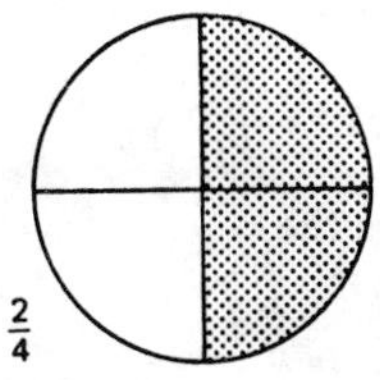

$\frac{2}{4}$

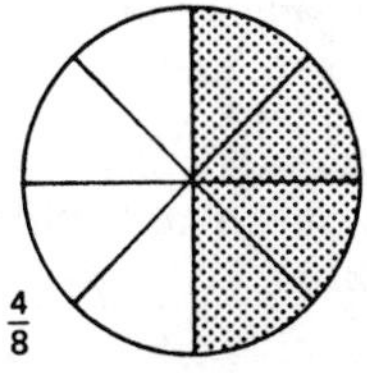

$\frac{4}{8}$

The fraction is given by

$$\frac{\textit{Number of shaded parts}}{\textit{Total number of parts}}$$

These are three different fractions, but the areas of the shaded parts are equal. So

$$\frac{1}{2} = \frac{2}{4} = \frac{4}{8}$$

Fractions which are equal but have different numerators and denominators are called *equivalent fractions*.

Equivalent fractions can be found for any given fraction by multiplying the numerator and denominator by the *same* whole number.

EXAMPLE

Find some equivalent fractions for $\frac{1}{3}$.

SOLUTION

$\frac{2 \times 1}{2 \times 3} = \frac{2}{6}$

$\frac{4 \times 1}{4 \times 3} = \frac{4}{12}$

$\frac{4 \times 2}{4 \times 6} = \frac{8}{24}$

Multiply top and bottom by the ***same*** *whole number.*

$\frac{4 \times 2}{4 \times 6} = \frac{8 \times 1}{8 \times 3} = \frac{8}{24}$

$\frac{1}{3}, \frac{2}{6}, \frac{4}{12}, \frac{8}{24}$ are all equivalent fractions.

Equivalent fractions can also be obtained by *dividing* the numerator and denominator by the same whole number.

Divide top and bottom by the ***same*** *whole number.*

$\frac{24}{32} = \frac{24 \div 4}{32 \div 4} = \frac{6}{8} = \frac{6 \div 2}{8 \div 2} = \frac{3}{4}$

$\frac{3}{4}$ is the simplest fraction equivalent to $\frac{24}{32}$ because there is no whole number which divides both 3 and 4. Finding the simplest equivalent fraction for a given fraction is called *reducing the fraction to its simplest form.*

A fraction is in its simplest form when there is no whole number which divides ***both*** *numerator and denominator exactly.*

A number which divides a given number exactly is said to be a *factor* of that number. A number which divides two given numbers exactly is a *common factor* of these two numbers. So to reduce a fraction to its simplest form it's necessary to look for common factors of the numerator and denominator.

For example, 4 is a factor of 24 but 7 is not.
4 is a ***common*** *factor of 24 and 32.*

EXAMPLE

Reduce $\frac{48}{60}$ to its simplest form.

SOLUTION

4 is a common factor of 48 and 60. So

$\frac{48}{60} = \frac{48 \div 4}{60 \div 4} = \frac{12}{15}$

Look for common factors of 48 and 60. 4 divides 48 exactly and it divides 60 exactly.

But we are not yet finished since 3 is a common factor of 12 and 15. Thus

$\frac{48}{60} = \frac{12}{15} = \frac{12 \div 3}{15 \div 3} = \frac{4}{5}$

4 and 5 have no common factors so we can go no further; $\frac{4}{5}$ is the simplest form of $\frac{48}{60}$.

TRY SOME YOURSELF

2(i) Which of the following are equivalent fractions for $\frac{2}{5}$:
(a) 4/5 (b) 4/10 (c) 4/15 (d) 10/25?

4/5 is an alternative way of writing $\frac{4}{5}$.

(ii) Which of the following are pairs of equivalent fractions:
(a) $\frac{2}{7}; \frac{4}{28}$ (b) $\frac{3}{16}; \frac{6}{8}$ (c) $\frac{4}{6}; \frac{2}{3}$?

(iii) Reduce each of the following fractions to its simplest form:
(a) $\frac{15}{30}$ (b) $\frac{15}{60}$ (c) $\frac{25}{100}$ (d) $\frac{54}{72}$.

1.4(iii) MANIPULATING FRACTIONS: ADDING AND SUBTRACTING

To add two fractions with the same denominator is fairly straightforward; you just add the numerators. For example,

$$\frac{3}{8} + \frac{2}{8} = \frac{3+2}{8} = \frac{5}{8}$$

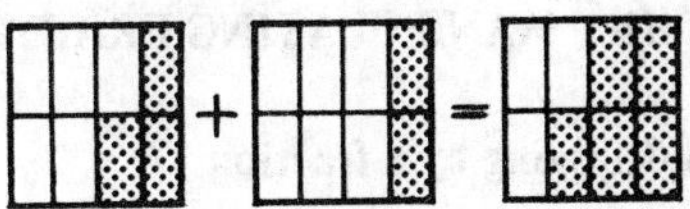

Similarly, if the denominators are the same, then to subtract one fraction from another you need only subtract one numerator from the other. For example,

$$\frac{5}{12} - \frac{3}{12} = \frac{5-3}{12} = \frac{2}{12}$$

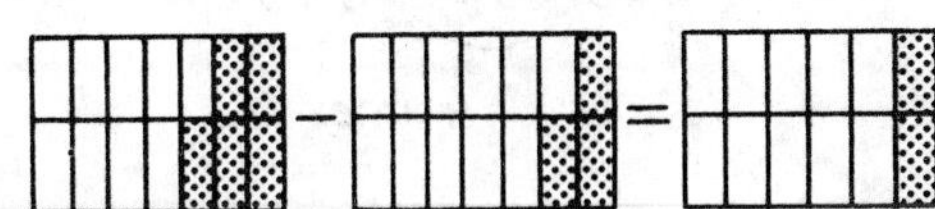

However, if the denominators are different then addition and subtraction are a bit more complicated.

The first thing to do is to make the denominators the same. This is called *finding a common denominator.* It can be done by considering equivalent fractions, as the following examples illustrate.

EXAMPLE

Evaluate $\frac{1}{4} + \frac{3}{8}$.

SOLUTION

We must first make the denominators the same. Since $\frac{1}{4} = \frac{2}{8}$, we get

$$\frac{1}{4} + \frac{3}{8} = \frac{2}{8} + \frac{3}{8} = \frac{2+3}{8} = \frac{5}{8}.$$

8 is a common denominator. When the fractions have a common denominator we can add or subtract them as before.

The next example involves subtraction.

EXAMPLE

Evaluate $\frac{9}{16} - \frac{2}{5}$.

SOLUTION

We must first make the denominators the same. Some equivalent fractions are

$$\frac{9}{16} = \frac{18}{32} = \frac{27}{48} = \frac{36}{64} = \mathbf{\frac{45}{80}} = \frac{54}{96}$$

$$\frac{2}{5} = \frac{4}{10} = \frac{8}{20} = \mathbf{\frac{32}{80}} = \frac{40}{100}$$

In this case, 80 is a common denominator, so

$$\frac{9}{16} - \frac{2}{5} = \frac{45-32}{80} = \frac{13}{80}$$

It's not often easy to find the common denominator. Sometimes you can just spot it; occasionally it's a question of trial and error. Alternatively you can find a common denominator by multiplying all the denominators together. This always works but it does mean that the numbers can become rather complicated.

TRY SOME YOURSELF

3 Evaluate each of the following. In each case give the answer in its simplest form.

(i) $\frac{3}{10}+\frac{5}{10}$ (ii) $\frac{11}{32}-\frac{7}{32}$ (iii) $\frac{7}{16}-\frac{2}{16}$ (iv) $\frac{1}{3}+\frac{1}{6}$ (v) $\frac{8}{9}+\frac{1}{5}$
(vi) $\frac{5}{16}+\frac{1}{3}-\frac{1}{6}$

1.4(iv) MANIPULATING FRACTIONS: MULTIPLYING AND DIVIDING

Multiplying by a fraction

Multiplication by a whole number is really just repeated addition:

$$4 \times 3 = \underbrace{3+3+3+3}_{\text{4 times}}$$

In the same way

$$4 \times \frac{3}{5} = \underbrace{\frac{3}{5}+\frac{3}{5}+\frac{3}{5}+\frac{3}{5}}_{\text{4 times}} = \frac{12}{5}$$

Just add the numerators since the denominators are the same.

You may have noticed that the same answer is obtained by multiplying:

$$4 \times \frac{3}{5} = \frac{4 \times 3}{5} = \frac{12}{5}$$

In fact *any* two fractions can be multiplied together by multiplying the numerators and multiplying the denominators. For example,

$$\frac{3}{4} \times \frac{2}{3} = \frac{3 \times 2}{4 \times 3} = \frac{6}{12}$$

Also

$$\frac{2}{7} \times \frac{3}{5} = \frac{2 \times 3}{7 \times 5} = \frac{6}{35}$$

Sometimes you may be asked to find, for example,

$\frac{1}{2}$ of 16 or $\frac{3}{4}$ of $\frac{1}{2}$

The 'of' just means multiply.

So $\frac{1}{2}$ of 16 means

$\frac{1}{2} \times 16 = 8$

and $\frac{3}{4}$ of $\frac{1}{2}$ means

$\frac{3}{4} \times \frac{1}{2} = \frac{3}{8}$

The principle is quite straightforward; now we introduce a quick way to multiply fractions which makes the multiplication easier. For example, consider

$$\frac{6}{15} \times \frac{5}{9} = \frac{6 \times 5}{15 \times 9} = \frac{30}{135} = \frac{2}{9}$$

Multiplying 15 by 9 is not the easiest of calculations. However, instead of multiplying out the numerator and denominator we could have looked for some common factors first. For example, 5 is a common factor of both numerator and denominator, so that

$$\frac{6 \times \cancel{5}^{1}}{{}_{3}\cancel{15} \times 9} = \frac{6 \times 1}{3 \times 9}$$

3 is also a common factor so the calculation can be simplified still further:

$$\frac{{}^{2}\cancel{6} \times 1}{{}_{1}\cancel{3} \times 9} = \frac{2 \times 1}{1 \times 9} = \frac{2}{9}$$

This makes the multiplication much easier at the end of the calculation.

This process, of simplifying the numbers first by finding common factors, is called *cancelling*.

EXAMPLE

Evaluate $\frac{4}{49} \times \frac{7}{16}$.

SOLUTION

Cancelling gives

$$\frac{\overset{1}{\cancel{4}}}{_7\cancel{49}} \times \frac{\overset{1}{\cancel{7}}}{\cancel{16}_4} = \frac{1}{7 \times 4} = \frac{1}{28}$$

4 is a common factor; so is 7.

TRY SOME YOURSELF

4 Evaluate each of the following. In each case give the answer in its simplest form.
(i) $\frac{11}{12} \times \frac{3}{11}$ (ii) $\frac{5}{6} \times \frac{9}{10}$ (iii) $\frac{17}{33} \times \frac{12}{34} \times \frac{3}{4}$
(iv) Find $\frac{3}{4}$ of 2. (v) Find $\frac{4}{5}$ of $\frac{1}{2}$.
(vi) A school contains 720 pupils. $\frac{5}{8}$ are boys and $\frac{3}{8}$ are girls. Give the number of boys and girls.

Dividing by a fraction

What is the meaning of $6 \div \frac{1}{2}$?

Do you expect the answer to be bigger, or smaller than 6?

Again there is an analogy with division by whole numbers.

$6 \div 2$ asks for the number of twos in 6; $6 \div 2 = 3$, since $2 \times 3 = 6$.

In the same way, $6 \div \frac{1}{2}$ is asking for the number of $\frac{1}{2}$s in 6. Think of 6 as 6 whole rectangles.

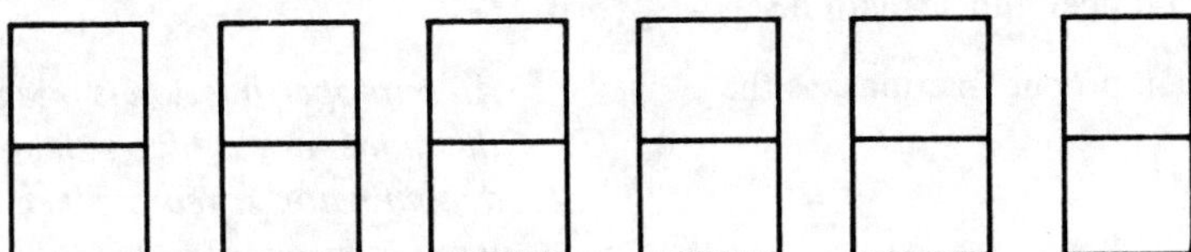

Each rectangle contains two half-rectangles, and 6 rectangles contain 6×2 half-rectangles. So

$$6 \div \tfrac{1}{2} = 6 \times 2 = 12$$

$1 = 2 \times \frac{1}{2}$.
$6 = 6 \times 1 = 6 \times 2 \times \frac{1}{2}$.

Similarly $\frac{1}{2} \div \frac{1}{4}$ is asking for the number of $\frac{1}{4}$s in $\frac{1}{2}$. Think of half a rectangle:

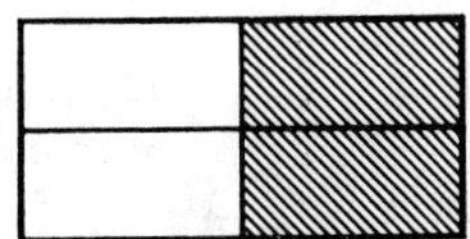

This contains two $\frac{1}{4}$ rectangles. So

$$\tfrac{1}{2} \div \tfrac{1}{4} = 2.$$

In fact

$$\tfrac{1}{2} \div \tfrac{1}{4} = \tfrac{1}{2} \times 4 = 2.$$

Each whole rectangle contains 4 ($\frac{1}{4}$ rectangles).
So $\frac{1}{2}$ a whole rectangle contains $\frac{1}{2} \times 4$($\frac{1}{4}$ rectangles).

This suggests a rule for dividing by a fraction.

To divide by a fraction, turn the fraction upside down and multiply.

EXAMPLE

Evaluate $\frac{1}{3} \div \frac{1}{6}$.

SOLUTION

Using the rule above

$$\frac{1}{3} \div \frac{1}{6} = \frac{1}{3} \times \frac{6}{1} = 2.$$

Turning the fraction upside down is called *finding its reciprocal.*

The reciprocal of 6 is $\frac{1}{6}$.
The reciprocal of $\frac{5}{6}$ is $\frac{6}{5}$.

TRY SOME YOURSELF

5 Evaluate each of the following. In each case give the answer in its simplest form.
(i) $10 \div \frac{1}{2}$ (ii) $\frac{2}{5} \div \frac{5}{7}$ (iii) $\frac{16}{9} \div \frac{4}{3}$ (iv) $\frac{14}{8} \div \frac{7}{16}$

1.4(v) MANIPULATING MIXED NUMBERS

So far we've concentrated on simple fractions (fractions where the numerator is smaller than the denominator; fractions less than one). We now consider how to manipulate mixed numbers, numbers like $3\frac{3}{4}$, a combination of a whole number and a fraction.

Mixed numbers can be turned into improper fractions, as the following example illustrates.

An improper fraction is one where the numerator is bigger than the denominator; it represents a number bigger than 1.

EXAMPLE

Rewrite $3\frac{3}{4}$ as an improper fraction.

SOLUTION

We first look at the whole number 3.

$$3 = 3 \times 1 = 3 \times \frac{4}{4} = \frac{12}{4}$$

We want to write 3 as a fraction with denominator 4.

Now

$$3\frac{3}{4} = 3 + \frac{3}{4} = \frac{12}{4} + \frac{3}{4} = \frac{15}{4}$$

15 is bigger than 4, so $\frac{15}{4}$ is an improper fraction.

Improper fractions can also be converted to mixed numbers. For example, consider $\frac{12}{7}$. Remember that $\frac{12}{7}$ is another way of writing $12 \div 7$. Now $12 \div 7 = 1$ remainder 5. This gives

$$\frac{12}{7} = 1\frac{5}{12}$$

The remainder gives the fraction part of the mixed number.

TRY SOME YOURSELF

6(i) Convert each of the following to improper fractions:
(a) $4\frac{1}{8}$ (b) $2\frac{2}{5}$ (c) $3\frac{3}{11}$ (d) $2\frac{15}{32}$.

(ii) Convert each of the following to mixed numbers:
(a) $\frac{14}{3}$ (b) $\frac{19}{5}$ (c) $\frac{39}{6}$ (d) $\frac{48}{42}$.

You might find it useful to reduce the fractions to their simplest form first.

The next example illustrates how to add or subtract mixed numbers.

EXAMPLE

Evaluate $3\frac{3}{4} + 2\frac{5}{8}$.

SOLUTION

$$\begin{aligned} 3\tfrac{3}{4} + 2\tfrac{5}{8} &= 3 + 2 + \tfrac{3}{4} + \tfrac{5}{8} \\ &= 5 + \tfrac{3}{4} + \tfrac{5}{8} \\ &= 5 + \tfrac{6}{8} + \tfrac{5}{8} \\ &= 5 + \tfrac{11}{8} \\ &= 5 + 1\tfrac{3}{8} \\ &= 6\tfrac{3}{8} \end{aligned}$$

First add or subtract the whole number part

Add or subtract the fraction parts, reducing the answer to its simplest form.

If necessary, change the improper fraction to a mixed number before recombining the parts.

TRY SOME YOURSELF

7 Use the method outlined above to evaluate each of the following:
(i) $1\frac{3}{8} + 2\frac{1}{4}$ (ii) $2\frac{2}{3} + 4\frac{3}{5}$ (iii) $2\frac{5}{16} - 1\frac{1}{8}$ (iv) $4\frac{1}{8} - 1\frac{3}{4}$

For multiplication or division, mixed numbers must *always* be converted to improper fractions, as shown by the next example.

EXAMPLE

Evaluate $1\frac{4}{7} \times 1\frac{2}{5} \div 1\frac{1}{10}$.

SOLUTION

$$1\tfrac{4}{7} = \tfrac{11}{7},\ 1\tfrac{2}{5} = \tfrac{7}{5},\ 1\tfrac{1}{10} = \tfrac{11}{10}.$$

First convert the mixed numbers to improper fractions.

Now

$$\begin{aligned} 1\tfrac{4}{7} \times 1\tfrac{2}{5} \div 1\tfrac{1}{10} &= \tfrac{11}{7} \times \tfrac{7}{5} \div \tfrac{11}{10} \\ &= \tfrac{\cancel{11}}{\cancel{7}} \times \tfrac{\cancel{7}}{\cancel{5}} \times \tfrac{\cancel{10}^{2}}{\cancel{11}} \\ &= 2. \end{aligned}$$

To divide by a fraction, turn it upside down and multiply.

Cancel out the common factors and multiply.

TRY SOME YOURSELF

8 Use the method outlined above to evaluate each of the following:
(i) $1\frac{1}{2} \times 2\frac{2}{3}$ (ii) $4\frac{1}{5} \div \frac{7}{10}$ (iii) $\frac{4}{5} \times 1\frac{7}{8}$ (iv) $2\frac{1}{4} \times 1\frac{1}{5} \div 1\frac{1}{8}$.

Calculations with brackets which involve fractions are treated in the same way as those with whole numbers; the brackets must be evaluated first.

EXAMPLE

Evaluate $(1\frac{7}{8} \div \frac{5}{6}) + 1\frac{2}{3}$.

SOLUTION

$$1\tfrac{7}{8} \div \tfrac{5}{6} = \tfrac{15}{8} \div \tfrac{5}{6} = \tfrac{15}{8} \times \tfrac{6}{5} = \tfrac{9}{4} = 2\tfrac{1}{4}$$

$$2\tfrac{1}{4} + 1\tfrac{2}{3} = (2 + 1) + \tfrac{1}{4} + \tfrac{2}{3}$$

$$= 3 + \tfrac{3+8}{12} = 3\tfrac{11}{12}$$

First work out the brackets, then complete the calculation.

TRY SOME YOURSELF

9 Evaluate each of the following:
(i) $(1\frac{1}{2} + \frac{3}{5}) \div 1\frac{2}{5}$ (ii) $2\frac{1}{4} + (4\frac{1}{4} \times \frac{2}{17})$ (iii) $(3\frac{1}{4} - 2\frac{1}{2}) \div (4\frac{1}{2} + 1\frac{1}{4})$.

1.4(vi) CONVERTING FRACTIONS TO DECIMALS

When you evaluate $6 \div 2$ you are trying to find the number which will give 6 when multiplied by 2.

$6 \div 2 = 3$ and $3 \times 2 = 6$

Now

$$\tfrac{4}{5} \times 5 = 4$$

This indicates that $\frac{4}{5}$ can be thought of as $4 \div 5$. According to the calculator (which works in decimals)

[4] [÷] [5] [=] [0·8]

$4 \div 5$ asks for the number which gives 4 when multiplied by 5. $\frac{4}{5} \times 5 = 4$, so $4 \div 5 = \frac{4}{5}$

So the decimal 0·8 represents the same number as the fraction $\frac{4}{5}$. (You can see that this is right if you recall that the place immediately after the decimal point stands for tenths: thus $0{\cdot}8 = \frac{8}{10} = \frac{4}{5}$.)

Every fraction has a decimal equivalent, although the decimal may not always be as short as 0·8. For example, although the decimal equivalent of $\frac{5}{6}$ is 0·8333333, according to a calculator with an 8 figure display, even this is not entirely accurate.

Try this yourself.

Converting fractions to decimals is easy when you use a calculator. The following example indicates how to convert a mixed number to a decimal.

EXAMPLE

Convert $1\frac{5}{6}$ to a decimal.

SOLUTION

$$1\tfrac{5}{6} = 1 + \tfrac{5}{6}$$

Rather than evaluate $1 + \frac{5}{6}$ it is easier to evaluate $\frac{5}{6} + 1$ to give

Try keying

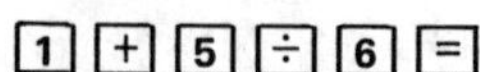

Does your calculator work the division out first to give 1·833, or does it work from left to right to give $(1 + 5) \div 6 = 1$?

It is very unlikely that you will need to convert decimals to fractions, since the quickest way to carry out a calculation is to use a calculator, which works in decimals anyway.

TRY SOME YOURSELF

10 Convert each of the following fractions to decimals:
(i) $\frac{1}{2}$ (ii) $\frac{1}{4}$ (iii) $\frac{1}{8}$ (iv) $\frac{3}{4}$ (v) $\frac{1}{3}$ (vi) $2\frac{2}{3}$ (vii) $3\frac{3}{8}$.

After you have worked through this section you should be able to

a Interpret the shaded part of a given area as a fraction
b Given any fraction, find a series of equivalent fractions
c Reduce a fraction to its simplest form
d Add, subtract, multiply and divide simple fractions
e Add, subtract, multiply and divide mixed numbers
f Convert a given fraction to a decimal

Finally, here are some exercises if you want more practice.

TRY SOME MORE YOURSELF

11(i) Write each of the following shaded areas as fractions of the given shape:

(a)

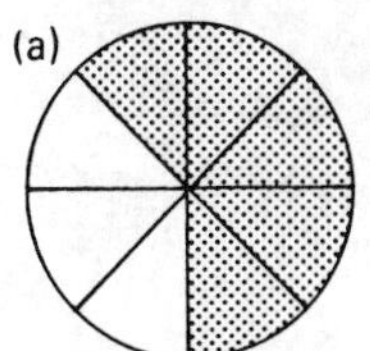

(b)

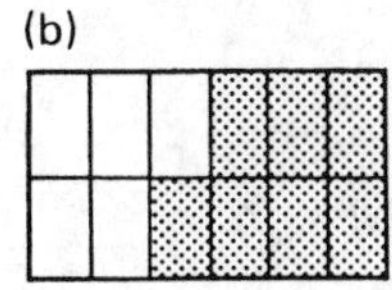

(c)

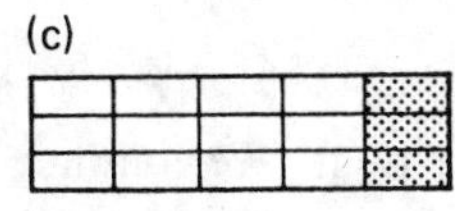

(ii) Reduce each of the following fractions to its simplest form:
(a) $\frac{5}{15}$ (b) $\frac{63}{72}$ (c) $\frac{115}{200}$ (d) $\frac{32}{128}$.

(iii) Evaluate each of the following:
(a) $\frac{4}{7}-\frac{1}{7}$ (b) $\frac{4}{9}+\frac{1}{3}$ (c) $\frac{8}{15}-\frac{1}{3}$ (d) $\frac{8}{15}+\frac{1}{9}-\frac{2}{5}$
(e) $\frac{14}{25}-\frac{7}{10}+\frac{1}{4}$.

(iv) Evaluate each of the following:
(a) $\frac{4}{5}\times\frac{15}{28}$ (b) $\frac{2}{7}\times\frac{14}{25}\times\frac{15}{24}$ (c) $\frac{3}{10}\times\frac{5}{13}\times\frac{39}{42}$.

(v) Evaluate each of the following:
(a) $\frac{16}{21}\div\frac{4}{7}$ (b) $\frac{12}{17}\div\frac{24}{34}$ (c) $\frac{4}{9}\times\frac{3}{16}\div\frac{7}{12}$ (d) $\frac{2}{5}\times\frac{1}{12}\div\frac{4}{15}$.

(vi) Evaluate each of the following:
(a) $2\frac{3}{5}+1\frac{5}{6}$ (b) $4\frac{6}{7}+1\frac{2}{3}-2\frac{1}{21}$ (c) $4\frac{1}{6}-5\frac{3}{4}+2\frac{3}{8}$.

(vii) Evaluate each of the following:
(a) $2\frac{3}{4}\times1\frac{5}{11}$ (b) $3\frac{7}{15}\div5\frac{1}{5}$ (c) $2\frac{1}{3}\times2\frac{4}{7}\div7$ (d) $4\frac{2}{5}\div\frac{11}{12}\times1\frac{7}{8}$.

(viii) Evaluate each of the following:
(a) $(1\frac{1}{9}\div1\frac{2}{3})-\frac{1}{6}$ (b) $(\frac{1}{4}\times\frac{2}{3})\div(1\frac{5}{7}-\frac{6}{7})$.

(ix) Convert each of the following fractions to decimals:
(a) $2\frac{3}{4}$ (b) $2\frac{7}{8}$ (c) $4\frac{2}{3}$ (d) $3\frac{2}{5}$.

1.5 Ratios and Percentages, and More Calculator Practice

TRY THESE QUESTIONS FIRST

1 In a certain constituency people voted Conservative, Labour and Liberal in the ratio 6 : 5 : 1. How many voted for each party if 48,000 people voted altogether?

2 A garage sells tyres at £20 plus VAT at 15%. A discount of 12% is offered to AA members. How much would the tyres cost an AA member?

3 Use the memory on your calculator to evaluate

$$\frac{14{\cdot}6 + 0{\cdot}79}{16{\cdot}7 + 4{\cdot}32}$$

4 Find $\sqrt{0{\cdot}4}$.

1.5(i) RATIOS

In the last section we discussed fractions. We continue the theme in this section by looking at ratios. Ratios give exactly the same information as fractions. The only difference is that whereas fractions give the denominator, with ratios you have to work it out for yourself. And although problems often give information as a ratio you will usually have to convert it to fractions in order to find the solution.

But what is ratio? Well, suppose a cocktail is made by mixing three parts vermouth to one part gin. Then the *ratio* of vermouth to gin is 3:1. Alternatively, we could say the *proportion* of vermouth to gin is 3:1.

3:1 is pronounced three to one.

The cocktail is made up from 3 parts vermouth to 1 part gin. There are 4 parts altogether, and

$\frac{3}{4}$ of the cocktail is vermouth

$\frac{1}{4}$ is gin

For example to make one litre of the cocktail we would need $\frac{3}{4}$ litre of vermouth and $\frac{1}{4}$ litre of gin.

This indicates that adding both sides of the ratio together gives the denominator of the fractions.

The next example illustrates how to solve a problem involving ratios.

EXAMPLE

Concrete is mixed from cement and aggregate (a mixture of sand and gravel) in the ratio 2 parts cement to 7 parts aggregate. How much cement and aggregate are needed to make 1400 kg of concrete?

SOLUTION

The ratio of cement to aggregate is 2:7. So $\frac{2}{9}$ is cement and $\frac{7}{9}$ is aggregate.

The total number of parts is $2 + 7 = 9$.

To make 1400 kg of concrete we need

$(\frac{2}{9} \times 1400)$ kg cement ($\triangleq$ 311 kg cement)

and

$(\frac{7}{9} \times 1400)$ kg aggregate ($\triangleq$ 1089 kg aggregate)

Notice that $311 + 1089 = 1400$. Add the separate parts together again to check your answer.

The next example indicates how a ratio can be built up.

EXAMPLE

A primary school class contained 20 girls and 16 boys. What was the ratio of girls to boys?

SOLUTION

The ratio of girls to boys was 20:16. However, the ratio can be simplified quite considerably. This can be done by dividing both sides by a common factor. So 20:16 is the same as 10:8 or 5:4. The ratio of girls to boys was therefore 5:4.

Check this: there are 36 children altogether. If the ratio of girls to boys is 5:4, how many girls and boys are there?

Thus, ratios can be simplified rather like fractions.

TRY SOME YOURSELF

1(i) Purple paint is mixed from red and blue in the ratio 3 parts red to 2 parts blue. How much of each colour is needed to make 15 litres of purple paint?

(ii) In a plant breeding experiment, germinated seeds can end up as short or tall plants. 100 germinated seeds produced 25 short plants and 75 long plants. What was the ratio of short plants to long plants?

(iii) A cocktail was made from gin, vermouth and orange juice in the ratio 2:3:1. What quantity of each was required to make up 2 pints of the cocktail?

The denominator of the fractions is given by adding all the parts together.

Scales on maps

The scale on a map is often indicated as a ratio. The scale on this map is 1″:4 miles.

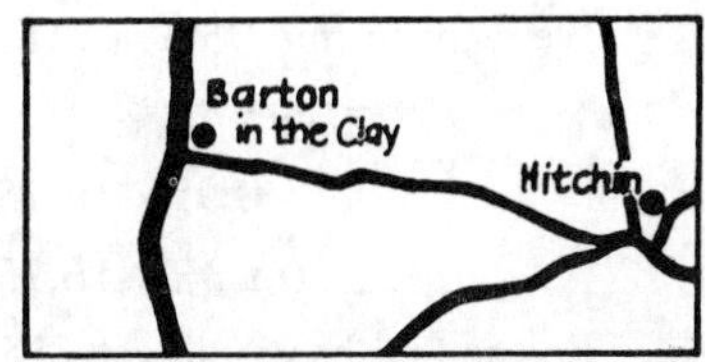

The ratio has to be interpreted slightly differently here. For example, on the map, the distance from Barton in the Clay to Hitchin is $1\frac{1}{4}$″. The actual distance is $(1\frac{1}{4} \times 4)$ miles, i.e. 5 miles.

> *CHECK YOUR ANSWERS*
>
> 1 A ratio of 6:5:1 means $\frac{6}{12}$ voted Conservative, $\frac{5}{12}$ voted Labour and $\frac{1}{12}$ voted Liberal. So 24,000 voted Conservative, 20,000 voted Labour and 4000 voted Liberal. — *Section 1.5(i)*
>
> 2 Total cost = 1·15 x 0·88 x £20 = £20·24. — *Section 1.5(ii)*
>
> 3 Estimate first then calculate to get 0·7321598. — *Section 1.5(iii)*
>
> 4 Using the [$\sqrt{x}$] key, $\sqrt{0{\cdot}4}$ = 0·6324555. — *Section 1.5(iii)*

TRY SOME YOURSELF

2 The scale on the map opposite is 1:50,000. What is the actual distance from Great Beccott to Yarde? (Take this ratio to mean 1 cm:50,000 cm.)

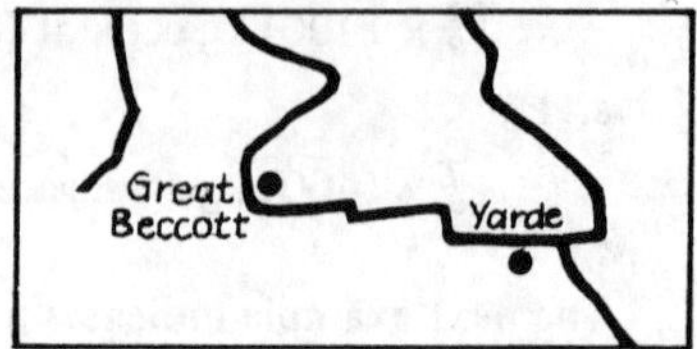

1.5(ii) PERCENTAGES

Percentages also indicate proportions. For example, labels in clothes indicate the various proportions of different yarns in the fabric.

% means 'per cent' or per 'one hundred'.

100% cotton indicates that the fabric is made entirely from cotton

80% cotton means that $\frac{80}{100}$ (or $\frac{4}{5}$) of the garment is cotton

20% polyester means that $\frac{20}{100}$ (or $\frac{1}{5}$) is polyester

100% cotton

80% cotton
20% polyester

The percentages on the label always total 100%, just as the corresponding fractions add up to 1, because the total (100%) refers to the whole garment.

80% + 20% = 100%
$\frac{4}{5} + \frac{1}{5} = 1$

Percentages can be manipulated either as fractions or decimals. The latter approach may be easier since we have already shown that it is quite easy to divide by 100, and

80% = $\frac{80}{100}$ = 0·8

3% = $\frac{3}{100}$ = 0·03 etc.

EXAMPLE

A building society offers 95% mortgages to first time buyers. How much would the Smiths get on a house valued at £16,750?

SOLUTION

We want to find 95% of £16,750.

95% = 0·95 or $\frac{95}{100}$

0·95 x 16,750 = 15912·5

(or $\frac{95}{100}$ x 16,750 = 15912·5)

So the Smiths would get £15,913.

We've rounded off the answer to the nearest pound.

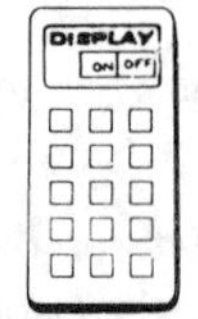

Your calculator may have a [%] key. Check in the maker's handbook.

You can of course always convert to a decimal by dividing by 100, and then multiply.

TRY SOME YOURSELF

3(i) Express each of the following percentages as fractions:
(a) 50% (b) 33% (c) 4% (d) 90%.

(ii) Express each of the following percentages as decimals:
(a) 14% (b) 25% (c) 3% (d) 0·5%.

(iii) A survey was carried out on 540 married couples to investigate family income. 65% of the wives worked; 50% of the husbands had annual salaries over £6000; 14% of the couples had a joint income of less than £50 per week.
(a) How many wives worked?
(b) How many husbands earned over £6000 per year?
(c) How many couples had a joint income of less than £50 per week?

Fractions and decimals can also be converted to percentages.

We have already shown that $80\% = \frac{80}{100} = 0{\cdot}8$. This indicates that to change a decimal to a percentage, we need only multiply by 100.

For example

0·14 = 14% *0·14 x 100 = 14*

0·4 = 40% *0·4 x 100 = 40*

To convert a fraction to a percentage it's necessary to change the fraction to a decimal first. So

$\frac{4}{5} = 0{\cdot}8 = 80\%$

and

Use your calculator.

$\frac{2}{3} \simeq 0{\cdot}6666667 \simeq 0{\cdot}67$ or 67%

We've rounded the decimal so that the percentage is a whole number.

Decimals or fractions bigger than 1 correspond to percentages greater than 100%. For example

$1\frac{2}{5} = 1{\cdot}4 = 140\%$ *1·4 x 100 = 140*

TRY SOME YOURSELF

4 Convert each of the following to percentages:
(i) (a) 0·2 (b) 0·14 (c) 0·90 (d) 2·5
(ii) (a) $\frac{1}{10}$ (b) $\frac{1}{5}$ (c) $\frac{1}{3}$ (d) $1\frac{1}{4}$.

Round off the percentages to whole numbers if necessary.

Percentage increase and decrease

Our everyday experience of percentages tends to centre on percentage increases (like VAT at 15%, or service charges) and percentage decreases (such as discount of 10%).

For example, £7 + 15% VAT means we actually have to pay

£7 + (15% of £7).

This means that we actually pay £8·05.

15% = 0·15
15% of £7 = £(0·15 x 7)
= £1·05
£7 + £1·05 = £8·05

Alternatively the actual amount can be calculated as follows. We pay (100% + 15%) of £7, that is 115% of £7.

Again this gives £8·05.

115% of £7 = £(1·15 x 7)
= £8·05

Discount can be calculated in the same way.

£7 with 10% discount means we actually pay £7 − (10% of £7) = £6·30.

Again the actual amount can be evaluated by calculating (100% − 10%) of £7 or 90% of £7.
This also gives £6·30.

10% = 0·1
10% of £7 = £(0·1 x 7)
= £0·7
£7 − £0·7 = £6·30

90% of £7 = £0·9 x 7
= £6·30

EXAMPLE

A restaurant bill comes to £12. In addition VAT is added at 15% and the restaurant also adds a service charge of 12%. Does it make any difference if the VAT is added first *then* the service charge or vice versa?

In practice, VAT must be paid last.

SOLUTION

15% = 0·15 (VAT)
12% = 0·12 (service charge)

Adding VAT first gives

115% of £12 = 1·15 x £12 = £13·80

115% = 100% + 15%

Adding the service charge on to £13·80 gives

112% of £13·80 = 1·12 x £13·80 = £15·46

112% = 100% + 12%

So the total bill is £15·46.

Adding these extras the other way round,

112% of £12 = 1·12 x £12 = £13·44
115% of £13·44 = 1·15 x £13·44 = £15·46

The total bill is the same. The order does not matter.

We've actually calculated 1·15 x 1·12 x £12 or 1·12 x 1·15 x £12. Since we're multiplying the order doesn't matter.
1·15 x 1·12 = 1·12 x 1·15

TRY SOME YOURSELF

5

£20 only

(i) How much will this tennis racquet cost if VAT at 15% has to be added?
(ii) How much will it cost if the customer gets 20% discount?
(iii) If the customer pays VAT *and* gets a discount how much does he pay? Does it make any difference whether the VAT is added first then the discount subtracted, or vice versa?

1.5(iii) CALCULATOR PRACTICE

The rest of this section is designed to help you find out more about how your calculator works.

Using the memory

Your calculator may have one or more memories. The memory can be used to store parts of a calculation to be used at a later stage. We are going to assume that you can *store* numbers in the memory and then *recall* the number some time later.

Check in the maker's handbook.

What do the memory keys look like? How are they used?

We are going to use the symbol [STO] to indicate that you should store the number and [RCL] when we want you to recall it. Check how to store and recall numbers using your calculator.

This information will be given in the maker's handbook.

Whatever method your calculator uses you will also need to find out how to clear the memory. Make sure you know how to do this. Check that you can do it by carrying out the following steps.

Store the number 2 in the memory.

Key [2] [STO].

Clear the display and press [RCL].

The display should indicate [2].

Now clear the memory and press [RCL] again. The display should now indicate [0].

If the display continues to show [2] you have not cleared the memory–so try again.

It's always a good idea to clear both the display and the memory before starting any new calculation, particularly since many calculators now can continue to store a number in the memory after the calculator is switched off!

Using the memory in calculations

If your calculator has a memory you can often use the memory when a calculation involves brackets. The brackets indicate which part to work out first. Instead of using brackets you can work out that part first and store it in the memory, recalling it again when needed.

EXAMPLE

Evaluate 3 + (2 − 1).

As in earlier sections, we're interested in the principles involved so we use easy numbers. This means that you can check what your calculator does by working out the answers in your head.

SOLUTION

Work out the brackets first and store the result in the memory.

A key sequence is

[2] [−] [1] [=] [STO]

Now the calculation is completed by keying

[3] [+] [RCL] [=] [4]

3 + (2 − 1) = 4

When the brackets are reached the result is recalled from the memory.

The complete key sequence is

[2] [−] [1] [=] [STO] [3] [+] [RCL] [=] [4]

Try this key sequence on your calculator. If you don't get the answer 4, check in the maker's handbook again how to use the memory.

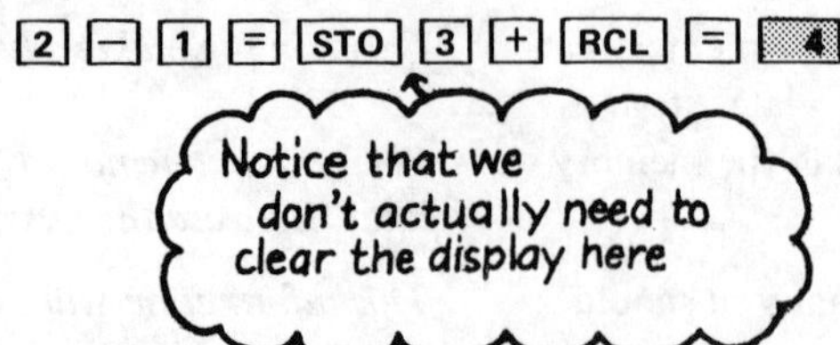

TRY SOME YOURSELF

6 Use the memory on your calculator to evaluate each of the following:
(i) (a) $6 + (7 - 3)$ (b) $7 + (2 \times 6)$ (c) $2 + (6 \div 2)$
(ii) (a) $6{\cdot}321 + (17{\cdot}6 - 24{\cdot}3)$ (b) $14{\cdot}7 \div (0{\cdot}2 + 10{\cdot}62)$
(c) $0{\cdot}12 + (0{\cdot}001 \times 14{\cdot}6)$.

Check the answers by some other method.
Estimate the answer first just in case you make a mistake.

We have indicated that $\frac{2}{3}$ is another way of writing $2 \div 3$. In the same way $\frac{32}{3 + 5}$ is another way of writing $32 \div (3 + 5)$. Evaluate $(3 + 5)$ first and then divide to get $32 \div 8 = 4$. It's essential to include the brackets in the key sequence so that the key sequence to evaluate $\frac{32}{3 + 5}$ is

[32] [÷] [(] [3] [+] [5] [)] [=] [4]

It's no good keying
[32] [÷] [3] [+] [5] [=]*!*
See what happens if you ***do*** *key this sequence!*

Alternatively you can use the memory like this:

[3] [+] [5] [=] [STO] [32] [÷] [RCL] [=] [4]

TRY SOME YOURSELF

7 Use the memory on your calculator to evaluate each of the following:
(i) $\frac{15}{2 + 3}$ (ii) $\frac{149{\cdot}6}{31{\cdot}2 + 18{\cdot}7}$ (iii) $\frac{64 - 28}{16 + 53}$

For more complicated calculations estimate the answer first, just in case you make a mistake.

Reciprocal key

Your calculator may have a reciprocal key, [1/x].

Check in the maker's handbook.

In Section 1.4(iii) we indicated that the reciprocal of a fraction is found by turning the fraction upside down. The reciprocal of any number can be found by dividing one by that number.

For example, the reciprocal of $\frac{2}{5}$ is $\frac{5}{2}$.

For example, the reciprocal of 0·25 is $\frac{1}{0{\cdot}25}$ (= 4).

$0{\cdot}25 = \frac{0{\cdot}25}{1}$

Turning this upside down we get $\frac{1}{0{\cdot}25}$.

Use the reciprocal key on your calculator to find the reciprocals of 2, 10 and 0·25.

The reciprocal key can also be used to evaluate expressions like $\frac{32}{3 + 5}$ because

$$\frac{32}{3 + 5} = 32 \times \left(\frac{1}{3 + 5}\right)$$
$$= 32 \times \text{reciprocal of } (3 + 5)$$

Try keying

[3] [+] [5] [=] [1/x] [×] [32] [4]

Notice that the calculation has been rearranged so that (3 + 5) is evaluated first.

$32 \times \frac{1}{(3+5)} = \frac{1}{(3+5)} \times 32$

We've evaluated

$\frac{1}{3+5} \times 32$

TRY SOME YOURSELF

8 Use the reciprocal key on your calculator to evaluate each of the following:
(i) $\frac{1}{6{\cdot}7}$ (ii) $\frac{4}{5+3}$ (iii) $\frac{149{\cdot}3}{159{\cdot}6 + 13{\cdot}8}$.

Estimate the answer first, just in case you make a mistake.

To evaluate calculations involving division, like Exercise 8(iii) above, you can use either the memory or the reciprocal key. It doesn't matter which method you use as long as you feel confident about the answer. In more complex calculations, though, you may need to use both these facilities.

Square and square root keys

In Section 1.3(iii) we showed that $10^2 = 10 \times 10$.

10^2 is the square of 10.

In the same way

$2^2 = 2 \times 2$

$4^2 = 4 \times 4$

$(-3)^2 = (-3) \times (-3)$ and so on

2^2 is the square of two.

4^2 is pronounced 'four squared'.

Your calculator might have a square key, [x^2].

Check in the maker's handbook.

Using this key you can quickly find the square of any number. For example, try keying [4] [x^2]. You should get [16]. What would you expect to get for $1^2, 2^2, 3^2, (-1)^2$? Try them on your calculator. Of course, if your calculator does not have a [x^2] key you can just multiply the numbers together.

$4^2 = 4 \times 4 = 16$

$(-1)^2 = (-1) \times (-1) = 1$

Key [4] [×] [4] [=] etc.

TRY THESE YOURSELF

9 Use your calculator to find each of the following:
(i) 14^2 (ii) $(-4)^2$ (iii) $(-147{\cdot}3)^2$ (iv) $(\frac{7}{8})^2$.

Change the fraction to a decimal first.

The number whose square gives a given number is called the *square root* of that number.

For example, the square root of 4 is 2 since $2^2 = 4$. This is written as

$\sqrt{4} = 2$

$\sqrt{4}$ is pronounced 'the square root of 4'.

If your calculator has an [x^2] key then it probably also has a key which finds the square root, [$\sqrt{\ }$].

Try keying [4] [$\sqrt{\ }$] on your calculator. You should get [2]. What would you expect to get for $\sqrt{16}, \sqrt{9}, \sqrt{1}$?

Try them and see!

Numbers less than one often cause difficulty because it's hard to anticipate the answer. Scientific notation may make the process clearer. For example

Would you expect 0·1 x 0·1 to be bigger or smaller than 0·1?

$$0{\cdot}1 \times 0{\cdot}1 = 10^{-1} \times 10^{-1} = 10^{-2} = 0{\cdot}01$$

So $(0{\cdot}1)^2 = 0{\cdot}01$.

Similarly $\sqrt{0{\cdot}01} = 0{\cdot}1$.

Since 0·1 x 0·1 = 0·01

You may have found these examples rather odd, if you expect that the square of a given number should be a larger number than the original number and the square root of a number should be smaller than the original number.

In fact the square of a number between 0 and 1 is smaller than the original number and the square root of such a number is bigger than the original number.

For example $(0{\cdot}1)^2 = 0{\cdot}1$ *and* $\sqrt{0{\cdot}01} = 0{\cdot}1$.

TRY THESE YOURSELF

10 Use your calculator to evaluate each of the following:
(i) $\sqrt{81}$ (ii) $\sqrt{9}$ (iii) $\sqrt{90}$ (iv) $\sqrt{0{\cdot}9}$ (v) $\sqrt{0{\cdot}09}$

You can check the answers to these exercises by squaring your answer. You should get the number you started with.

This concludes our investigation on how to use some of the keys on your calculator. We hope that you now have confidence in using your calculator when you need to evaluate numerical calculations. As long as you estimate the order of magnitude of the answer you will be able to judge for yourself whether the answer given by your calculator is accurate or not. Of course we also hope that you are now tempted to explore other processes on your calculator. The maker's handbook should outline the facilities available, but if you do investigate some unfamiliar processes, remember that it is most helpful to use easy numbers at first, so that you can check the answers in your head.

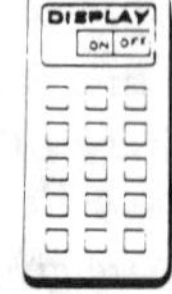

Finally, we assume that you have a calculator available when you work through each module of this book. Whenever we introduce a new key we will include some detailed notes, but you will usually find that we just indicate a calculator in the margin (without showing the key sequences). So if you do get into difficulties later on, you might find it useful to come back to this module and refresh your memory by working through some of the exercises again.

After you have worked through this section you should be able to

a Express a pair of numbers as a ratio in the simplest form
b Given a ratio, work out corresponding proportions of a fixed quantity
c Convert a fraction or decimal to a percentage and vice versa

d Work out percentage increases and decreases
e Use the store and recall facilities on your calculator
f Use the reciprocal key on your calculator
g Use the square and square root keys on your calculator

Finally, here are some exercises if you want more practice.

TRY SOME MORE YOURSELF

11(i) (a) What is the ratio of 80 to 16 in its simplest form?
(b) What is the ratio of 14 to 21 in its simplest form?
(c) A shade of paint is obtained by mixing red, orange and white paint in the ratio 4:2:1. How much of each colour is needed to make up 14 litres of the required shade?

(ii) (a) Express 65% as a fraction.
(b) Express 0·1% as a decimal.
(c) Express $\frac{3}{5}$ as a percentage.
(d) Express 0·6 as a percentage.

(iii) Students are offered 15% discount at a certain restaurant. Four students went for a meal and the bill came to £20. VAT was also added at 15% and a service charge of 10%. What was the final cost?

(iv) Use the memory on your calculator to evaluate each of the following. Round the answers to two decimal places:
(a) $4+(6\div 7)$ (b) $2{\cdot}142-(3{\cdot}6\times 0{\cdot}13)$ (c) $\frac{14{\cdot}7+16{\cdot}8}{15{\cdot}2\times 11{\cdot}4}$

(v) Use the reciprocal key on your calculator to evaluate each of the following. Round the answers to two significant figures.
(a) $\frac{14{\cdot}81}{16{\cdot}3+3{\cdot}2}$ (b) $\frac{193{\cdot}2-16{\cdot}3}{(0{\cdot}1\times 0{\cdot}03)}$ (c) $\frac{196\times(18{\cdot}4-0{\cdot}032)}{(18{\cdot}3-0{\cdot}134)}$

(vi) Use your calculator to find each of the following:
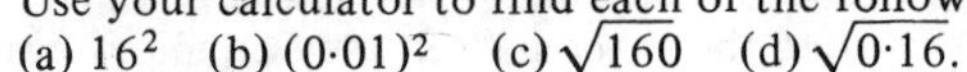
(a) 16^2 (b) $(0{\cdot}01)^2$ (c) $\sqrt{160}$ (d) $\sqrt{0{\cdot}16}$.

Section 1.1 Solutions

1
(i) 43 rounds *down* to 40 (3 is *less than 5*).
(ii) 58 rounds *up* to 60 (8 is *more than 5*).
(iii) 204 rounds *down* to 200 (4 is *less than 5*).
(iv) 96 rounds *up* to 100 (6 is *more than 5*).
(v) 1005 rounds *up* to 1010 (5 *is half-way*).
Notice that when rounded to the nearest 10, numbers always end in a zero.

2
(i) 1*57* rounds *up* to 200 (57 is over half-way).
(ii) 1*19* rounds *down* to 100 (19 is less than half-way).
(iii) 11*32* rounds *down* to 1100.
(iv) 4979 rounds *up* to 5000.
(v) 31,3*50* rounds *up* to 31,400.
Notice that when rounded to the nearest 100, numbers always end in at least two zeros. We've italicised the last two digits. Round down if they are less than 50 and round up if they are 50 or more.

3
(i) (a) We want to round to the nearest 100. Look at the last two digits: 7,183,5*30* rounds down to 7,183,500.
(b) Look at the last 3 digits: 7,183,*530* rounds up to 7,184,000 since 530 is bigger than 500. (Notice that when rounded to the nearest 1000, the number ends with three zeros.)
(c) 7,18*3,530* (look at the last 4 digits). Round down, to 7,180,000 (the number ends in 4 zeros).
(d) 7,1*83,530* (look at the last 5 digits). Round up, to 7,200,000 (the number ends in 5 zeros).
(ii) 8,000,000
(iii) 60 miles
(iv) 30,000

4

(i) Round off the numbers to 50 + 20 + 30 + 60 + 50 + 90 = 300 pence or £3. (In fact the exact total is £2·94.)

(ii) 384 rounds to 400. So in a year Dave earns about £400 × 12 which is £4800 (in fact he earns exactly £4608 a year).

5

(i) Round the numbers to get 30 × 4.
So the area $\simeq$ 120 square feet. (The exact area is 104 square feet.)

(ii) Round the numbers to get 20 × 10.
Area $\simeq$ 20 × 10 = 200 square feet. (The exact amount required is 204 square feet. You'd be in a bit of a mess here if you just bought 200 square feet. This example illustrates that in practice it is usually safer to overestimate rather than underestimate.)

6 Key sequences for each of the calculations are

(i) [5] [+] [3] [=] [8]
(ii) [6] [−] [2] [=] [4]
(iii) [5] [×] [3] [=] [15]
(iv) [8] [÷] [4] [=] [2]
(v) [6] [×] [3] [=] [18]
(vi) [9] [÷] [3] [=] [3]

7

(i) (a) Estimate: 300 + 100 = 400. Calculate: 354.
(b) Estimate: 70,000 − 20,000 = 50,000. Calculate: 48,398.
(c) Estimate: 800,000 + 800,000 = 1,600,000. Calculate: 1,577,426.
(d) Estimate: 11,000,000 − 9,000,000 = 2,000,000. Calculate: 1,937,538.
In each case the exact answer is close to the estimate, so we conclude that there are probably no mistakes.

(ii) (a) Estimate: 500 − 500 = 0. So the answer is wrong. (In fact 1006 = 507 + 499.)
(b) Estimate: 1300 + 900 = 2200. Again the answer is wrong. (In fact 3077 = 2176 + 901.)
(c) Estimate: 2,000,000 − 1,000,000 = 1,000,000. So the answer is probably right (in fact this is the correct answer).
(d) Estimate: 800,000 − 500,000 = 300,000. So the answer is wrong (in fact 49,245 = 792,564 − 743,319).
The mistakes which produced these wrong answers all consisted of keying errors—which are very easy to make. For example, it's easy to press the wrong operation key, or to enter the digits in the wrong order. That is why it's useful to be able to estimate—then these mistakes are likely to be eliminated.

8 In each case round both numbers to the same level of accuracy.

(i) Estimate by rounding off to the nearest 10: 560 − 10 = 550. Calculate: 548.
(ii) Estimate by rounding to the nearest 100: 2000 − 200 = 1800. Calculate: 1764.
(iii) Estimate by rounding to the nearest 1000: 15,000 + 0 = 15,000. Calculate: 14,991.
(iv) Estimate by rounding to the nearest 1000: 20,000 + 2000 = 22,000. Calculate: 21,166.
Again, the estimates give a good indication of the accuracy of the answer. You may not have obtained such good estimates if you rounded the numbers less accurately. Nevertheless you should still have been able to discover whether you made a mistake when using your calculator since your estimate should at least have indicated the order of magnitude.

9

(i) Estimate: 300 × 200 = 3 × 2 × 100 × 100 = 60,000.
Calculate: 59,800.
(ii) Estimate: 6000 × 3000 = 18,000,000.
Calculate: 15,888,012.
(iii) Estimate: 10,000 × 600 = 6,000,000.
Calculate: 9,155,968.
(iv) Estimate: 600 × 100 = 60,000.
Calculate: 94,168.
(v) Estimate: 3000 × 200 = 600,000.
Calculate: 453,887.
Although the estimates are not particularly close to the exact answer, they do indicate the order of magnitude and provide some guidelines as to whether or not you made a mistake when keying the calculation on your calculator.

10

(i) What do we have to calculate?
The annual mortgage repayments are £12 × 131. The rates are £310. We need to calculate the sum of these two.
Estimate: 10 × 100 = 1000 (mortgage);
1000 + 300 = 1300 (mortgage + rates).
Calculate: 12 × 131 = £1572 (mortgage).
Total = £1572 + £310 = £1882.
The estimate is not particularly good—but it does indicate that the answer should be a 4 digit number between 1000 and 2000.

(ii) We need to calculate 360 × 240.
Estimate: 400 × 200 = 80,000.
Calculate: 360 × 240 = 86,400.
The area is therefore 86,400 sq. ft.

(iii) We need to calculate £367 × 5312.
Estimate: 400 × 5000 = 2,000,000.
Calculate: 367 × 5312 = 1,949,504.
The annual salary bill is therefore £1,949,504.

11

(i) (a) 20 (b) 10 (c) 90 (d) 100 (e) 30 (f) 120 (g) 1020

(ii) (a) 200 (b) 300 (c) 400 (d) 1000 (e) 1100 (f) 10,100 (g) 1,001,100

(iii) (a) 2000 (b) 0 (c) 2000 (d) 20,000 (e) 18,000 (f) 1,349,000

(iv) (a) 1,000,000 (b) 700 (c) 100 (d) 900

(v) (a) £1000 (b) 10,000 sq. yds

(vi) (a) 32 (b) 337 (c) 128,540 (d) 100,633 (e) 93,233

(vii) (a) 57 (b) 51 (c) 1591 (d) 4391 (e) 45,120

(viii) (a) 637 (b) 9933 (c) 6,335,848 (d) 3,092,250 (e) 135,407

(ix) (a) £1144 (b) 11,684 sq. yds

Section 1.2 Solutions

1

(i) 3 left

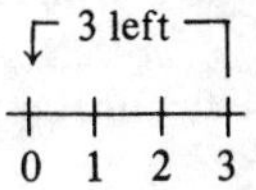

0 1 2 3

$3 - 3 = 0$

(ii) 6 left

(−2) (−1) 0 1 2 3 4 5

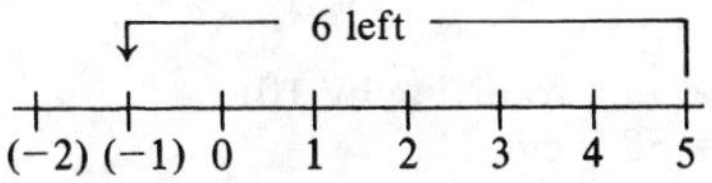

$5 - 6 = (-1)$

(iii) 9 left

(−7) (−6) (−5) (−4) (−3) (−2) (−1) 0 1 2 3

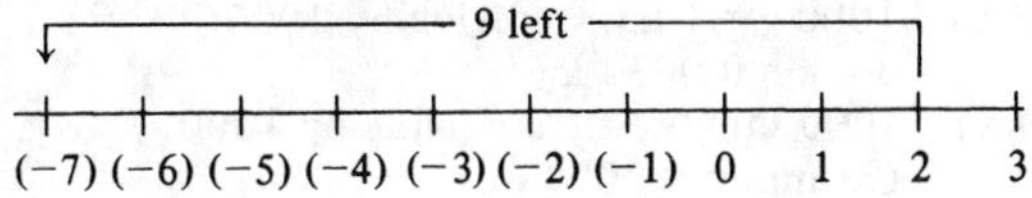

$2 - 9 = -7$

2 Try keying

(i) [4] [+/−]

(ii) [6] [+/−]

(iii) [1782] [+/−]

(iv) [2941] [+/−]

3

(i) $(-3) - 7 = -10$

(ii) $(-2) - 9 = -11$

(iii) $2 + (-6) = 2 - 6 = -4$

(iv) $4 + (-7) = 4 - 7 = -3$

(v) $(-3) + (-8) = (-3) - 8 = (-11)$

(vi) $(-1) + (-2) = (-1) - 2 = (-3)$

4

(i) $(-3) + 6 = 3$

(ii) $(-7) + 4 = -3$

(iii) $3 - (-2) = 3 + 2 = 5$

(iv) $5 - (-7) = 5 + 7 = 12$

(v) $(-2) - (-6) = (-2) + 6 = 4$

(vi) $(-1) - (-1) = (-1) + 1 = 0$

5

(i) Sarah has £(−26). So she must add £26 to get £0 (then she'll be out of debt).

(ii) Jim started at (−51) metres and ended at 20 metres.
We want $20 - (-51)$.
$$20 - (-51) = 20 + 51 = 71$$
So Jim climbed 71 metres.

(iii) Temperature started at $(-2)^\circ$ C and ended at $(-12)^\circ$ C.
We want $(-12) - (-2) = -10$.
The temperature rose $(-10)^\circ$ C. In other words it fell 10° C.

(iv) We want £12 − £30 = £(−18) and £(−18) − £16 = £(−34).
So Dave had £(−34). (He was £34 overdrawn.)

6

(i) $(+) \times (-) = (-)$, so $6 \times (-2) = (-12)$.

(ii) $(-) \times (-) = (+)$, so $(-3) \times (-4) = 12$.

(iii) $(+) \times (-) = (-)$, so $2 \times (-10) = (-20)$.

(iv) $(+) \div (-) = (-)$, so $20 \div (-4) = (-5)$.

(v) $(-) \div (-) = (+)$, so $(-36) \div (-12) = 3$.

(vi) $(-) \div (+) = (-)$, so $(-42) \div 14 = (-3)$.

7 In each case evaluate the brackets first.

(i) $(5 + 3) = 8$
So $(5 + 3) - 14 = 8 - 14 = (-6)$

(ii) $(2 - 9) = (-7)$
So $7 \times (2 - 9) = 7 \times (-7) = (-49)$.

(iii) $(6 - 2) = (-4)$
So $4 - (6 - 2) = 4 - (-4) = 4 + 4 = 8$

(iv) $(2 - 3) = (-1)$ and
$(6 + 3) = 9$
So $(2 - 3) \times (6 + 3) = (-1) \times 9 = (-9)$

8 In each case evaluate the brackets first.

(i) $((-2) + 7) = 5$
So $((-2) + 7) \div 5 = 5 \div 5 = 1$

(ii) $((-3) - (-4)) = (-3) + 4 = 1$
So $((-3) - (-4)) \times 5 = 1 \times 5 = 5$

(iii) $(2 + (-2)) = 2 - 2 = 0$
$(6 \times (-2)) = (-12)$
and $(2 + (-2)) + (6 \times (-2)) = (-12)$

9 Evaluate innermost bracket first.

(i) $(5 - 7) = (-2)$
$\{(5 - 7) + 6\} = \{(-2 + 6\} = 4$
$\{(5 - 7) + 6\} - 3 = 4 - 3 = 1$

(ii) $(3 - 8) = (-5)$
$[(3 - 8) - 4] = [(-5) - 4] = (-9)$
$16 + [(3 - 8) - 4] = 16 + (-9) = 16 - 9 = 7$

(iii) $(6 + 2) = 8$
$[5 - (6 + 2)] = [5 - 8] = (-3)$
$(2 + 3) = 5$
$(2 + 3) \times [5 - (6 + 2)] = 5 \times (-3) = (-15)$

10

(i) 2 x (3 + 4) = (2 x 3) + (2 x 4)
= 6 + 8 = 14
(Check: 3 + 4 = 7, 2 x 7 = 14.)

(ii) 3 x (2 − 6) = (3 x 2) + (3 x (−6))
= 6 + (−18) = (−12)

(iii) 4 x (3 − 2 + 1) = (4 x 3) + (4 x (−2)) + (4 x 1)
= 12 − 8 + 4 = 8

11

(i) 2(3 + 4) = (2 x 3) + (2 x 4)
= 6 + 8 = 14

(ii) 3(2 − 6) = (3 x 2) + (3 x (−6)) = (−12)

(iii) −(6 + 3) = −6 − 3 = −9

(iv) −(2 − 4) = −2 + 4 = 2

(v) −(3 + 6 − 4) = −3 −6 + 4 = −5

12

(i) (a) 10 (b) (−4) (c) (−6) (d) 6 (e) (−10) (f) 6 (g) (−5)

(ii) (a) 1 (b) (−5) (c) (−11) (d) (−11)

(iii) (a) 15 (b) 14 (c) 9 (d) (−2) (e) (−4)

(iv) (a) (−32) (b) (−14) (c) 30 (d) 2 (e) (−108)

(v) (a) (−9) (b) (−5) (c) 9 (d) 3

(vi) (a) 9 (b) (−15) (c) (−25)

(vii) (a) 27 (b) 30 (c) (−14) (d) 9

(viii) Remember to insert multiplication signs when multiplying brackets on a calculator, and remember to estimate the answer first, just in case you *do* make a mistake.
(a) (−1485) (b) 7,462,598 (c) 169,260 (d) (−241,668)

Section 1.3 Solutions

1

(i)

U	1/10	1/100	1/1000
3	7	0	1
3	0	7	1

3·701 is the bigger
(3 units, 7 tenths, 1 thousandth)

(ii)

U	1/10	1/100
0	3	
0	0	8

0·3 is the bigger
(0 units, 3 tenths)

(iii)

U	1/10	1/100
7	0	6
7	6	

7·6 is the bigger
(7 units, 6 tenths)

2

(i) 6·34 x 10 = 63·4 (Move decimal point one place right.)

(ii) 14·752 x 100 = 1475·2 (Move decimal point two places right.)

(iii) 0·674 x 10 = 6·74 (Move decimal point one place right.)

(iv) 0·00042 x 100 = 0·042 (Move decimal point two places right.)

(v) 1·03 x 100 = 103 (Move decimal point two places right.) Notice that 103 is the same as 103·0

3

(i) 58·9 ÷ 10 = 5·89 (Move decimal point one place left.)

(ii) 589 = 589·0, so 589·0 ÷ 10 = 58·9 (Move decimal point one place left.)

(iii) 58·9 ÷ 1000 = 0·0589 (Move decimal point three places left.)

(iv) 0·589 ÷ 1000 = 0·000589 (Move decimal point three places left.)

(v) 0·0589 ÷ 10 = 0·00589 (Move decimal point one place left.)

4

(i) 10 mm = 1 cm, so divide by 10:
589 mm = 58·9 cm.

(ii) 1 km = 1000 m, so multiply by 1000:
59·5 km = 59500 m.

(iii) 1000 g = 1 kg, so divide by 1000:
3·4 g = 0·0034 kg.

(iv) 1000 mm = 1 m, so divide by 1000:
0·3 mm = 0·0003 m.

5

(i) 4942·1 = 4·9421 x 10^3

(ii) 32 = 3·2 x 10^1

(iii) 734,000 = 7·34 x 10^5

(iv) 200,000 = 2 x 10^5

6

(i) 0·49 = 4·9 x 10^{-1}

(ii) 0·0062 = 6·2 x 10^{-3}

(iii) 0·0004 = 4 x 10^{-4}

(iv) 0·0000101 = 1·01 x 10^{-5}

7

(i) 3·4 x 10^2 = 340

(ii) 7·06 x 10^{-4} = 0·000706

(iii) 1·49 x 10^3 = 1490

(iv) 1·3 x 10^{-5} = 0·000013

8

(i) $10^1 \times 10^4 = 10^{(1+4)} = 10^5$
(ii) $10^2 \times 10^3 = 10^{(2+3)} = 10^5$
(iii) $10^0 \times 10^5 = 10^{(0+5)} = 10^5$
(iv) $10^3 \div 10 = 10^3 \div 10^1 = 10^{(3-1)} = 10^2$
(v) $10^4 \div 10^6 = 10^{(4-6)} = 10^{-2}$
(vi) $10^2 \div 10^2 = 10^{(2-2)} = 10^0 = 1$

9

(i) $10^3 \times 10^{-4} = 10^{(3+(-4))} = 10^{(3-4)} = 10^{-1}$
(ii) $10^2 \div 10^{-5} = 10^{(2-(-5))} = 10^{(2+5)} = 10^7$
(iii) $10^{-3} \div 10^{-4} = 10^{(-3-(-4))} = 10^{(-3+4)} = 10^1 = 10$

10

(i) $100 \times 1000 = 10^2 \times 10^3 = 10^{(2+3)} = 10^5$
$= 100{,}000$
(ii) $10 \div 1000 = 10^1 \div 10^3 = 10^{(1-3)} = 10^{-2}$
$= 0{\cdot}01$
(iii) $1000 \times 0{\cdot}01 = 10^3 \times 10^{-2} = 10^{(3+(-2))}$
$= 10^{3-2} = 10^1 = 10$
(iv) $0{\cdot}01 \times 0{\cdot}01 = 10^{-2} \times 10^{-2} = 10^{-4} = 0{\cdot}0001$
(v) $0{\cdot}01 \div 0{\cdot}01 = 10^{-2} \div 10^{-2} = 10^{(-2-(-2))}$
$= 10^{((-2))+2)} = 10^0 = 1$

11

(i) 3·1 (ii) 3·06 (iii) 3·057 (iv) 3 (v) 3·05710
Notice that we leave the zero at the end of 3·05710 even though we'd normally just write 3·0571. Writing the number as 3·05710 means that we're certain that the number is nearer to 3·05710 than to 3·05711 or 3·05709.

12

(i) $45{,}901 = 4{\cdot}5901 \times 10^4$. Rounding to two significant figures gives $4{\cdot}6 \times 10^4$ or 46,000.
(ii) $0{\cdot}049 = 4{\cdot}9 \times 10^{-2}$. To one significant figure it's 5×10^{-2} or 0·05.
(iii) $0{\cdot}4003 = 4{\cdot}003 \times 10^{-1}$. To three significant figures it's $4{\cdot}00 \times 10^{-1}$ or 0·400. Again we leave the zeros at the end to indicate that we are certain of the accuracy to this level.

13

(i) Estimate: $(4 \times 10^0) + (3 \times 10^0) - (1 \times 10^0)$
$= (4 + 3 - 1) \times 10^0 = 6 \times 10^0 = 6$.
Calculate: 5·3283.
In Section 1.2(v) we showed that, for example,
$(3 \times 6) + (2 \times 6) = (3 + 2) \times 6$
The same principle has been used here:
$(4 \times 10^0) + (3 \times 10^0) - (1 \times 10^0)$
$= (4 + 3 - 1) \times 10^0$ etc.
(ii) Estimate: $(5 \times 10^2) + (6 \times 10^2) - (2 \times 10^2)$
$= 9 \times 10^2 = 900$.
Calculate: 878·13.
(iii) Estimate: $(6 \times 10^{-1}) - (2 \times 10^{-1}) + (4 \times 10^{-1})$
$= 8 \times 10^{-1} = 0{\cdot}8$.
Calculate: 0·86
Notice that the estimates give reasonable approximations to the exact answers.

14

(i) Estimate: $(3 \times 10^{-2}) + (2 \times 10^1)$.
Consider only the largest power, so the answer should be about 2×10^1 or 20.
Calculate: 21·432 = 21. (Rounded to two significant figures.)
(ii) Estimate: $(4 \times 10^{-3}) - (4 \times 10^{-4})$.
Consider only the largest power, (10^{-3}), so the answer should be about 4×10^{-3} or 0·004.
Calculate: 0·00387 = 0·0039. (Rounded to two significant figures.)
(iii) Estimate: $(1 \times 10^2) - (2 \times 10^1) + (2 \times 10^2)$.
Consider only the largest power, which gives $(1 \times 10^2) + (2 \times 10^2)$. So the answer should be about 3×10^2 or 300.
Calculate: 323·9 = 320. (Rounded to two significant figures.)

15

(i) Estimate: $(6 \times 10^2) \div (3 \times 10^1)$
$= (6 \div 3) \times (10^2 \div 10^1) = 2 \times 10 = 20$.
Calculate: 21·962963 = 22·0. (Rounded to three significant figures.)
(ii) Estimate: $(5 \times 10^0) \times (8 \times 10^2)$
$= (5 \times 8) \times (10^0 \times 10^2) = 40 \times 10^2 = 4000$.
Calculate: 4397·1831 = 4400. (Rounded to three significant figures.)
(iii) Estimate: $(1 \times 10^3) \div (3 \times 10^1)$
$= (1 \div 3) \times (10^3 \div 10^1) \doteq 0{\cdot}3 \times 10^2 = 30$.
Calculate: 40·941176 = 40·9. (Rounded to three significant figures.)
(iv) Estimate: $(1 \times 10^{-2}) \times (9 \times 10^{-3})$
$= (1 \times 9) \times (10^{-2} \times 10^{-3}) = 9 \times 10^{-5} = 0{\cdot}00009$.
Calculate: 0·0001288 = 0·000129. (Rounded to three significant figures.)
Notice that $0{\cdot}0001288 \doteq 1 \times 10^{-4}$ and 9×10^{-5} is not far off 10×10^{-5} which is just 1×10^{-4}, so the answer does agree reasonably well with the estimate.
In each of these exercises, even if the estimate is not particularly close to the exact answer, it does indicate the order of magnitude.

16 In each case you must work out the brackets first.

(i) Estimate: $6{\cdot}32 \times 4{\cdot}6 \doteq (6 \times 10^0) \times (5 \times 10^0)$
$= (3 \times 10^1)$.
Complete the estimate: $(7 \times 10^2) + (3 \times 10^0)$
$\doteq 7 \times 10^2 = 700$.
Calculate: 699·072. (To three decimal places.)
(ii) Estimate: $0{\cdot}013 + 0{\cdot}004 \doteq (1 \times 10^{-2}) + (4 \times 10^{-3}) \doteq 1 \times 10^{-2}$. (Only consider the highest power.)
Complete the estimate: $(1 \times 10^{-2}) \div (1 \times 10^{-2}) = 1$.
Calculate: 0·7058824 = 0·706. (Rounded to three decimal places.)

(iii) Estimate: $14{\cdot}6 + 15{\cdot}21$
$\hat{=} (1 \times 10^1) + (2 \times 10^1) = 3 \times 10^1$.
$17{\cdot}02 - 206 \hat{=} (2 \times 10^1) - (2 \times 10^2)$
$\hat{=} (-2) \times 10^2$.
Complete the estimate: $(3 \times 10^1) \times ((-2) \times 10^2)$
$= (3 \times (-2)) \times (10^1 \times 10^2) = (-6) \times 10^3$
$= -6000$.
Calculate: $-5633{\cdot}4938 = -5633{\cdot}494$.
(Rounded to three decimal places.)

17

(i) (a) 0·2 (b) 36·721 (c) 0·0045
(ii) (a) 54, 163·2 (b) 72·9 (c) 0·31 (d) 0·01
(e) 5·41632 (f) 0·729 (g) 0·000031
(h) 0·000342
(iii) (a) 0·549 kg (b) 0·67 m (c) 0·0145 m
(d) 3710 g
(iv) (a) 10^{-1} (b) 10^3 (c) 10^0 (d) 10^1
(e) 10^{-5} (f) $0{\cdot}00100 = 0{\cdot}001 = 10^{-3}$
(v) (a) $1{\cdot}4952 \times 10^1$ (b) $1{\cdot}78 \times 10^2$
(c) $3{\cdot}4 \times 10^{-2}$ (d) $1{\cdot}00 \times 10^{-4}$ (e) $1{\cdot}3 \times 10^{-2}$
(vi) (a) 30,000 (b) 0·021 (c) 1970
(d) 0·0000102
(vii) (a) 10^5 (b) $10^0 = 1$ (c) 10^{-6} (d) 10^6
(e) 10^2 (f) 10^2 (g) 10^4 (h) 10
(viii) (a) Estimate: 30. Calculate: 32·2.
(b) Estimate: 100. Calculate: 127·591.
(c) Estimate: 100. Calculate: 151·04.
(d) Estimate: 0·01. Calculate: 0·0112.
(ix) (a) Estimate: 2000. Calculate: 2216·199.
(b) Estimate: 0·5. Calculate: 0·9121965.
(c) Estimate: 20. Calculate: 12·932886.
(d) Estimate: 0·001. Calculate: 0·001716.
(e) Estimate: 0·001. Calculate: 0·0019219.
(x) (a) 32·2 (b) 127·6 (c) 151·0 (d) 0
(xi) (a) $2{\cdot}2 \times 10^3$ or 2200 (b) $9{\cdot}1 \times 10^{-1}$ or 0·91
(c) $1{\cdot}3 \times 10^1$ or 13 (d) $1{\cdot}7 \times 10^{-3}$ or 0·0017
(e) $1{\cdot}9 \times 10^{-3}$ or 0·0019

Section 1.4 Solutions

1

(i) (a) $1\frac{1}{2}''$ (b) $2\frac{1}{8}''$
(ii) (a) $\frac{3}{4}$ square (b) $\frac{5}{12}$ rectangle (c) $\frac{4}{10}$ rectangle
(d) $\frac{3}{16}$ circle

2

(i) (a) $\frac{4}{5}$ is already in its simplest form. It is *not* the same as $\frac{2}{5}$.
(b) $\frac{4}{10} = \frac{4 \div 2}{10 \div 2} = \frac{2}{5}$. So $\frac{4}{10}$ and $\frac{2}{5}$ are equivalent fractions.
(c) $\frac{4}{15}$ is already in its simplest form. It is *not* the same as $\frac{2}{5}$.
(d) $\frac{10}{25} = \frac{10 \div 5}{25 \div 5} = \frac{2}{5}$. So $\frac{10}{25}$ and $\frac{2}{5}$ are equivalent fractions.

(ii) (a) $\frac{2}{7}$ is in its simplest form; $\frac{4}{28} = \frac{1}{7}$. So $\frac{4}{28}$ and $\frac{2}{7}$ are not equivalent.
(b) $\frac{3}{16}$ is in its simplest form; $\frac{6}{8} = \frac{3}{4}$. So $\frac{3}{16}$ and $\frac{6}{8}$ are not equivalent.
(c) $\frac{4}{6} = \frac{4 \div 2}{6 \div 2} = \frac{2}{3}$. So $\frac{4}{6}$ and $\frac{2}{3}$ are equivalent.
(iii) (a) $\frac{15}{30} = \frac{1}{2}$ (b) $\frac{15}{60} = \frac{1}{4}$ (c) $\frac{25}{100} = \frac{1}{4}$
(d) $\frac{54}{72} = \frac{6}{8} = \frac{3}{4}$

3

(i) $\frac{3}{10} + \frac{5}{10} = \frac{3+5}{10} = \frac{8}{10} = \frac{4}{5}$
(ii) $\frac{11}{32} - \frac{7}{32} = \frac{11-7}{32} = \frac{4}{32} = \frac{1}{8}$
(iii) $\frac{7}{16} - \frac{2}{16} = \frac{7-2}{16} = \frac{5}{16}$
(iv) $\frac{1}{3} + \frac{1}{6} = \frac{2}{6} + \frac{1}{6} = \frac{2+1}{6} = \frac{3}{6} = \frac{1}{2}$
(v) $\frac{8}{9} + \frac{1}{5} = \frac{40+9}{45} = \frac{49}{45}$
(vi) $\frac{5}{16} + \frac{1}{3} - \frac{1}{6} = \frac{15+16-8}{48} = \frac{23}{48}$

4

(i) $\frac{\cancel{11}}{{}_{4}\cancel{12}} \times \frac{\cancel{3}^{1}}{\cancel{11}} = \frac{1}{4}$
(ii) $\frac{\cancel{5}}{{}_{2}\cancel{6}} \times \frac{\cancel{9}^{3}}{\cancel{10}_{2}} = \frac{3}{2 \times 2} = \frac{3}{4}$
(iii) $\frac{\cancel{17}}{{}_{11}\cancel{33}} \times \frac{\cancel{12}^{3}}{\cancel{34}_{2}} \times \frac{\cancel{3}}{\cancel{4}_{1}} = \frac{3}{11 \times 2} = \frac{3}{22}$
(iv) $\frac{3}{4}$ of $2 = \frac{3}{4} \times 2 = \frac{3 \times \cancel{2}}{\cancel{4}_{2}} = \frac{3}{2}$
(v) $\frac{4}{5}$ of $\frac{1}{2} = \frac{4}{5} \times \frac{1}{2} = \frac{2}{5}$
(vi) $\frac{5}{8}$ of 720 = 450, so there are 450 boys.
$\frac{3}{8}$ of 720 = 270, so there are 270 girls.

5

(i) $10 \div \frac{1}{2} = 10 \times \frac{2}{1} = 20$
(ii) $\frac{2}{5} \div \frac{5}{7} = \frac{2}{5} \times \frac{7}{5} = \frac{14}{25}$
(iii) $\frac{16}{9} \div \frac{4}{3} = \frac{\cancel{16}^{4}}{{}_{3}\cancel{9}} \times \frac{\cancel{3}^{1}}{\cancel{4}_{1}} = \frac{4}{3}$
(iv) $\frac{14}{8} \div \frac{7}{16} = \frac{\cancel{14}^{2}}{{}_{1}\cancel{8}} \times \frac{\cancel{16}^{2}}{\cancel{7}_{1}} = 4$

6

(i) (a) $\frac{33}{8}$ (b) $\frac{12}{5}$ (c) $\frac{36}{11}$ (d) $\frac{79}{32}$
(ii) (a) $4\frac{2}{3}$ (b) $3\frac{4}{5}$ (c) $\frac{39}{6} = \frac{13}{2} = 6\frac{1}{2}$
(d) $\frac{48}{42} = \frac{8}{7} = 1\frac{1}{7}$

7

(i) $1\frac{3}{8} + 2\frac{1}{4} = 3 + \frac{3}{8} + \frac{1}{4} = 3 + \frac{5}{8} = 3\frac{5}{8}$
(ii) $2\frac{2}{3} + 4\frac{3}{5} = 6 + \frac{2}{3} + \frac{3}{5} = 6 + \frac{19}{15} = 7\frac{4}{15}$
(iii) $2\frac{5}{16} - 1\frac{1}{8} = 1 + \frac{5}{16} - \frac{1}{8} = 1 + \frac{3}{16} = 1\frac{3}{16}$
(iv) $4\frac{1}{8} - 1\frac{3}{4} = 3 + \frac{1}{8} - \frac{3}{4} = 3 + \frac{(-5)}{8} = 2\frac{3}{8}$
Notice that in this last exercise the fraction has a negative numerator, so we must evaluate $3 - \frac{5}{8}$. This can be done by considering

$$2 + 1 - \frac{5}{8} = 2 + \frac{8-5}{8} = 2\frac{3}{8}$$

(We've taken 1 from the whole number part and changed it to $\frac{8}{8}$.)

8

(i) $1\frac{1}{2} \times 2\frac{2}{3} = \frac{3}{2} \times \frac{8}{3} = 4$

(ii) $4\frac{1}{5} \div \frac{7}{10} = \frac{21}{5} \div \frac{7}{10} = \frac{21}{5} \times \frac{10}{7} = 3 \times 2 = 6$

(iii) $\frac{4}{5} \times 1\frac{7}{8} = \frac{4}{5} \times \frac{15}{8} = \frac{3}{2} = 1\frac{1}{2}$

(iv) $2\frac{1}{4} \times 1\frac{1}{5} \div 1\frac{1}{8} = \frac{9}{4} \times \frac{6}{5} \div \frac{9}{8} = \frac{9}{4} \times \frac{6}{5} \times \frac{8}{9} = \frac{12}{5}$
$= 2\frac{2}{5}$

9

(i) Work out the brackets first.
$1\frac{1}{2} + \frac{3}{5} = 1 + \frac{11}{10} = 2\frac{1}{10}$
$(1\frac{1}{2} + \frac{3}{5}) \div 1\frac{2}{5} = 2\frac{1}{10} \div 1\frac{2}{5} = \frac{21}{10} \div \frac{7}{5} = \frac{21}{10} \times \frac{5}{7}$
$= \frac{3}{2} = 1\frac{1}{2}$

(ii) $4\frac{1}{4} \times \frac{2}{17} = \frac{17}{4} \times \frac{2}{17} = \frac{1}{2}$
$2\frac{1}{4} + (4\frac{1}{4} \times \frac{2}{17}) = 2\frac{1}{4} + \frac{1}{2} = 2\frac{3}{4}$

(iii) $3\frac{1}{4} - 2\frac{1}{2} = (3 - 2) + \frac{1}{4} - \frac{1}{2} = 1 - \frac{1}{4} = \frac{3}{4}$
$4\frac{1}{2} + 1\frac{1}{4} = 5 + \frac{1}{2} + \frac{1}{4} = 5\frac{3}{4}$
So
$(3\frac{1}{4} - 2\frac{1}{2}) \div (4\frac{1}{2} + 1\frac{1}{4}) = \frac{3}{4} \div 5\frac{3}{4}$
$= \frac{3}{4} \div \frac{23}{4} = \frac{3}{4} \times \frac{4}{23} = \frac{3}{23}$

10 (i) 0·5 (ii) 0·25 (iii) 0·125 (iv) 0·75
(v) 0·33 (Rounded to two decimal places.)
(vi) 2·67 (Rounded to two decimal places.)
(vii) 3·375

11

(i) (a) $\frac{5}{8}$ circle (b) $\frac{7}{12}$ rectangle
(c) $\frac{1}{5}$ or $\frac{3}{15}$ rectangle

(ii) (a) $\frac{1}{3}$ (b) $\frac{7}{8}$ (c) $\frac{23}{40}$ (d) $\frac{1}{4}$

(iii) (a) $\frac{3}{7}$ (b) $\frac{7}{9}$ (c) $\frac{1}{5}$ (d) $\frac{11}{45}$ (e) $\frac{11}{100}$

(iv) (a) $\frac{3}{7}$ (b) $\frac{1}{10}$ (c) $\frac{3}{28}$

(v) (a) $\frac{4}{3}$ or $1\frac{1}{3}$ (b) 1 (c) $\frac{1}{7}$ (d) $\frac{1}{8}$

(vi) (a) $4\frac{13}{30}$ (b) $4\frac{10}{21}$ (c) $\frac{19}{24}$

(vii) (a) 4 (b) $\frac{2}{3}$ (c) $\frac{6}{7}$ (d) 9

(viii) (a) $\frac{1}{2}$ (b) $\frac{7}{36}$

(ix) (a) 2·75 (b) 2·875 (c) 4·67 (Rounded to two decimal places.) (d) 3·4

Section 1.5 Solutions

1

(i) The ratio of red to blue is 3:2 (3 + 2 = 5; this gives the denominator). So $\frac{3}{5}$ is red and $\frac{2}{5}$ is blue. To make 15 litres we need
$\frac{3}{5} \times 15 = 9$ litres of red paint
and
$\frac{2}{5} \times 15 = 6$ litres of blue

(ii) There are 25 short plants and 75 long plants. The ratio is therefore 25:75 or 1:3.

(iii) 2 + 3 + 1 = 6; so there were 6 parts altogether, and $\frac{2}{6}$ or $\frac{1}{3}$ was gin, $\frac{3}{6}$ or $\frac{1}{2}$ was vermouth and $\frac{1}{6}$ was orange juice.
2 pints of the cocktail required
$\frac{1}{3} \times 2 = \frac{2}{3}$ pint of gin
$\frac{1}{2} \times 2 = 1$ pint of vermouth
and
$\frac{1}{6} \times 2 = \frac{1}{3}$ pint orange juice

2 On the map the distance is about 2 cm. So the actual distance is 2 x 50,000 cm = 100,000 cm or 1 km.

3

(i) (a) $\frac{50}{100} = \frac{1}{2}$ (b) $\frac{33}{100}$ (c) $\frac{4}{100} = \frac{1}{25}$
(d) $\frac{90}{100} = \frac{9}{10}$

(ii) (a) 14 ÷ 100 = 0·14 (b) 25 ÷ 100 = 0·25
(c) 3 ÷ 100 = 0·03 (d) 0·5 ÷ 100 = 0·005

(iii) (a) 65% of 540 = 0·65 x 540 = 351, so 351 wives worked.
(b) 50% of 540 = 0·5 x 540 = 270, so 270 husbands earned over £6000 a year.
(c) 14% of 540 = 0·14 x 540 = 75·6. But 75·6 is not a whole number. We'd probably say 75 or 76 couples had a joint income of less than £50 per week.

4

(i) (a) 0·2 x 100 = 20% (b) 0·14 x 100 = 14%
(c) 0·90 = 90% (d) 2·5 = 250%

(ii) (a) $\frac{1}{10}$ = 0·1 = 10% (b) $\frac{1}{5}$ = 0·2 = 20%
(c) $\frac{1}{3}$ = 0·333 ≏ 33% (d) $1\frac{1}{4}$ = 1·25 = 125%

5

(i) 1·15 x £ 20 = £23

(ii) 0·80 x £20 = £16 (without adding VAT)

(iii) 1·15 x 0·80 x £20 = £18·40 = 0·80 x 1·15 x £20
This shows that the order doesn't matter; it still costs the same.

6

(i) (a) [7] [−] [3] [=] [STO] [6] [+] [RCL] [=] [10]
Check: 6 + (7 − 3) = 6 + 4 = 10.
(b) [2] [x] [6] [=] [STO] [7] [+] [RCL] [=] [19]
Check: 7 + (2 x 6) = 7 + 12 = 19.
(c) [6] [÷] [2] [=] [STO] [2] [+] [RCL] [=] [5]
Check: 2 + (6 ÷ 2) = 2 + 3 = 5.

(ii) (a) Estimate: 17·6 − 24·3 ≏ (2 x 10) − (2 x 10) = 0.
To complete the estimate: (1 x 10) + 0 = (1 x 10).
Calculate: [17·6] [−] [24·3] [=] [STO] [6·321] [+] [RCL] [=] [−0·379].

(b) Estimate:
$(1 \times 10^1) \div \{(2 \times 10^{-1}) + (1 \times 10^1)\} \simeq 1$.
Calculate: [0·2] [+] [10·62] [=] [STO] [14·7] [÷] [RCL] [=] [1·359]. (Rounded to three decimal places.)

(c) Estimate:
$(1 \times 10^{-1}) + \{(1 \times 10^{-3}) \times (1 \times 10^1)\} \simeq 1 \times 10^{-1} = 0{\cdot}1$.
Calculate: [0·001] [x] [14·6] [=] [STO] [0·12] [+] [RCL] [=] [0·135]. (Rounded to three decimal places.)

7

(i) $\frac{15}{2+3} = 15 \div (2+3)$
Calculate [2] [+] [3] [=] [STO] [15] [÷] [RCL] [=] [3].

(ii) Estimate:
$(1 \times 10^2) \div \{(3 \times 10^1) + (2 \times 10^1)\} \simeq 2$.
Calculate: [31·2] [+] [18·7] [=] [STO] [149·6] [÷] [RCL] [=] [2·998]. (Rounded to three decimal places.)

(iii) Estimate:
$\{(6 \times 10^1) - (3 \times 10^1)\} \div \{(2 \times 10^1) + (5 \times 10^1)\} \simeq 0{\cdot}5$.
Calculate: [16] [+] [53] [=] [STO] [64] [−] [28] [=] [÷] [RCL] [=] [0·522]. (Rounded to three decimal places.)
You may have used different key sequences to the ones given here. That doesn't matter— as long as you got the right answers.

8

(i) 0·149 (Rounded to three decimal places.)

(ii) 0·5

(iii) Estimate:
$(1 \times 10^2) \div \{(2 \times 10^2) + (1 \times 10^1)\} \simeq 0{\cdot}5$.
Calculate: [159·6] [+] [13·8] [=] [1/x] [x] [149·3] [=] [0·861]. (Rounded to three decimal places.)

9 (i) 196 (ii) 16 (iii) 21697·29 (iv) 0·765625
Notice that from part (iii), the square of a negative number is always positive, since $(-) \times (-) = (+)$.

10 (i) 9 (ii) 3 (iii) 9·48683 (iv) 0·948683 (v) 0·3
Did you expect $\sqrt{0{\cdot}9}$ to be 0·3? Well, remember that 0·3 is less than 1, so we'd expect that $(0{\cdot}3)^2$ would be less than 0·3 whereas 0·9 is bigger than 0·3. In fact part (iv) shows that $\sqrt{0{\cdot}09} = 0{\cdot}3$.

11

(i) (a) 5:1 (b) 2:3
(c) 8 litres red, 4 litres orange, and 2 litres white

(ii) (a) $\frac{13}{20}$ (b) 0·001 (c) 60% (d) 60%

(iii) $0{\cdot}85 \times 1{\cdot}15 \times 1{\cdot}10 \times £20 = £21{\cdot}51$

(iv) (a) 4·86 (b) 1·67 (c) 0·18

(v) (a) 0·76 (b) 59,000 (c) 200

(vi) (a) 256 (b) 0·0001 (c) 12·649 (d) 0·4

MODULE 2

2.1 Symbols

2.2 Algebraic Simplification

2.3 Brackets

2.4 Solving Equations

2.5 Equations, Formulas and Proportion

2.1 Symbols

TRY THESE QUESTIONS FIRST

1 Express the following sequence of instructions using a symbol to represent the unknown number:

Think of a number. Add 3. Multiply by (−2). Add 4. Divide by 3. Add the number you first thought of.

2 (i) Write down the coefficient of s in the expression below:

$2b - 6s + 3$

(ii) Evaluate the following expression when $b = 7$

$$\frac{2(b-2) - 3b}{11}$$

3 (i) Write down a formula which gives the perimeter of a rectangle with sides l and w.

(ii) In a country hospital a certain number of boys were born one week. The total number of births in that week was 11. Write down an equation which gives the number of girls born that week.

2.1(i) SYMBOLS

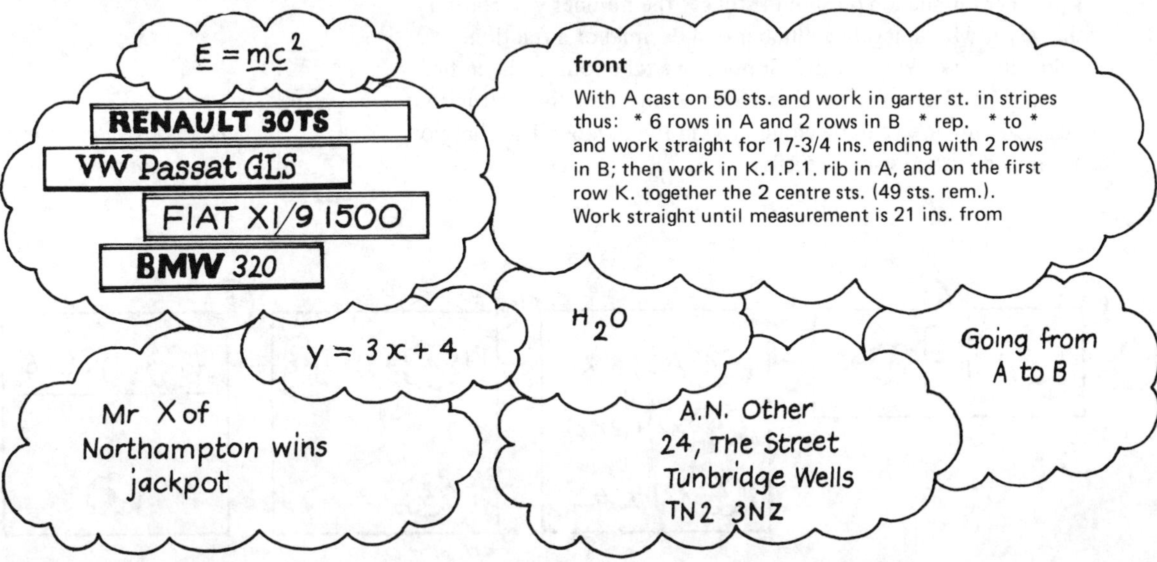

CHECK YOUR ANSWERS

1 We've used a to stand for the unknown number. The sequence becomes *Section 2.1(i)*

$$\frac{[(a+3)\times(-2)]+4}{3}+a$$

2 (i) The coefficient of s is -6. *Section 2.1(ii)*

(ii) $\frac{2(7-2)-21}{11}=-1$

3 (i) Perimeter $= P = 2l + 2w$ *Section 2.1(iii)*

(ii) Let b be the number of boys and g the number of girls born that week.
Then $g = 11 - b$.

All the examples in the diagram above contain abbreviations or symbols. You may recognise some of these, such as the postal code or the knitting instructions. In certain expressions a symbol is used to stand for an unknown quantity, for example, in

Other expressions (such as $E = mc^2$) may not be so familiar

'Mr X' or 'From A to B'.

In this case the symbols stand for unknown names.

In this section we are going to concentrate on symbols which stand for unknown numbers. First of all, here is a puzzle for you to try. It is better if you can do it in your head if possible.

Otherwise write down the intermediate steps.

1 Think of a number–any number you like.
2 Add three to the number you thought of.
3 Multiply the result by 2.
4 Now subtract 6.
5 Finally divide by 2.

Did you end up with the number you started with?

If not, you did something wrong so try again.

Try another number. You should still get the number you started with. Try it with a negative number or a decimal or a fraction. It should still work. Why is this? Suppose we take four as the initial number. The diagram below traces what happens as the instructions are carried out. We've put a cloud around the number 4 so that you can follow its progress more easily.

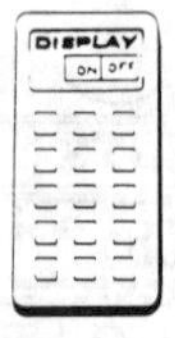

Use your calculator if necessary.

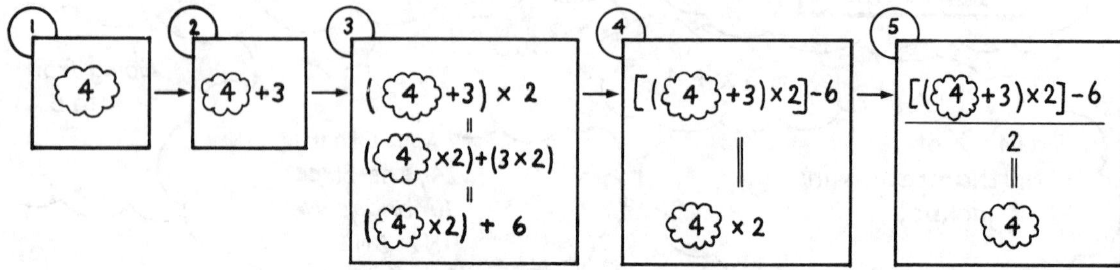

At stage 3 we multiplied out the brackets to make calculation easier at stage 4. The same process applies to any number. Choose a number, and follow through the diagram below, putting the number you choose in the clouds.

Remember that

$(4 + 3) \times 2 = 2 \times (4 + 3)$

and

$= (2 \times 4) + (2 \times 3)$
$= (4 \times 2) + (3 \times 2)$

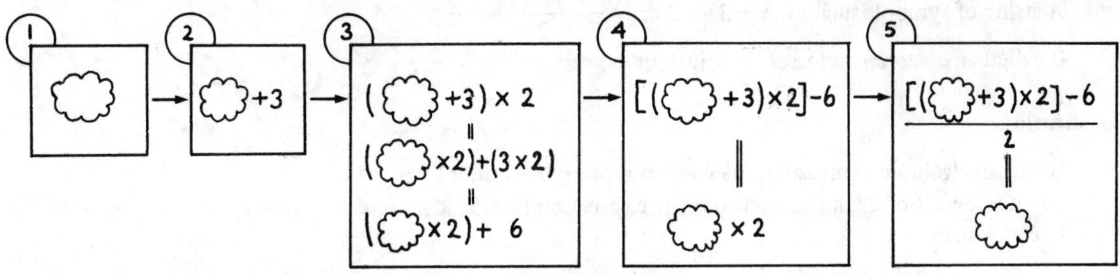

In this diagram the cloud stands for the unknown number. But clouds can be cumbersome and it is more usual to use a letter. In the diagram below we've used the letter *A* to stand for the unknown number.

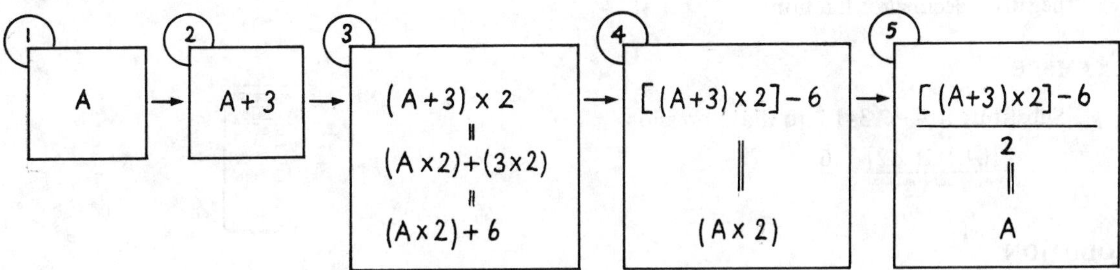

The next example indicates how to write down a sequence of instructions for a similar puzzle.

EXAMPLE

Express the following sequence of instructions using a symbol to stand for the unknown number.
Think of a number. Add 5. Multiply the result by 2.

SOLUTION

We choose x to stand for the unknown number. The sequence then becomes:

Think of a number:	x
Add 5:	$x + 5$
Multiply by 2:	$(x + 5) \times 2$

TRY SOME YOURSELF

1 Express each of the following sequences of instructions using a symbol to stand for the unknown number:

(i) Think of a number. Add 2. Multiply by 3. Subtract 4.

(ii) Think of a number. Multiply by 6. Add 2. Divide by 3. Add 1.

(iii) Think of a number. Multiply by 3. Subtract 2. Multiply by the number you first thought of.

A string of symbols such as $(x + 3) \times 2$ or $\dfrac{[(A + 3) \times 2] - 6}{2}$ is called an *algebraic expression* or just an *expression.*

Substitution

Given an algebraic expression, it's easy to replace the symbol with any number. For example, consider the expression $(x + 5) \times 2$. If $x = 3$, then

$$(x + 5) \times 2 = (3 + 5) \times 2 = 8 \times 2 = 16$$

We have just replaced the symbol x with the number 3, and evaluated the resulting expression.

This process is called *substitution.* In fact any number can be substituted into an algebraic expression, whether it's positive or negative, decimal or fraction.

EXAMPLE

Substitute $A = -2{\cdot}341$ into the expression

$$\frac{[(A + 3) \times 2] - 6}{2}$$

SOLUTION

Putting $A = -2{\cdot}341$ gives

$$\frac{[(-2{\cdot}341 + 3) \times 2] - 6}{2} = -\,2{\cdot}341$$

Notice that this is the example from the beginning of the section —so the answer ***should*** *be the number we first started with— as indeed it is.*

TRY SOME YOURSELF

2(i) Substitute $r = 4$ into the expression

$$(4 + r) - 6$$

(ii) Substitute $m = (-1)$ into the expression

$$[(m + 3) \times 2] + 7$$

(iii) Substitute $d = 0{\cdot}12$ into the expression

$$[(2 \times d) + 16] \div 8$$

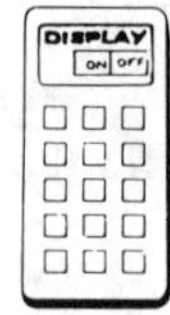

2.1(ii) ABBREVIATING THE NOTATION

Expressions like $[(2 \times m) + 16] \div 8$ and $[(A + 3) \times 2] - 6$ are very cumbersome to manipulate, so we now introduce a neater way of writing them down.

For example, instead of writing

$2 \times x$

it is shorter and neater to write

$2x$

We just omit the multiplication sign.

$x \times 2$ is also abbreviated to $2x$.

You may have thought that $x \times 2$ should be written $x2$, but it is usual in mathematics to put the number first. The same applies if the number is negative. Here are some examples:

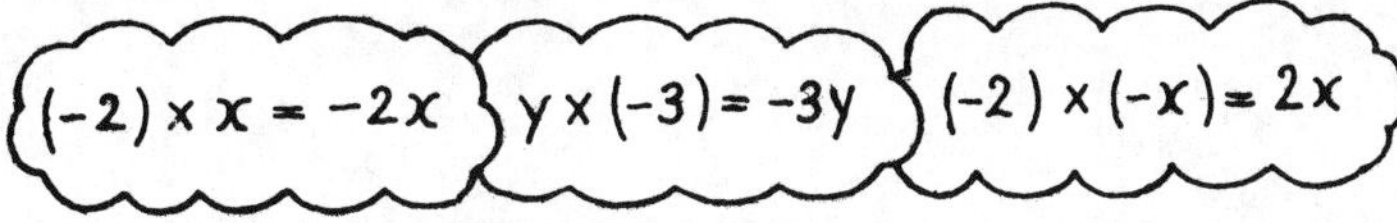

Remember that

$(-) \times (-) = (+)$

So that

$(-2) \times (-x) = 2x.$

Using this system we should perhaps write

$1 \times a$ as $1a$ and $(-1) \times a$ as $-1a$

However, in practice the number 1 is omitted and

$1 \times a$ is just written as a

$(-1) \times a$ is written as $-a$

The multiplication sign is also omitted when multiplying brackets, so that

$3 \times (x + 2)$ is written as $3(x + 2)$

$(y + 2) \times 3$ is written as $3(y + 2)$

Compare this with brackets involving only numbers:

$(1 + 2) \times 3 = 3 \times (1 + 2)$
$= 3(1 + 2)$

Again, it is usual to put the number first. Thus

$[(A + 3) \times 2] - 6$ is written as $2(A + 3) - 6$.

EXAMPLE

Write $2 \times d \times (-4)$ in its most abbreviated form.

SOLUTION

We first rewrite the expression putting the numbers first.

$2 \times d \times (-4) = 2 \times (-4) \times d = -8d$

The order should always be sign–number–symbol.

TRY SOME YOURSELF

3 Write each of the following expressions in its most abbreviated form:
(i) $4 \times a$ (ii) $(-2) \times a$ (iii) $y \times 3$
(iv) $b \times (-3) \times (-2)$ (v) $(x + 5) \times 3$ (vi) $(b + 2) \times (-4)$.

When the expression is in its most abbreviated form, the number in front of the symbol is called its *coefficient*. For example, the coefficient of a in the expression $4a$ is 4.

Here are some more examples:

The coefficient of t in $-6t$ is -6

The coefficient of s in $143s$ is 143

The coefficient of x in $-1980x$ is -1980

Although the multiplication sign can be omitted when writing down the expression, it's important to remember that it *is* intended. This is particularly important when substituting numerical values into an expression.

For example, in the expression 4a, if we substitute a = 7 we get 28 not 47.

EXAMPLE

Substitute $x = 3$ into the expression

$$\frac{2(x+3)-7x}{4}$$

SOLUTION

We need to substitute $x = 3$. This gives

$$\frac{2(x+3)-7x}{4} = \frac{2(3+3)-21}{4} = \frac{-9}{4}$$

Remember that 7x means 7 × x.

TRY SOME YOURSELF

4(i) Substitute $x = 5$ into the expression

$$3(x+5)$$

(ii) When $x = 4$, what is the value of the expression below?

$$\frac{3(x+5)-3x+1}{2} - x$$

2.1(iii) FORMULAS AND EQUATIONS

The perimeter of a square is the total length of its sides. This square, whose sides are each 3, has perimeter

$$3 + 3 + 3 + 3 = 4 \times 3$$

The perimeter of any square can be derived in the same way. If each side has length l the perimeter is

$$l + l + l + l = 4 \times l \text{ or } 4l$$

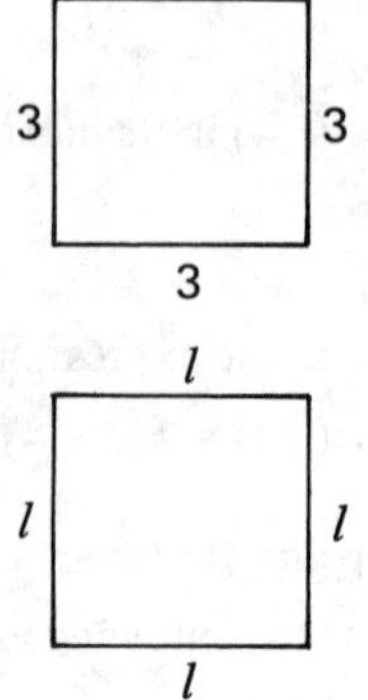

Now, if P stands for the perimeter, then

$$P = 4l$$

This is a *formula* for the perimeter of a square.

A formula can also be derived for the area of a rectangle:

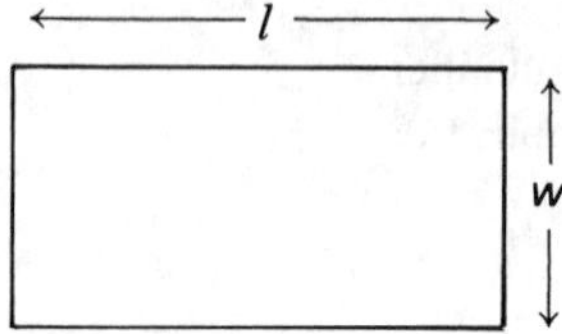

l stands for the unknown length and w for the unknown width. A stands for the area.

Area = length × width

$A = lw$

Formulas can turn up in lots of different contexts. They provide an easy way of remembering a general result—such as the area of a rectangle.

More generally if one expression is equated with another the result is usually called an *equation.* Here are some equations:

We now introduce some more terminology. Consider the equation

$P = 2l - 3w$

$2l$ and $-3w$ are called *terms.* A *term* consists of a *symbol, or symbols,* and a *coefficient.*

A term may involve two or more symbols. For example, in the term 6tz the symbols tz have coefficient 6.

The term involving l is $2l$. It has coefficient 2. The coefficient of w in the term involving w is -3.

EXAMPLE

Write down the coefficient of the term involving x in

$y = 2x + 3z$

SOLUTION

The term involving x is $2x$, and the coefficient is 2.

TRY SOME YOURSELF

5(i) Write down the term involving x in

$T = 14 - 6x$

(ii) Write down the term involving pq in

$5p + 6pq + 4q$

(iii) Write down the coefficients of each of the terms in

$x - y + xy + z$

Deriving an equation from a problem

Problems can often be solved by deriving an equation and solving the equation. For the moment we're going to discuss how to derive equations. Later in this module, in Section 2.4, we will discuss how to *solve* some equations. The following example illustrates all the ideas we have discussed so far.

EXAMPLE

Think of a number. Double it. Add 3. Call the number you started with x and the result y. Express these instructions as an equation and find the value of y when $x = -4$.

SOLUTION

The instructions become

Think of a number:	x
Double it:	$2x$
Add 3:	$2x + 3$

The answer is y, so that

$$y = 2x + 3$$

When $x = -4$,

$$y = (2 \times (-4)) + 3 = -5$$

To find the value of y when $x = -4$ we must substitute $x = -4$ into the right hand side of the equation.

Since the right hand side is equal to the left hand side this gives the corresponding value of y.

TRY SOME YOURSELF

6 In each of the following let x stand for the number you choose and y for the result. Express the instructions as an equation and find the corresponding value of y for the given value of x.

(i) (a) Think of a number. Multiply it by 3. Subtract 2.
(b) What is the value of y when $x = -2$?

(ii) (a) Think of a number. Multiply it by -2. Add 1. Multiply by 2.
(b) What is the value of y when $x = 3$?

(iii) (a) Think of a number. Multiply it by 3. Subtract 2. Divide by 4. Add 4. Multiply by the number you first thought of.
(b) What is the value of y when $x = 2$?

In each of these exercises you were told that the unknown quantities were x and y. However, problems are often expressed in words and in order to solve such a problem you will need to identify the unknown quantities and choose suitable symbols for yourself. The next example illustrates how such an equation might be derived.

EXAMPLE

Oranges cost 8 pence each and lemons 7 pence. A mixed bag of oranges and lemons costs 46 pence. Express this information as an equation.

SOLUTION

We don't know how many oranges were in the bag or how many lemons. These are the unknown quantities so we must now choose suitable symbols.

First identify the unknown quantities.
The choice of symbols is arbitrary.

Let m stand for the number of oranges and n stand for the number of lemons.

Most letters would be OK here– but you should beware of choosing the letter o since it looks like a zero.

1 orange costs 8 pence, so m oranges cost $8m$ pence. Similarly, 1 lemon costs 7 pence, so n lemons cost $7n$ pence. The total cost was therefore $(8m + 7n)$ pence.

But the total cost of the mixed bag was 46 pence.

If we now equate these two pieces of information we get an equation

$$46 = 8m + 7n$$

We equate the total cost giving

$46 = 8m + 7n$

Usually the most difficult part in solving a problem is choosing the symbols and deriving an equation. Having obtained an equation it is often relatively easy to find the solution. The following exercises give you some practice in choosing symbols and deriving the equation. Try them now, but don't worry if you find them difficult. After you have worked through the remaining sections in this module you may like to have another go, when, hopefully you will find they're much easier.

This equation involves the unknowns m and n.

TRY SOME YOURSELF

7(i) A gallon of petrol costs £1·70. Write down an expression for the cost in pounds of x gallons of petrol.

(ii) A hotel employed a certain number of people, each earning £50 per week. There were a number of guests staying in the hotel one week, each paying £80 a week.
  (a) What was the cost in pounds of the wages bill for a week?
  (b) What was the total income from the guests that week?
  (c) The total profit that week was £1100. Derive an equation which relates the profit to the number of guests and the number of employees.

The profit is the difference between income and wages.

After you have worked through this section you should be able to

a Write down a sequence of instructions using a symbol to stand for the unknown quantity
b Write down an algebraic expression in its most abbreviated form
c Identify the coefficient of a given symbol in an expression
d Substitute a numerical value for the unknown in an algebraic expression
e Choose symbols to stand for unknown quantities in a given situation and write down an equation to summarise the information

Finally, here are some exercises if you want more practice.

TRY SOME MORE YOURSELF

8(i) Write down each of the following sets of instructions using a symbol for the unknown number:
  (a) Think of a number. Multiply it by 4. Subtract 2. Multiply the result by 6. Add the number you first thought of.
  (b) Think of a number. Subtract 7. Multiply the result by (−1). Add 14. Subtract the number you first thought of.

(ii) Write each of the following in its shortest form:
  (a) $7 \times a$ (b) $3 \times b : (-4)$ (c) $(-z) \times (-3)$ (d) $0{\cdot}12 \times (-r)$.

(iii) (a) Substitute $l = 1{\cdot}4$ into the expression

$$\frac{(5l + 7) - l}{2}$$

(b) Substitute $f = -3{\cdot}2$ into the expression

$$7[(1{\cdot}4f - 2{\cdot}3) - 4{\cdot}6f]$$

(iv) (a) Write down the coefficient of b in

$$14a - 7b + 8c$$

(b) Write down the coefficients of z and w in

$$3t - w + z$$

(v) A gallon of petrol costs £1·70 and a can of oil £0·40.
Let the total cost be £c.

(a) Derive an equation for the total cost if you bought x gallons of petrol and one can of oil.
(b) Derive an equation for the total cost if you bought x gallons of petrol and y cans of oil.
(c) What is the value of c if $x = 9{\cdot}8$ and $y = 2$?

(vi) A hotel employs a number of people each of whom earns £30 a week. One week there were the same number of guests every night, each paying £17 a night.

(a) What was the wages bill that week?
(b) What was the total income that week?
(c) Suppose the week's profit was £P. Derive an equation for P in terms of the number of employees and the number of guests.
(d) Now suppose that all the guests stayed for four weeks. How much profit was made over the four week period?

2.2 Algebraic Simplification

TRY THESE QUESTIONS FIRST

1 Simplify $-5x + 2y + 7x - 3y$.

2 (i) Rewrite $3 \times f \times (-2) \times l$ in standard form.
(ii) Simplify $2 \times a \times 3b \times a^2$.

3 (i) Rewrite $(2x + 3y) \div (3p \times 2q)$ as an algebraic fraction.
(ii) Find the numerical value of this expression when $x = 1, y = 2, p = 2, q = 3$.
(iii) Write $\frac{1}{y^2}$ using a negative power.

4 Simplify $2ab + b^2 + 7b^2 - 3ab + a^2$.

5 A body's momentum is the product of its mass and its velocity. What are the units of measurement of momentum if mass is measured in grams and velocity in centimetres per second?

LIKE TERMS

simplifying an algebraic expression are, perhaps, ıplifying the expression makes it shorter and easier ɔn you will find that it's not just a matter of making look neater: algebraic simplification is essential. simplification is called algebraic manipulation and ɔften a tedious process it is worth mastering.

Suppose you had to evaluate $5d + 2d$ when $d = 12$. You could do it by just substituting $d = 12$ as follows:

$$(5 \times 12) + (2 \times 12) = 60 + 24 = 84$$

However, it's simpler to say

$$(5 \times 12) + (2 \times 12) = (5 + 2) \times 12 = 7 \times 12 = 84$$

In fact you could make the same simplification to the original expression to get

$$5d + 2d = (5 + 2)\, d = 7d$$

Now if $d = 12$, $7 \times 12 = 84$.

Terms such as $5d$ and $2d$ are called *like terms* because although they involve different coefficients the symbol involved is the same (d).

The numerical expression

$$(5 \times 12) + (2 \times 20)$$

cannot be combined in this way; there is no alternative to evaluating the brackets separately. Similarly

$$5d + 2s$$

cannot be simplified any further. $5d$ and $2s$ are *unlike terms*, since $5d$ involves the symbol d whereas $2s$ involves the symbol s, a different symbol.

That is, unless we know numerical values for d and s.

EXAMPLE

Simplify $7d + 4d + d$.

SOLUTION

$7d + 4d + d = 12d$

The terms are combined by adding the coefficients:

$$7 + 4 + 1 = 12$$

The coefficient of the term d is 1 since although we just write d we mean $1 \times d$.

TRY SOME YOURSELF

1 Simplify each of the following expressions:
(i) $8x - 6x$ (ii) $3y + 4y - 6y$ (iii) $4p + 7p + 18p$.

Combine the terms by adding or subtracting the coefficients.

CHECK YOUR ANSWERS

1 $-5x + 2y + 7x - 3y = 2x - y$ *Section 2.2(i)*

2 (i) $3 \times f \times (-2) \times l = -6fl$ *Section 2.2(ii)*

(ii) $2 \times a \times 3b \times a^2 = 6a^3b$

3 (i) $(2x + 3y) \div (3p \times 2q) = (2x + 3y) \div (6pq) = \frac{2x + 3y}{6pq}$ *Section 2.2(iii)*

(ii) $\frac{2x + 3y}{6pq} = \frac{(2 \times 1) + (3 \times 2)}{6 \times 2 \times 3} = \frac{8}{36} = \frac{2}{9}$

(iii) $\frac{1}{y^2} = y^{-2}$

4 $2ab + b^2 + 7b^2 - 3ab + a^2 = a^2 + 8b^2 - ab$ *Section 2.2(iv)*

5 Momentum = mass × velocity. *Section 2.2(v)*

Velocity is measured in $\frac{\text{cm}}{\text{s}} = \text{cm s}^{-1}$.

The units of momentum are therefore g × cm s^{-1} or g cm s^{-1}.

Expressions often contain some like terms and some unlike terms. Such expressions are simplified by combining the different like terms together. This process is called *collecting like terms.*

EXAMPLE

Simplify $8x + 4y - 2x + 6y$ by collecting like terms together.

SOLUTION

$$\begin{aligned} 8x + 4y - 2x + 6y &= (8x - 2x) + (4y + 6y) \\ &= (8 - 2)x + (4 + 6)y \\ &= 6x + 10y \end{aligned}$$

First rearrange the expression to put like terms next to each other then add or subtract the coefficients.

TRY SOME YOURSELF

2(i) Simplify each of the following expressions by collecting like terms:
(a) $6a + 2b + 7a + 3b$ (b) $-2r + 4p + 2p + 6r$
(c) $2e + f - 3e + 5e$ (d) $3 + 4a - 2 - a$.

(ii) (a) Write down an expression for the perimeter of this figure. Simplify the expression by collecting like terms.
(b) Evaluate this expression when $a = 0{\cdot}25$ cm, $b = 1{\cdot}36$ cm and $l = 4{\cdot}27$ cm.

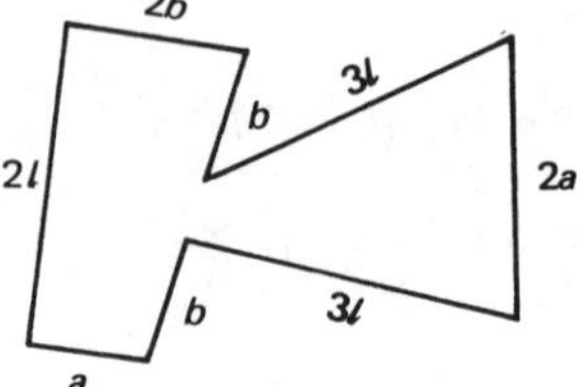

2.2(ii) SYMBOLS MULTIPLIED BY SYMBOLS

The area of a rectangle is given by the formula

$A = lw$

where l is the length of the rectangle and w is the width.

This formula involves two symbols multiplied together. Often more than two symbols are multiplied together. In such cases the multiplication signs are always omitted. The standard form for writing down such expressions is illustrated by

When several symbols are multiplied together they should always be combined into this standard form. For example

$2 \times b \times a \times 3$ is written as $6ab$

$a \times p \times (-2) \times r$ is written $-2apr$

$z \times (-3) \times t \times (-2)$ is written $6tz$

The numbers are multiplied to give the coefficient. Remember that (−) × (−) = (+).

TRY SOME YOURSELF

3 Write each of the following terms in its shortest form:

(i) (a) $7 \times e \times d \times c$
(b) Evaluate the term when $c = 2{\cdot}4$, $d = 4{\cdot}1$ and $e = 2{\cdot}5$.

(ii) (a) $5 \times (-x) \times z \times y$
(b) Evaluate the term when $x = -4{\cdot}25$, $y = -1{\cdot}26$ and $z = -4{\cdot}25$.

(iii) (a) $a \times 4 \times c \times 3 \times d$
(b) What happens if $a = 2{\cdot}9$ and $c = -1$?

(iv) (a) $r \times (-3) \times (-q) \times 2 \times p$
(b) What happens if $p = \frac{1}{2}$?

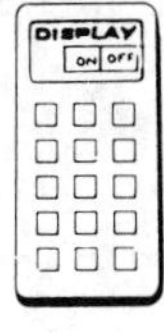

Powers

You've already seen power notation used for numbers. For example

$$10 \times 10 = 10^2$$
$$\text{and } 2 \times 2 = 2^2$$

The same notation is used for symbols, so that

$a \times a = a^2$ — *Pronounced 'a squared'.*

$b \times b \times b = b^3$ — *Pronounced 'b cubed'.*

$z \times z \times z \times z = z^4$ — *Pronounced 'z to the fourth'.*

$d \times d \times d \times d \times d = d^5$ — *Pronounced 'd to the fifth'.*

and so on.

Powers are used to simplify algebraic expressions as illustrated by the next example.

EXAMPLE

Simplify $2 \times a \times 3a \times 4a$.

SOLUTION

$$2 \times a \times 3a \times 4a = 24(a \times a \times a)$$
$$= 24a^3$$

$a \times a \times a = a^3$ — Three as

TRY SOME YOURSELF

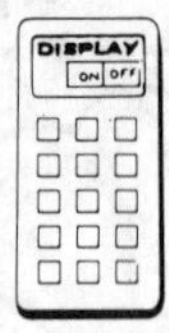

4 Simplify each of the following terms then evaluate the expression by substituting the given value:

(i) $2 \times b \times 3b \times 4b$. Evaluate the expression when $b = 2$.

(ii) $3a \times (-4a)$. Evaluate the expression when $a = -3$.

(iii) $6z \times 3 \times z \times (-4z)$. Evaluate the expression when $z = 2{\cdot}5$.

The next example shows how a term containing more than one symbol is simplified using power notation.

EXAMPLE

Simplify $2 \times a \times 2b \times 3a \times 4b$.

SOLUTION

$$\begin{aligned} 2 \times a \times 2b \times 3a \times 4b &= 2 \times (a \times 3a) \times (2b \times 4b) \\ &= 2 \times 3a^2 \times 8b^2 \\ &= 48a^2b^2 \end{aligned}$$

First rearrange the expression grouping the letters together. Work out the powers then rewrite the term putting the coefficient first and the symbols in alphabetical order.

TRY SOME YOURSELF

5 Simplify each of the following expressions. In each case evaluate the expression for the given substitutions.

(i) (a) $3 \times a \times c \times a \times c$
(b) Evaluate the expression when $a = 2$ and $c = 3$.

(ii) (a) $6aba$
(b) Evaluate the expression when $a = 1$ and $b = 0$.

(iii) (a) $2a \times (-3b) \times 4a \times b^2$
(b) Evaluate the expression when $a = 1$ and $b = -4{\cdot}59$.

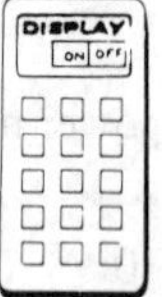

2.2(iii) ALGEBRAIC FRACTIONS

Numerical fractions can be obtained by dividing one number by another. Division of one symbol by another can be written in exactly the same way. Just as

$3 \div 7$ is written as $\frac{3}{7}$,

so $a \div b$ is written as $\dfrac{a}{b}$

Fractions like $\dfrac{a}{b}$ are called *algebraic fractions.* They are manipulated just like numerical fractions. In particular, two fractions are multiplied together by multiplying the numerators and multiplying the denominators. For example

$\dfrac{a}{b}$ is often written as a/b to save space.

$$\frac{a}{b} \times \frac{c}{d} = \frac{ac}{bd}$$

and

$$a \times \frac{1}{b} = \frac{a}{b} = a \div b$$

Just as

$$\frac{3}{4} \times \frac{5}{7} = \frac{3 \times 5}{4 \times 7}$$

and

$$3 \times \frac{1}{4} = \frac{3}{4} = 3 \div 4$$

Thus to divide by a symbol multiply by the reciprocal.

Here, the reciprocal of b is $\frac{1}{b}$ and $a \div b = a \times \frac{1}{b}$.

The next example indicates how an algebraic fraction is simplified and evaluated for given numerical values.

EXAMPLE

(i) Rewrite $(3a \times 2b \times a) \div (a + b)$ as an algebraic fraction.

(ii) Evaluate the fraction when $a = 1{\cdot}2$ and $b = 3{\cdot}5$.

SOLUTION

(i)
$$(3a \times 2b \times a) \div (a + b) = (3a \times 2b \times a) \times \frac{1}{(a+b)}$$
$$= ((3a \times a) \times 2b) \times \frac{1}{(a+b)}$$
$$= 6a^2b \times \frac{1}{(a+b)}$$
$$= \frac{6a^2b}{a+b}$$

To divide by an expression, multiply by the reciprocal–the reciprocal of $(a + b)$ is $\frac{1}{(a+b)}$.

(ii) When $a = 1{\cdot}2$, $b = 3{\cdot}5$,
$$\frac{6a^2b}{a+b} = \frac{6 \times (1{\cdot}2)^2 \times 3{\cdot}5}{1{\cdot}2 + 3{\cdot}5} = 6{\cdot}43$$

We have rounded the answer to two decimal places. Try this yourself on your calculator just to get some practice.

(Of course the expression $(3a \times 2b \times a) \div (a + b)$ could have been evaluated directly. This should give the same answer and acts as a check as to whether or not the simplification is correct.)

You could try substituting $a = 1$ and $b = 2$ to check that both methods give the same answer.

TRY SOME YOURSELF

6 Rewrite each of the following as algebraic fractions. In each case evaluate the fraction for the given substitutions.

(i) (a) $(a + b) \div (c \times d)$
(b) Evaluate the fraction when $a = 1$, $b = 3$, $c = 2$ and $d = 1$.

(ii) (a) $(2x \times 3y) \div (5p \times 7q)$
(b) Evaluate the fraction when $x = 8{\cdot}4$, $y = 7{\cdot}3$, $p = 2{\cdot}2$ and $q = 3{\cdot}0$.

(iii) (a) $(a \div b) \times (c \div d) \times (a \div d)$
(b) Evaluate the fraction when $a = 2$, $b = 1$, $c = 3$ and $d = 1$.

Round off your answer to one decimal place.

Negative powers

The notation for negative powers of numbers can also be extended to symbols. In the same way that

$$\frac{1}{10 \times 10} \text{ is written as } 10^{-2},$$

$$\text{so } \frac{1}{b \times b} \text{ is written as } b^{-2}$$

10^{-2} is pronounced ten to the minus two, b^{-2} is pronounced b to the minus two.

Some other examples are:

$$\frac{1}{a} = a^{-1}$$ (Pronounced *a* to the minus one.)

$$\frac{1}{y \times y} = y^{-2}$$

$$\frac{1}{d \times d \times d} = d^{-3}$$

Compare this with $\frac{1}{10} = 10^{-1}$

Thus the expression $\frac{1}{y \times y}$ can either be written as

y^{-2} or $\frac{1}{y^2}$ or $1 \div (y \times y)$

So there are many different ways of writing down the same information.

Algebraic fractions are often expressed using negative powers.

EXAMPLE

Express $\frac{m}{s}$ using the notation s^{-1}.

SOLUTION

$$\frac{m}{s} = m \times \frac{1}{s} = m \times s^{-1} = ms^{-1}$$

TRY SOME YOURSELF

7(i) Rewrite $\frac{1}{z \times z \times z \times z}$ as a power of z.

(ii) Rewrite the fraction $\frac{l}{p}$ using power notation.

(iii) Rewrite $\frac{s}{t^2}$ as a product of powers.

2.2(iv) MORE SIMPLIFICATION

In Section 2.2(i) you saw that an expression such as

$8x + 4y - 2x + 6y$

is simplified by collecting like terms. The same process applies if the terms are more complex, involving powers, or products of different symbols.

$8x + 4y - 2x + 6y$
$= (8x - 2x) + (4y + 6y)$
$= 6x + 10y$

The main difficulty lies in identifying the like terms. For example,

$2a^2$ and $4a^2$ are like terms

but

$2a^2$ and $4b^2$ are not

One term involves a^2; the other involves b^2.

Similarly,

$4b^2c$ and $-9cb^2$ are like terms

but

$4b^2c$ and $-9bc^2$ are not

One term involves b^2c; the other involves bc^2.

In order to be like terms, the symbols *and* the powers of each symbol must be the same.

TRY SOME YOURSELF

8 Identify like terms in each of the following lists of terms. Watch out—there may be more than one collection of like terms in any given list!

(i) $a, ab, c, 5ab, 5a^2b$

(ii) $x^2, 3xy, y^2, -xy, y, 2xy, x$

(iii) $c^3, 4c^2d, 2d^2c, -3c^2d, 4c^3, -3cd^2$

(iv) $-t, t, t^2, t^3, t^4$

*Look for the same symbols **and** the same powers.*

The next example illustrates how to simplify an expression involving more complex terms. As with simpler expressions, this involves collecting like terms.

EXAMPLE

Simplify $2bc + 3d^2 + 3b^2c + 6d^2 - 4bc$.

SOLUTION

The first thing is to identify the like terms.

$$2bc + 3d^2 + 3b^2c + 6d^2 - 4bc = (2bc - 4bc) + (3d^2 + 6d^2) + 3b^2c$$

like (marking $3d^2$ and $6d^2$)

like (marking $2bc$ and $4bc$)

$$= -2bc + 9d^2 + 3b^2c$$

Rearrange the expression putting like terms next to each other. The coefficients can then be added or subtracted.

TRY SOME YOURSELF

9 Simplify each of the following expressions by collecting like terms:
(i) $5t^2 + 3t^2$ (ii) $8y^2 - 7y^2 + 6y^2$
(iii) $4p^2 + pq - 3p^2 - 3pq + 5p^2$ (iv) $2c^3 + 4c^2d - 3dc^2 + 2$.

2.2(v) UNITS: AN APPLICATION

All the symbols we have used so far have been used to represent unknown numbers. Since they represent numbers they can be manipulated in a number-like fashion. However, there is one important application of symbols in which the symbols do not stand for numbers: symbols are also used as abbreviations for units of measurement. Nevertheless, such symbols can also be manipulated like numbers.

Units of measurement are essential in practical situations. It's no good ordering from a shop a piece of material 2 by 3. Would this mean 2 metres by 3 metres, 2 feet by 3 feet, 2 yards by 3 yards or what? The unit of measurement is needed in order to make the situation clear.

The table below lists some typical units of measurement, together with the symbol usually used as an abbreviation.

Quantity	Metric	Symbol	Imperial	Symbol
Distance	millimetre	mm	inch	in
	centimetre	cm	foot	ft
	metre	m	yard	yd
	kilometre	km		
Weight	gram	g	ounce	oz
	kilogram	kg	pound	lb
Time	millisecond	ms	second	s
	second	s	hour	hr
	hours	hr		

This table gives common metric and imperial units of measurement.

Notice that most of the symbols used as abbreviations are not single letters. In this case cm does not mean c x m.

In manipulating the symbol cm we must therefore treat it as a single letter.

Other units of measure can be derived from these units. For example, area is measured in square units. Consider the rectangle opposite. The formula

$$A = lw$$

gives its area. Substituting $l = 4$ and $w = 3$ gives $A = 12$. To find the units of measurement, the formula can be used again. l is measured in metres, as is w, so the unit of measurement for the area is

$$\text{m} \times \text{m} \ (= \text{m}^2).$$

Thus

$$A = 12 \text{ m}^2$$ (pronounced 12 square metres.)

Had the rectangle been 4 cm by 3 cm then the area would be

$$A = 12 \text{ cm} \times \text{cm}$$

or

$$A = 12 \text{ cm}^2$$

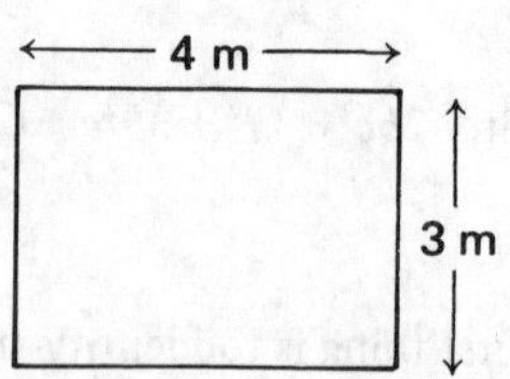

When manipulating units, the units of measurement must be the same. For example, it's no good multiplying m by cm.

cm is treated as a single letter. Thus cm x cm = cm², pronounced square centimetres.

EXAMPLE

A vehicle moves 12 metres in 2 seconds. Find its speed.

SOLUTION

Speed (v) is defined by the formula

$$v = \frac{d}{t}$$

where d is distance and t is time. Putting $d = 12$ and $t = 2$ we get $v = 6$. To find the units of measurement we put the units of distance and time into the formula to get m/s or m s^{-1}. So the velocity is 6 ms^{-1} (pronounced 6 metres per second).

The formula provides an easy way of remembering this general result.

$$v = \frac{d}{t} = \frac{12}{2} = 6$$

TRY SOME YOURSELF

10(i) Find the volume of a box of sides 5 cm, 2 cm and 3 cm.
(ii) A vehicle travels 5 kilometres in 2 hours. What is its speed?
(iii) Density is equal to mass divided by volume. If mass is measured in grams and volume in cubic centimetres, what is the unit of measurement of density?

The volume is given by length x width x height.

After you have worked through this section you should be able to

a Simplify an algebraic expression by identifying and collecting like terms
b Convert a string of symbols multiplied together into standard notation
c Simplify an algebraic expression involving multiplication and powers
d Convert an expression involving division of two expressions into an algebraic fraction, simplifying the result
e Substitute numerical values into any of the expressions listed above
f Use algebraic manipulation to find the units of measurement of area, velocity, volume, etc.

Finally, here are some exercises if you want more practice.

TRY SOME MORE YOURSELF

11(i) Simplify each of the following expressions:
(a) $3y + 4y + 9y - 8y$ (b) $8r + 3d - d + r$ (c) $2 + 3s - 2s + 5$.

(ii) Write each of the following expressions in standard form:
(a) $4 \times q \times f \times (-2)$ (b) $3 \times r \times 2s \times 4r$
(c) $(-2) \times (-a) \times 3b \times 4a$ (d) $r \times r \times r \times s \times s \times r$.

(iii) Simplify each of the following and write it as an algebraic fraction:
(a) $9 \times x \times y \div (3x + 4y)$ (b) $9 + x + y \div (3x \times 4y)$
(c) $9x + y \div (3 \times (x + y))$.

(iv) Rewrite each of the following as a product of symbols, using negative powers:
(a) $\frac{c}{s}$ (b) $\frac{m}{v}$ (c) $\frac{m \times m \times m}{t^2}$.

(v) Simplify each of the following expressions:
(a) $2x^2y + 4x - 9y^2x + yx^2 - 3x$
(b) $4 + r^2 - 3sr - 3r^2 + 4rs$
(c) $3st + 6s^2 - 5ts^2 + st + st^2$

(vi) (a) Simplify the expression
$$4x - 3yx + 7x + 2yx$$
(b) Evaluate the expression by substituting $x = 1$ and $y = 3$.
(c) Evaluate the expression by substituting $x = 2$ and $y = -4$.
(d) Evaluate the expression by substituting $x = 1{\cdot}32$ and $y = 0{\cdot}41$.

(vii) (a) A cube has side 4 ft. What is its volume?
(b) A plane travelled 14 kilometres in 6 seconds. What was its speed?

2.3 Brackets

TRY THESE QUESTIONS FIRST

1 Expand $-2(x + 7y - 2z)$.

2 Simplify $p(x - 2q) - 6q(x + p)$.

3 (i) Evaluate $4t(t + 3) - 2(4 - 2t^2)$ when $t = 2$.
(ii) Expand $(a - 2b)^2$.

2.3(i) EXPANDING BRACKETS

In Module 1, Section 1.2, we discussed how brackets are used in numerical calculations. For example, negative numbers such as (–3) and (–5) are sometimes written in brackets, and brackets indicate which part of a calculation should be carried out first.

For example, to calculate (3 + 2) x (7 + 5) the brackets must be worked out first to give 5 x 12 = 60.

Brackets were also used in the 'think of a number' sequences at the beginning of Section 2.1 in this module. Again, the brackets indicated which part should be done first. Expressions often involve one set of brackets inside another, for example as in

$$[(4 + 3) \times 2] - 6$$

Here, different notations are used to differentiate between the brackets.

The following exercises are intended to refresh your memory about how to handle brackets in numerical calculations.

TRY SOME YOURSELF

1 Evaluate each of the following:
(i) $7 \times (8 - 3)$ (ii) $(3 - 5) \times 2$ (iii) $4 + [2 \times (6 + 3)]$
(iv) $-(3 + 4)$ (v) $-(2 - 5 + 6)$.

CHECK YOUR ANSWERS

1 $-2(x + 7y - 2z) = ((-2) \times x) + ((-2) \times 7y) - ((-2) \times 2z)$
$= -2x + (-14y) - (-4z)$
$= -2x - 14y + 4z$
$= 4z - 2x - 14y$ *Section 2.3(i)*

2 $p(x - 2q) - 6q(x + p) = px - 2pq - 6qx - 6pq$
$= px - 8pq - 6qx$ *Section 2.3(ii)*

3 (i) $4t(t + 3) - 2(4 - 2t^2) = 4t^2 + 12t - 8 + 4t^2$
$= 8t^2 + 12t - 8$ *Section 2.3(iii)*

When $t = 2$,
$8t^2 + 12t - 8 = 32 + 24 - 8$
$= 48$

(ii) $(a - 2b)^2 = a^2 - 4ab + 4b^2$

In algebraic expressions (involving symbols) brackets are used in much the same way. Here are some examples:

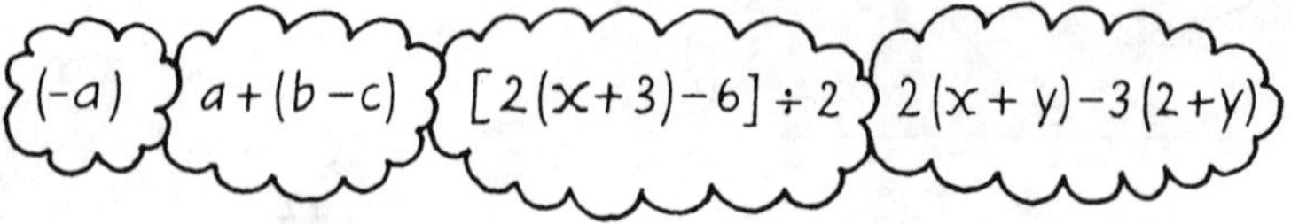

Brackets avoid ambiguity; they indicate exactly what is intended.

In numerical calculations, it is usually easier to work out the brackets first. However, in an algebraic expression it often doesn't make sense to work out the brackets. Consequently, the emphasis in algebraic manipulation is to replace an expression involving brackets by an equivalent one which doesn't. This process of removing the brackets is often called *expanding the expression.*

For example, in $2(\mathbf{a}+\mathbf{b})$ we can't work out the brackets unless we are given values for ***a*** *and* ***b***.

Removing brackets in algebraic expressions relies heavily on the rules of arithmetic, so we'll spend some time discussing how they can be applied to symbols; in particular we'll show how the rules for manipulating negative numbers can be used with symbols.

The rules for adding and subtracting are:

> ***Adding a negative number is the same as subtracting a positive number.***
> ***Subtracting a negative number is the same as adding a positive number.***

In the same way

$$a + (-b) = a - b$$

and

$$a - (-b) = a + b$$

Similarly

$$(-a) + (-b) = -a - b$$

and

$$(-a) - (-b) = -a + b$$

The rules for multiplication and division are:

> ***If the signs are the same the answer is positive.***
> ***If the signs are different the answer is negative.***

In the same way

$$a \times b = ab \quad \text{and} \quad (-a) \times (-b) = ab$$

$$a \div b = \frac{a}{b} \quad \text{and} \quad (-a) \div (-b) = \frac{a}{b}$$

Signs are the same.

whereas

$$(-a) \times b = -ab \quad \text{and} \quad a \times (-b) = -ab$$

$$(-a) \div b = -\frac{a}{b} \quad \text{and} \quad a \div (-b) = -\frac{a}{b}$$

Signs are different.

TRY SOME YOURSELF

2 Use the rules outlined above to simplify each of the following:
(i) $y - (-z)$ (ii) $(-u) + (-v)$ (iii) $(-t) \times r$
(iv) $x \times (-x) \times x$ (v) $(-z) \times (-z) \times x \times (-x)$.

A numerical calculation which involves multiplying brackets can be evaluated by working out the brackets first and then multiplying. Alternatively, the brackets can be multiplied out first. For example,

$$2(3 + 4) = 2 \times 7 = 14$$

or

$$2(3 + 4) = (2 \times 3) + (2 \times 4) = 6 + 8 = 14$$

Remember that $2(3 + 4) = 2 \times (3 + 4)$.

Similarly

$$2(3-4) = (2 \times 3) - (2 \times 4) = 6 - 8 = -2$$
$$-2(3+4) = (-2 \times 3) + (-2 \times 4) = -6 - 8 = -14$$
$$-2(3-4) = (-2 \times 3) - (-2 \times 4) = -6 + 8 = 2$$

If the expression involves symbols then the only way to proceed is to multiply out the brackets. For example

$$2(a+b) = 2a + 2b$$
$$2(a-b) = 2a - 2b$$
$$-2(a+b) = -2a - 2b$$
$$-2(a-b) = -2a + 2b$$

As with numbers,

$$2(a+b) = 2 \times (a+b).$$

In particular notice that

$$-(a+b) = (-1) \times (a+b) = -a - b$$

and that

$$-(a-b) = (-1) \times (a-b) = -a + b$$

In each case the algebraic expression which involves brackets has been replaced by an equivalent expression with no brackets; the expression has been expanded.

EXAMPLE

Expand $-7(x - 2y)$

SOLUTION

We need to multiply out the brackets.

$$-7(x-2y) = (-7 \times x) - (-7 \times 2y) = -7x - (-14y) = -7x + 14y$$

Multiply each term inside the brackets by the number outside and add or subtract accordingly.

The final expression could be written as $14y - 7x$, since it is neater to put the positive term first.

TRY SOME YOURSELF

3 Expand each of the following expressions by removing the brackets:
(i) $3(x+y)$ (ii) $5(2x-y)$ (iii) $-4(a+2b-c)$
(iv) $-5(a-b)$.

Remember that order doesn't matter when multiplying numbers together, so that

$$(3+4) \times 2 = 2 \times (3+4)$$

Similarly with symbols,

$$(a+b) \times 2 = 2 \times (a+b) = 2(a+b)$$

The easiest way to expand an expression like $(a+b) \times 2$ is to turn the expression round and expand $2(a+b)$.

Remember also that although we write 2(3 + 4) we ***don't*** *write (3 + 4)2, the number should always come before the brackets.*

TRY SOME YOURSELF

4 Remove the brackets from each of the following expressions:
(i) $x + (y + z)$ (ii) $(a + b) - (c + d)$ (iii) $(a + b) \times (-2)$
(iv) $(a - b + 2c) \times (-5)$.

2.3(ii) EXPANDING MORE COMPLICATED EXPRESSIONS

Often an algebraic expression can be simplified by first removing the brackets then collecting like terms.

EXAMPLE

Simplify $d - g - (3d - 4g)$.

SOLUTION

Since $-(3d - 4g) = -3d + 4g$,

$$\begin{aligned} d - g - (3d - 4g) &= d - g - 3d + 4g \\ &= d - 3d - g + 4g \\ &= -2d + 3g \\ &= 3g - 2d \end{aligned}$$

Remember

$$-(3d - 4g) = (-1) \times (3d - 4g)$$

Collect like terms together and add or subtract coefficients.
Put the positive term first so that the expression doesn't start with a minus sign.

TRY SOME YOURSELF

5 Simplify each of the following expressions by removing the brackets and collecting like terms:
(i) $r - (s - r)$ (ii) $3z + 2y - (y - 2z)$
(iii) $3 + 2a - b - (3b - 4)$.

Here's a more complicated example.

EXAMPLE

Simplify $x - 2(x + 3y - 4z^2)$.

SOLUTION

Since

$$\begin{aligned} -2(x + 3y - 4z^2) &= -2x - 6y + 8z^2 \\ x - 2(x + 3y - 4z^2) &= x - 2x - 6y + 8z^2 \\ &= -x - 6y + 8z^2 \\ &= 8z^2 - x - 6y \end{aligned}$$

Collect like terms together and rewrite the expression, putting the positive term first.

TRY SOME YOURSELF

6 Simplify each of the following expressions:
(i) $x + y + 2(x - 2y)$ (ii) $-(a + 2b) + 3(2c + b)$
(iii) $4t - 2(4 - t)$ (iv) $2(r + 6s) - 3(4s - r)$.

Expressions in which the brackets are multiplied by a symbol can be simplified in the same way. Just as

$$2(b + c) = 2b + 2c$$

so

$$a(b + c) = ab + ac$$

EXAMPLE

Expand $2b(x - y)$.

SOLUTION

$$\begin{aligned} 2b(x - y) &= (2b \times x) - (2b \times y) \\ &= 2bx - 2by \end{aligned}$$

EXAMPLE

Expand $-2x(3x - y)$.

SOLUTION

$$\begin{aligned} -2x(3x - y) &= ((-2x) \times 3x) - ((-2x) \times y)) \\ &= -6x^2 - (-2xy) \\ &= -6x^2 + 2xy \\ &= 2xy - 6x^2 \end{aligned}$$

TRY SOME YOURSELF

7 Expand each of the following expressions:
(i) $a(b - c)$ (ii) $3a(b - c)$ (iii) $2r(3r + s - t)$
(iv) $-3z(2y - z)$.

If the multiplication looks complicated, write down all the steps, as in the example above.

The following example shows how a complicated algebraic expression can be simplified by removing the brackets first.

EXAMPLE

Simplify $t(2t + 3) - 2t(1 - 3t)$

SOLUTION

$$\begin{aligned} t(2t + 3) - 2t(1 - 3t) &= 2t^2 + 3t - 2t + 6t^2 \\ &= 2t^2 + 6t^2 + 3t - 2t \\ &= 8t^2 + t \end{aligned}$$

$$\begin{aligned} t(2t + 3) &= 2t^2 + 3t \\ -2t(1 - 3t) &= -2t - ((-2t) \times 3t) \\ &= -2t - (-6t^2) \\ &= -2t + 6t^2 \end{aligned}$$

TRY SOME YOURSELF

8 Simplify each of the following expressions by expanding the brackets and collecting like terms:
(i) $x(2x - 3) - 3x(5 - 2x)$ (ii) $x(y - z) - y(x + z)$
(iii) $2a(b - 3a) - 2b(3a - 2b)$ (iv) $-2p(p + q) + 3p(2p - q)$.

2.3(iii) SUBSTITUTION

Substituting numbers

You've already seen how to evaluate an algebraic expression by substituting numerical values. Sometimes it is more convenient to simplify the algebraic expression first before substituting.

EXAMPLE

Evaluate $x + y + 2(x - y)$ when $x = 3$ and $y = -2$.

SOLUTION

Simplify the expression first to get

$$\begin{aligned} x + y + 2(x - y) &= x + y + 2x - 2y \\ &= x + 2x + y - 2y \\ &= 3x - y \end{aligned}$$

Expand the brackets then collect like terms together.

When $x = 3$, $y = -2$,

$$\begin{aligned} 3x - y &= (3 \times 3) - (-2) \\ &= 9 + 2 \\ &= 11 \end{aligned}$$

Here, it is quicker to substitute into the simplified expression rather than the original one.

TRY SOME YOURSELF

9(i) Evaluate $6t - 4s + 3(s - 4t)$ when $t = -1$ and $s = 4$.
(ii) Evaluate $2(r + 6s) - 3(4s - r)$ when $r = 10$ and $s = -5$.
(iii) Evaluate $x(2x - 3) - 3x(5 - 2x)$ when $x = 1$.

In the following exercises we ask you to substitute more complicated numbers and you will need to use your calculator. To check that you have the right key sequence for each expression, first try out an easier substitution, such as 1 or 2. Then you can compare the result you obtained using your calculator with that obtained on paper, or in your head.

TRY SOME YOURSELF

10(i) Evaluate $2(a + 2b) - 3(b - 2a)$ when
(a) $a = 0{\cdot}1$ and $b = 0{\cdot}7$
(b) $a = 143$ and $b = -91$.
(ii) Evaluate $x(2x - 3) - 3x(5 - 2x)$ when
(a) $x = 0{\cdot}9$
(b) $x = -0{\cdot}31$.
(iii) Evalaute $6y^2 - 7x(3 - 2y) + y(2x - y)$ when
(a) $x = 0{\cdot}2$ and $y = 0{\cdot}3$
(b) $x = -12$ and $y = -31$.

Substituting symbols

Symbols can be substituted into algebraic expressions in the same way. We're going to investigate what happens to

$$c(x + y)$$

if $c = (a + b)$.

First of all we expand the brackets:

$$c(x + y) = cx + cy$$

Now we substitute $c = (a + b)$ to get

$$cx + cy = (a + b)x + (a + b)y$$
$$= x(a + b) + y(a + b)$$

Rewrite these brackets the other way round, since

$$(a + b)x = (a + b) \times x$$
$$= x \times (a + b)$$
$$= x(a + b)$$

Now we can expand the brackets again:

$$x(a + b) + y(a + b) = xa + xb + ya + yb$$
$$= ax + bx + ay + by$$

What does this show? It indicates that

$$(a + b)(x + y) = ax + bx + ay + by$$

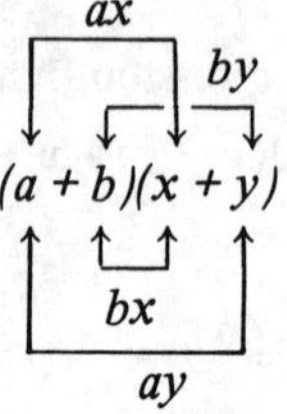

So, to multiply two brackets together you must multiply each symbol in the first brackets by each symbol in the second. Similarly

$$(x + 2)(x + 3) = x^2 + 2x + 3x + 6$$
$$= x^2 + 5x + 6$$

You might like to think of this in relation to a rectangle of length $(x + 3)$ and width $(x + 2)$. The area of such a rectangle is given by

length × width

which is

$$(x + 3)(x + 2)$$

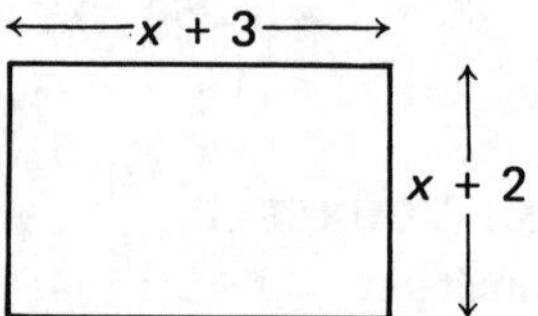

But the rectangle can also be split up into separate areas–A, B, C, and D. The total area is therefore

$$A + B + C + D$$
$$= x^2 + 3x + 2x + 6$$
$$= x^2 + 5x + 6$$

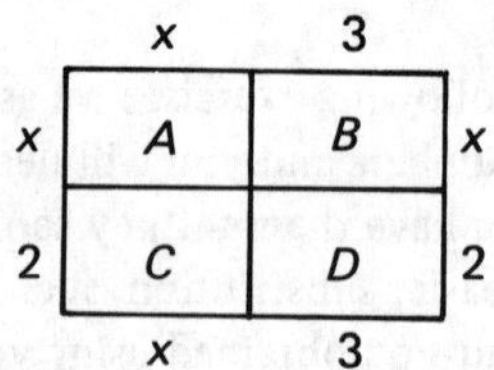

The expression $(r + s)(r + s)$ is usually written as $(r + s)^2$. From above, then,

Just as $a \times a = a^2$

$$(r + s)^2 = (r + s)(r + s)$$
$$= r^2 + rs + rs + s^2$$
$$= r^2 + 2rs + s^2$$

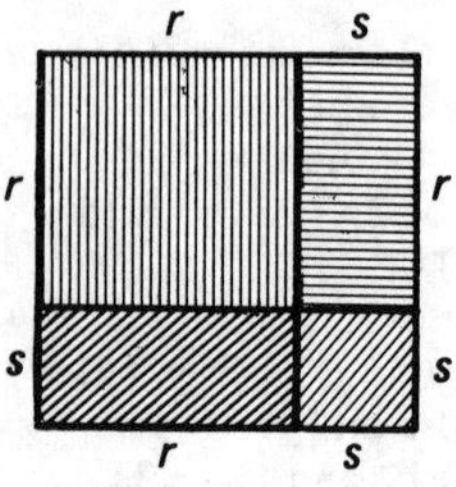

TRY SOME YOURSELF

11(i) Expand $x(y - z)$.

(ii) Substitute $x = y - z$ in (i) and simplify the result.

(iii) Hence write down the expansion of $(y - z)^2$

The following example illustrates that any two brackets can be multiplied together by multiplying each term in the first brackets by each term in the second.

EXAMPLE

Expand $(a + b)(a + 2b)$.

SOLUTION

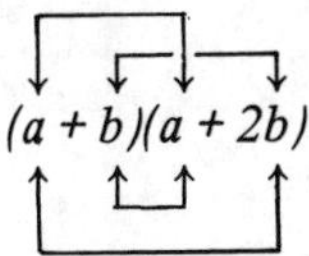

The resulting expression should contain four terms indicated by the arrows above, so that

$$\begin{aligned}(a + b)(a + 2b) &= a^2 + 2ab + ab + 2b^2 \\ &= a^2 + ab + 2b^2 \\ &= a^2 + 3ab + 2b^2\end{aligned}$$

Rewrite the terms with the symbols in alphabetical order and collect like terms.

TRY SOME YOURSELF

12 Simplify each of the following expressions by expanding the brackets and collecting like terms:
(i) $(x - 1)(x + 2)$ (ii) $(x + 5)(4 + 3x)$ (iii) $(a - 3)(a + 3)$
(iv) $(2a - b)^2$ (v) $(2p + q)(p - 3q)$.

After you have worked through this section you should be able to

a Remove the brackets from an algebraic expression and simplify the result, using the rules of addition, subtraction, and multiplication
b Evaluate an algebraic expression for given numerical substitutions by simplifying the expression first then substituting
c Expand brackets of the form $(a + x)(b + y)$, and simplify the result

Finally, here are some exercises if you want more practice.

TRY SOME MORE YOURSELF

13 Simplify each of the following:

(i) (a) $3(b + c)$ (b) $-2(s - t)$ (c) $-3(x - 4y + 2z)$

(ii) (a) $a + (b - 2c)$ (b) $4a - (3t + 5)$ (c) $29 - (t + 7)$

(iii) (a) $3z - (2y + z)$ (b) $4x - (6y + 2x)$ (c) $19t - 6 + (3 - 4t)$

(iv) (a) $6a + 2(a - 3b)$ (b) $4t - 3s - 2(s - 2t)$
(c) $6 - 2x + 4(2 - 6x)$

(v) (a) $4x(2x + 3) - x(1 - 2x)$ (b) $2y(3x + 2) - 3(1 - 2y)$
(c) $4p(q - 3r) + 2q(r + 3p)$

(vi) (a) $(r + t)^2$ (b) $(3 - x)^2$ (c) $(2 - 3x)(1 - x)$

(vii) (a) Simplify the expression $2x(3x + y) - y(x + 2y)$.
(b) Evaluate this expression for $x = 2$ and $y = 3$.
(c) Evaluate this expression for $x = 0$ and $y = 4$.
(d) Evaluate this expression for $x = 0{\cdot}19$ and $y = -31{\cdot}6$.

2.4 Solving Equations

TRY THESE QUESTIONS FIRST

1 Solve $-3(2-3x)=\frac{x}{2}+2$.

2 A quarterly telephone bill was £26·24 which consisted of rental of £12 and the metered units charge. If each metered unit cost 4 pence, how many metered units were used?

3 An electricity board has a fixed quarterly charge of £7 and charges 2·5 pence per unit of electricity consumed. If t denotes the total charge in pounds and n the number of units consumed in a quarter, write down an equation relating t and n. Evaluate n when $t = 10{\cdot}40$.

2.4(i) SOLVING EQUATIONS

At the beginning of Section 2.1(i) we invited you to try a 'think of a number' puzzle in which we asked you to follow a sequence of instructions starting with any number you liked.

Here is another puzzle. It, too, is based upon a sequence of instructions, but this time we give you the answer. The instructions are:

Think of a number. Multiply it by 2. Add 7 to the result.

Start with x: x
Multiply by 2: $2x$
Add 7: $2x + 7$

Now, what was the original number if the final answer is 13?

If the answer is 13 then
$$2x + 7 = 13$$

Perhaps you have guessed it already. In this particular example that's not too difficult. But if the final answer had been -3681, guessing the original number might have taken a bit longer.

What happens when
$$2x + 7 = -3681?$$

$$2x + 7 = 13$$

is an *equation*. The value of x which makes the left hand side equal to 13 is called the *solution*. There is a systematic approach to solving such an equation which can be applied no matter how complicated it is.

*x is the **unknown quantity**, sometimes just called the '**unknown**'.*

The overall strategy is to get x on one side of the equals sign and everything else on the other side. For the moment, we will always isolate x on the left hand side of the equation.

This choice is arbitrary; you can equally well isolate x on the right hand side.

Balancing an equation

The equation $2x + 7 = 13$ may be thought of as a balance with $2x + 7$ in the left hand pan and 13 in the right hand pan. In order to maintain the balance, whatever is done to one side of the equation must also be done to the other. This suggests the following rule for addition and subtraction:

> ***Any number or symbol can be added to or subtracted from one side of the equation provided the same number or symbol is added to or subtracted from the other side.***

We'll refer to this rule as the balancing rule.

So these operations are all valid:

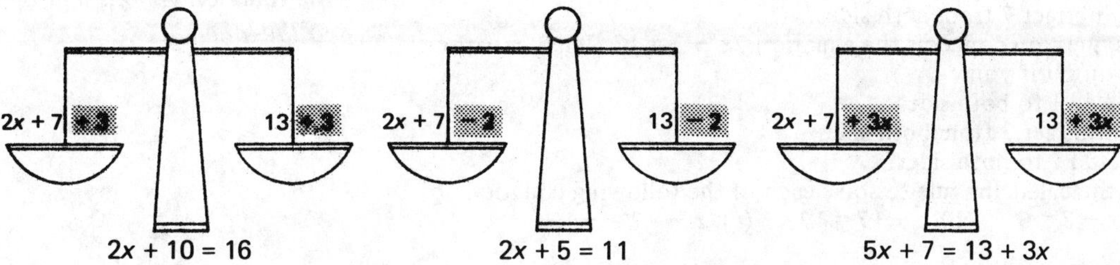

Whereas the following situation is not:

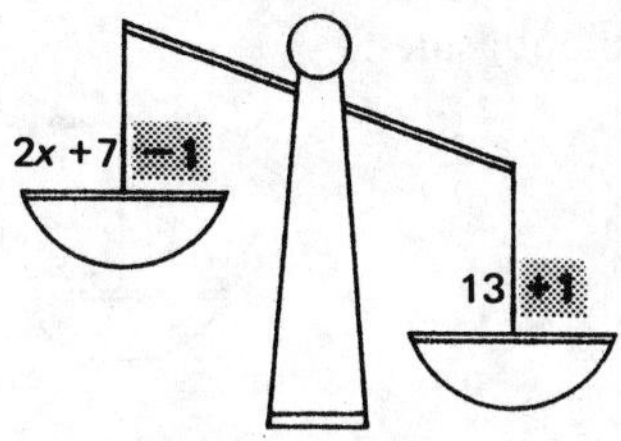

*In this case we've **subtracted** one from the left hand side but **added** one to the right hand side.*

Some equations can be solved completely using the balancing rule.

EXAMPLE

Solve $y + 7 = 13$.

In this equation y is the unknown.

SOLUTION

We need to isolate y on the left hand side. This means subtracting 7. In order to maintain the balance we must also subtract 7 from the right hand side. This gives

$$y + 7 - 7 = 13 - 7$$
$$\therefore y = 13 - 7 = 6$$

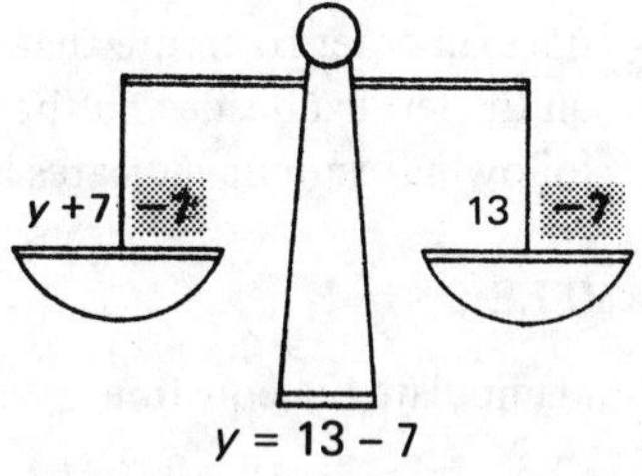

Check that $y = 6$ is the solution by substituting back into the original equation.

CHECK YOUR ANSWERS

1 $x = \frac{16}{17}$ *Section 2.4(i)*

2 If n denotes the number of metered units, then *Section 2.4(ii)*

$$26{\cdot}24 = 12 + \tfrac{4}{100}\, n$$

The solution is $n = 356$.

3 $t = 7 + 0{\cdot}025n$ *Section 2.4(iii)*

When $t = 10{\cdot}40$, $n = 136$.

TRY SOME YOURSELF

1(i) This question concerns the equation $2x + 7 = 13$. What equations do you get if you
(a) Add 1 to both sides
(b) Add 7 to both sides
(c) Subtract 7 from both sides?

(ii) This question concerns the equation $2x - 7 = 13$. What equations do you get if you
(a) Add 1 to both sides
(b) Subtract 7 from both sides
(c) Add 7 to both sides?

(iii) Use the balancing rule to solve each of the following equations:
(a) $t - 3 = 6$ (b) $z + 17 = 12$ (c) $x - 12 = -4$.

In order to maintain the balance, whatever you do to one side of the equation you must also do to the other.

The purpose of using the balancing rule is to simplify the equation. In some instances, as in part (iii) above, the equation can be solved completely by just adding or subtracting *numbers* from both sides. But this is not often the case.

Consider the equation

$$3x - 3 = 2x + 1$$

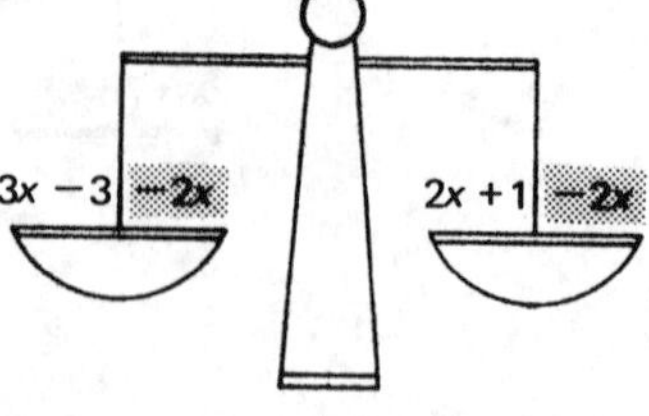

If we subtract $2x$ from each side we get

$$x - 3 = 1$$

In the original equation the unknown, x, appeared on both sides of the equation; now it appears only on the left hand side. Manipulating the equation so that the unknown appears only on the left hand side is called *taking the unknown to the left hand side.* The most difficult part is in deciding what to add or subtract to both sides, in order to ensure that the resulting equation has the unknown on the left hand side and the numbers on the right hand side. The following example indicates how it is done.

The equation $x - 3 = 1$ can be further simplified by adding 3 to both sides to get $x = 4$.

EXAMPLE

Manipulate the equation

$$2x + 5 = 4x + 8$$

so that the unknown, x, is on the left hand side and the numbers are on the right hand side.

SOLUTION

First of all we take the unknown to the left hand side.

$2x + 5 = 4x + 8$

Identify the unwanted term involving the unknown on the right hand side and add or subtract to get rid of it.

Subtract $4x$ from both sides to get

$2x + 5 - 4x = 4x + 8 - 4x$

that is

$-2x + 5 = 8$

We now take all the numbers to the right hand side:

$-2x + 5 = 8$

Identify the unwanted term on the left hand side and add or subtract to get rid of it.

Subtract 5 from both sides to get

$-2x + 5 - 5 = 8 - 5$

that is

$-2x = 3$

The resulting equation has the form

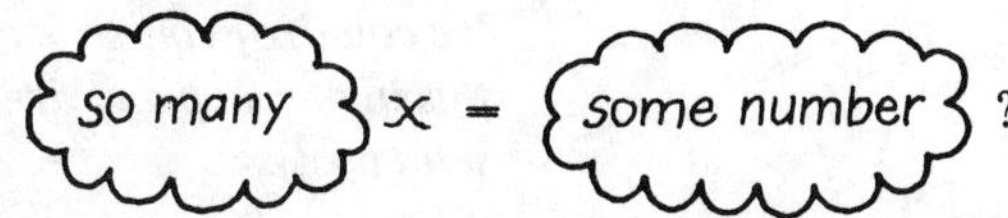

TRY SOME YOURSELF

2 What do you have to add or subtract to both sides of the following equations in order to get them in the form

Identify the unwanted terms then add or subtract accordingly.

so many x = some number ?

(i) $3x + 2 = 5$ (ii) $4x - 7 = 3$ (iii) $5 - 2x = 2$
(iv) $4x - 3 = 2x + 1$

Multiplying and dividing

The balance of an equation is also preserved if both sides of the equation are multiplied or divided by the same number or symbol.

We can divide by any positive or negative number, but not zero.

Thus, if $6x = 18$ we can divide both sides by 6 to get $x = 3$.

Less obvious is $-3x = 7$. Here, we can divide both sides by -3 to get

$$\frac{-3x}{-3} = \frac{7}{-3}$$

and so

$$x = -\tfrac{7}{3}$$

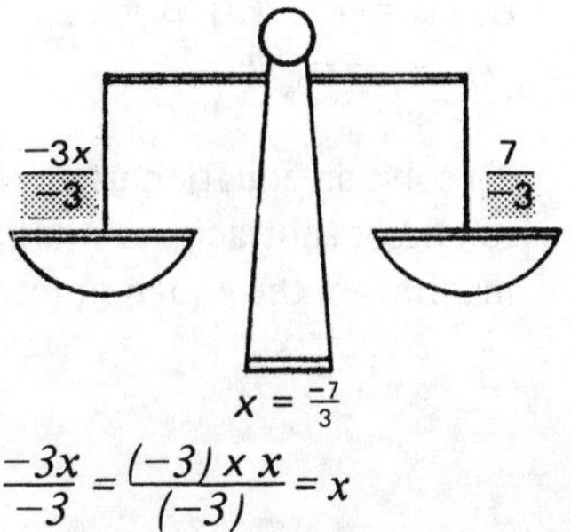

$$\frac{-3x}{-3} = \frac{(-3) \times x}{(-3)} = x$$

The objective is usually to solve the equation. Therefore, starting with an equation of the form

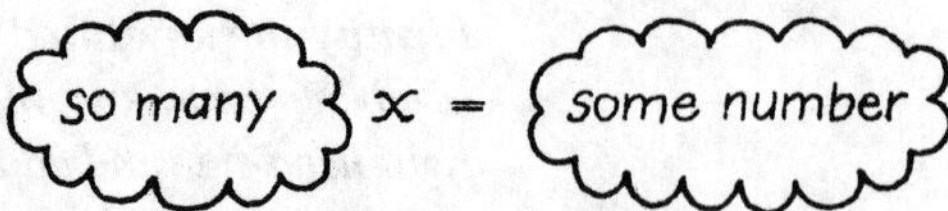

the equation must be manipulated into the form

where the coefficient of x is equal to 1. The resulting number is the solution of the equation. This process involves multiplying or dividing both sides of the equation by the same number. Again, the hardest part is deciding by what number to multiply or divide.

Remember that

$$1 \times x = x$$

EXAMPLE

Solve $\frac{2}{3}y = 24$

y is the unknown in this equation. The objective is to make the coefficient of y equal to 1.

SOLUTION

$\frac{2}{3}y = 24$

We **multiply** both sides by **3** to get

$$3 \times \tfrac{2}{3}y = 3 \times 24$$

$$2y = 72$$

We now **divide** both sides by **2** to get

$$\frac{2y}{2} = \frac{72}{2}$$

$$y = 36$$

We could, of course, have done this in one step and divided both sides by $\frac{2}{3}$. This is equivalent to multiplying both sides by $\frac{3}{2}$ (the reciprocal of $\frac{2}{3}$).

You can now check that $y = 36$ is the solution by substituting back into the original equation. Check that the left hand side is indeed equal to the right hand side.

$\frac{2}{3} \times 36 = 24 =$ right hand side

TRY SOME YOURSELF

3 Solve each of the following equations:
(i) $3x = 4$ (ii) $5x = -3$ (iii) $-2x = 11$ (iv) $-\frac{1}{4}y = 7$
(v) $-\frac{4}{9}z = -2$.

Check your answers by substituting into the original equation.

To solve an equation using the balancing rule you will usually need to add or subtract *and* multiply or divide. The first step is to manipulate the equation into the form

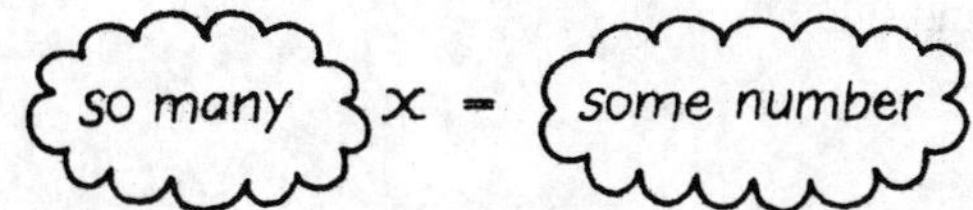

by adding or subtracting to get rid of the unwanted terms. Then multiply or divide to make the coefficient of x equal to 1. The resulting number is the solution. The following examples show how to put the steps together.

EXAMPLE

Solve $2x + 7 = 13$.

SOLUTION

We first isolate the unknown, x, on the left hand side:

$$2x + 7 = 13$$
$$2x = 6$$

*7 is the unwanted number term on the left hand side, so subtract 7 from **both** sides.*

We now divide by 2 to make the coefficient of x equal to 1:

$$x = \tfrac{6}{2} = 3$$

Check that $x = 3$ is the solution by substituting.

EXAMPLE

Solve $4x - 3 = 2x + 1$.

SOLUTION

$$4x - 3 = 2x + 1$$
$$2x - 3 = 1$$ *Subtract 2x from both sides.*
$$2x = 4$$ *Add 3 to both sides.*
$$x = 2$$ *Divide by 2.*

Check that $x = 2$ is the solution by substituting.

TRY SOME YOURSELF

4 Solve each of the following equations:
(i) $2x + 7 = -3681$ (ii) $1 - 3x = 19$ (iii) $5y + 3 = 2 + y$
(iv) $1 - 4t = 10 - 7t$

First isolate the unknown on the left hand side, then make the coefficient equal to 1. Check your solutions by substituting into the original equations.

Coping with fractions

In dealing with an equation like

$$3x + 3 = 8 - \frac{x}{3}$$

you may find it helpful to eliminate the fraction first by multiplying both sides of the equation by 3. Thus

This makes the resulting algebra less complicated.

$$3(3x + 3) = 3(8 - \tfrac{x}{3})$$
$$9x + 9 = 24 - x$$

It is probably easier to use brackets here and then multiply them out, remembering to multiply every term by 3.

The unknown, x, can now be isolated on the left hand side by adding x to both sides to get

$$10x + 9 = 24$$

and then subtracting 9 from both sides to get

$$10x = 15$$

Finally, the solution is found by dividing both sides of the equation by 10, so that

$$x = 1{\cdot}5$$

Check that $x = 1{\cdot}5$ is the solution by substituting into the original equation.

TRY SOME YOURSELF

5 Solve each of the following equations:

(i) $2x + 1 = \frac{x}{2} + 4$ (ii) $6x + 6 = 1 - \frac{x}{4}$ (iii) $1 - \frac{x}{3} = 2 + x$.

Remove the fractions first, then manipulate the equation into the form

so many x = some number

Check your solutions by substituting back into the original equation.

2.4(ii) SOLVING EQUATIONS WITH ONE UNKNOWN

Solving equations often involves all the techniques introduced in this module. The following examples illustrate the complete process. In each case you should carefully study the strategy used to solve the equation.

An equation with one unknown is an equation involving only one symbol which stands for the unknown quantity.

EXAMPLE

Solve $3(x - 2) = \frac{x}{2} + 4$.

SOLUTION

Strategy:

$3(x - 2) = \frac{x}{2} + 4$		
$6(x - 2) = x + 8$	(Multiply throughout by 2.)	*Remove fractions.*
$6x - 12 = x + 8$	(Multiply out brackets.)	*Remove brackets.*
$5x - 12 = 8$	(Subtract x from both sides.)	*Take x to the left.*
$5x = 20$	(Add 12 to both sides.)	*Take the numbers to the right.*
$x = 4$	(Divide by 5.)	*Make coefficient of x equal to 1.*

Check for yourself that $x = 4$ is the solution by substituting.

EXAMPLE

Solve $3(x - \frac{1}{2}) = x + 5$.

SOLUTION

Strategy:

$3(x - \frac{1}{2}) = x + 5$

$3x - \frac{3}{2} = x + 5$ (Multiply out brackets.) *Remove brackets.*

$6x - 3 = 2x + 10$ (Multiply throughout by 2.) *Remove fractions.*

$4x - 3 = 10$ (Subtract $2x$ from both sides.) *Take x to the left.*

$4x = 13$ (Add 3 to both sides.) *Take the numbers to the right.*

$x = \frac{13}{4}$ (Divide by 4.) *Make coefficient of x equal to 1.*

Check that $x = \frac{13}{4}$ is the solution.

Sometimes it is more convenient to take the unknown to the right hand side of the equation and the numbers to the left. Both methods are equally acceptable. As you get more practice you will be able to judge for yourself which is most appropriate. In the next example we leave the unknown on the right hand side.

In such cases the equation is manipulated into the form

some number = so many x

EXAMPLE

Solve $5 = \frac{(3x - 4)}{2}$

SOLUTION

Strategy:

$5 = \frac{(3x - 4)}{2}$

$10 = 3x - 4$ (Multiply by 2.) *Remove fractions.*

$14 = 3x$ (Add 4 to both sides.) *Leave x on the right. Take the numbers to the left.*

$\frac{14}{3} = x$ (Divide by 3.) *Make coefficient of x equal to 1.*

This is of course the same as saying $x = \frac{14}{3}$.

Check that $x = \frac{14}{3}$ is the solution.

The same procedures apply when the numbers are more complicated.

EXAMPLE

Solve $3 + \frac{2{\cdot}5x}{100} = 4 + \frac{2{\cdot}0x}{100}$.

SOLUTION

$$3 + \frac{2{\cdot}5x}{100} = 4 + \frac{2{\cdot}0x}{100}$$

Strategy:

First multiply throughout by 100, so that *Remove fractions.*

$$(100 \times 3) + (100 \times \tfrac{2\cdot5}{100}x) = (100 \times 4) + (100 \times \tfrac{2\cdot0}{100}x)$$

$$300 + 2{\cdot}5x = 400 + 2{\cdot}0x$$

$$300 + 0{\cdot}5x = 400$$ *Take x to the left.*

$$0{\cdot}5x = 100$$ *Take the numbers to the right.*

$$x = \tfrac{100}{0\cdot5} = 200$$ *Make coefficient of x equal to 1.*

Check that x = 200 is the solution.

The strategy for solving equations may be summarised as follows.

1 Remove fractions
2 Remove brackets
3 Take the unknown to one side of the equation
4 Take the numbers to the other side
5 Make the coefficient of the unknown equal to 1

It is sometimes better to carry out step (2) before step (1) as in the second example above. With practice, you will be able to judge for yourself which way is most appropriate. As a general rule though, it is a good idea to get rid of fractions as soon as possible.

TRY SOME YOURSELF

6 Solve each of the following equations:

(i) $5(2x - 3) = 30$ (ii) $\frac{x+4}{3} = x$ (iii) $b - \frac{3(b+1)}{5} = 1$

(iv) $a + 2 = \frac{5(1-a)}{4}$

Check your solutions by substituting back into the equation.

Word problems

Practical problems are usually expressed in words, in which case you first need to derive an algebraic equation. The solution can then be found using algebraic manipulation. In Module 2, Section 2.1(iii), we briefly considered how to derive equations from word problems. We now return to this subject, but this time, in addition to deriving the equation, we will go on to solve the problem as well. The first example is a 'think of a number' puzzle.

The hardest part in solving a problem lies in translating the words into symbols.

EXAMPLE

The following sequence of instructions resulted in a final answer of 4:

Think of a number. Add 4. Multiply by 3. Subtract the original number. Divide by 2. Subtract 5.

What was the original number?

SOLUTION

First, we must write the sequence of instructions using a symbol to stand for the unknown number. Let n stand for the unknown number. Then we get

$$\frac{3(n+4) - n}{2} - 5$$

The final answer is 4, so we get the equation

$$\frac{3(n+4) - n}{2} - 5 = 4$$

The equation can now be solved as follows:

$$3(n+4) - n - 10 = 8$$

$$3n + 12 - n - 10 = 8$$

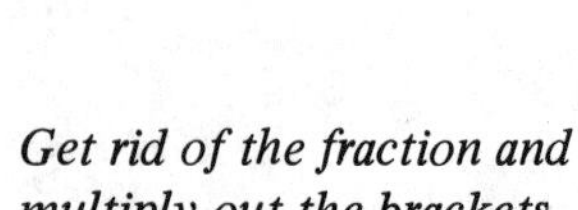

Get rid of the fraction and multiply out the brackets.

$$2n + 2 = 8$$

Simplify both sides.

$$2n = 6$$

$$n = 3$$

Take the numbers to the right hand side and divide by 2.

You can check that $n = 3$ is the solution by following through the original instructions.

The next example is not so straightforward.

EXAMPLE

An electricity board offers a choice between two tariffs:

Tariff 1: A fixed charge of £3 per quarter plus a charge of 2·5 pence per unit consumed.

Tariff 2: A fixed charge of £4 per quarter plus a charge of 2·0 pence per unit consumed.

How many units would need to be consumed in a quarter for the overall cost under both tariffs to be the same?

SOLUTION

We need to find the number of units consumed in a quarter satisfying certain conditions.

Identify the unknown quantity.

Let this unknown be x.

*Choose a symbol to stand for the unknown: **x** is the number of units consumed in a quarter.*

Under Tariff 1 the overall cost in pounds is given by

$$3 + \tfrac{2\cdot5}{100}x$$

Convert pence to pounds by dividing by 100. Then all the costs are given in pounds.

whereas under Tariff 2 it is

$$4 + \tfrac{2\cdot0}{100}x$$

If the overall cost is to be the same,

$$3 + \frac{2\cdot5x}{100} = 4 + \frac{2\cdot0x}{100}$$

In fact we have already solved this equation; we obtained

$$x = 200$$

See the example on page 105.
If 200 units are consumed in a quarter then the cost is the same under both tariffs.

You might like to consider when it is cheaper to use Tariff 1 and when it is cheaper to use Tariff 2. For example, if 250 units are consumed in a quarter, which tariff is cheaper? Which is cheaper if 150 units are consumed?

TRY SOME YOURSELF

7(i) The following sequence of instructions resulted in a final answer of $\frac{5}{2}$:

Think of a number. Multiply by 2. Subtract $\frac{1}{2}$. Multiply by 3. Add 2.

What was the original number?

(ii) This problem is similar to the one in the example above. If the following choice of tariffs apply, how many units would need to be consumed in a quarter for the overall quarterly costs to be the same? Give your answer to the nearest whole number of units.

Tariff 1: A fixed charge of £2·86 per quarter plus a charge of 2·51 pence per unit.

Tariff 2: A fixed charge of £3·77 per quarter plus a charge of 1·65 pence per unit.

2.4(iii) EQUATIONS WITH TWO UNKNOWNS

Often, problems involve not just one unknown, but two. It is still possible to derive an equation involving the two unknowns but it cannot be 'solved' in the sense that equations in one unknown can be solved.

Consider the following situation:

Soviet cellist Rostropovich and comedian Jimmy Edwards were both born on 23rd March but in different years. In fact Jimmy Edwards was seven when Rostropovich was born and so was always seven years older than the cellist. If r denotes Rostropovich's age in years and e denotes Jimmy Edwards's age in years, the relationship between the two ages can be expressed as

$$e = r + 7$$

This relationship is an equation in two unknowns. But there is no 'solution' as such; in fact there is a whole range of values of e and r such that the equation is valid. For example, here are a few:

For an equation in one unknown there is only one solution, but for an equation in two unknowns there are many ***pairs*** *of values which satisfy the equation.*

$$e = 8 \text{ and } r = 1$$

or

$$e = 20 \text{ and } r = 13$$

or

$$e = 57 \text{ and } r = 50$$

The equation is not valid for *any* values of e and r. If $e = 1$ and $r = 2$, then the left hand side is *not* equal to the right hand side. However, given any value of either e or r, then the other value can be calculated from the equation.

EXAMPLE

Find r when $e = 13$.

SOLUTION

We need to substitute $e = 13$ into the equation $e = r + 7$. This gives

This is an equation in one unknown, so we can find the solution by manipulation. Notice that we leave r on the right hand side.

$$13 = r + 7$$

$$6 = r \text{ or } r = 6 \quad \text{(Subtract 7 from both sides.)}$$

The following exercises give you some more practice in working with equations in two unknowns. The important thing to remember is that in such cases there are always many pairs of values which satisfy the equation: there is no such thing as '*the*' solution.

TRY SOME YOURSELF

8(i) This question concerns the equation $l = w + 10$.
  (a) Find l when $w = 5$.
  (b) Find l when $w = 35$.
  (c) Find w when $l = 5$.

(ii) This question concerns the equation $y = 3x - 5$.
  (a) Find y when $x = 3$.
  (b) Find y when $x = -2$.
  (c) Find x when $y = 10$.

(iii) A car salesman wishes to make a fixed profit of £250 from each customer who trades in an old car for a new one. Let P denote the profit (in pounds) that the salesman makes on the old car (when he eventually sells it), and N denote the profit (in pounds) on the new car.
  (a) Write down an equation which represents the salesman fulfilling his objective.
  (b) Find P when $N = 150$, and interpret your answer.
  (c) Find P when $N = 300$, and interpret your answer.

(iv) Let t denote the length of time in seconds that it takes for a radio signal from Earth to reach a space vehicle. Suppose it takes 10 seconds to understand such a signal and to respond to it.
  (a) Write down an equation which gives the total time in seconds (m) which elapses from the time the message is transmitted from Earth to the time the response is received back on Earth.
  (b) Find m when $t = 3$, and interpret your answer.
  (c) Find m when $t = 30$, and interpret your answer.
  (d) Find t when $m = 13$, and interpret your answer.
  (e) Find t when $m = 7$, and interpret your answer.

It actually takes about 3 seconds for a radio signal to reach the moon!

After you have worked through this section you should be able to

a Balance an algebraic equation in one unknown by adding, subtracting, multiplying or dividing
b Solve an algebraic equation in one unknown by balancing the equation and manipulating the algebra
c Given a word problem, derive an algebraic equation involving one unknown and consequently solve the problem
d Given an equation in two unknowns, find one of the unknowns given the value of the other
e Given a word problem, derive an algebraic equation involving two unknowns

Finally, here are some exercises if you want more practice.

TRY SOME MORE YOURSELF

9(i) Solve each of the following equations:
  (a) $2x - 7 = 5x - 9$ (b) $b + 5b = 12$ (c) $5z + 12 = z$
  (d) $7d = 24 - d$.

(ii) Solve each of the following equations:
  (a) $5(x + 12) = x$ (b) $2(a + 1) + 1 = a$
  (c) $3(y + 1) - 4(y - 2) = 5$.

(iii) Solve each of the following equations:

(a) $\frac{(x+1)}{2} - 1 = \frac{1}{4}$ (b) $\frac{(d-3)}{2} + 1 = \frac{3}{4}$

(c) $r + 1 = \frac{2(r-1)}{3}$.

(iv) The following sequence of instructions resulted in the original number:

Think of a number. Subtract 6. Multiply by 2. Divide by 3. Add 5.

What was the original number?

(v) A firm pays a fixed wage of £50 for a 40 hour week. In addition overtime is paid at £3·50 an hour. Bob earned £71 one week. How many hours overtime did he work?

(vi) (a) If $y = 7x + 12$, what is the value of y when $x = 3$?
(b) If $r + f = 2$, what is the value of f when $r = -0{\cdot}3$?
(c) If $D = 4f + 3$, what is the value of f when $D = 179$?

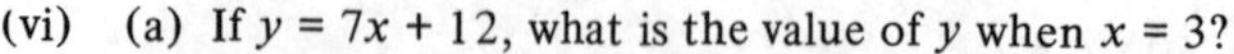

(vii) The price of a taxi ride consists of a fixed fare of 45 pence, plus a further charge of 5 pence for each tenth of a mile. Choose symbols to represent the distance covered and the total cost of the journey. What was the total cost if the distance travelled was 4 miles?

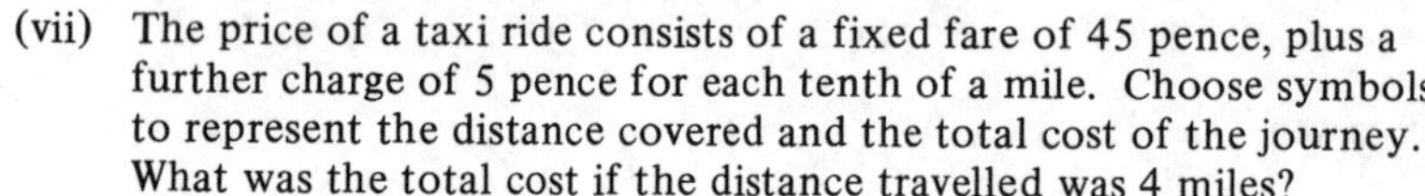

2.5 Equations, Formula and Proportion

TRY THESE QUESTIONS FIRST

1 (i) What is the subject of the equation $3y + 4 = x$?

(ii) Make x the subject of the equation $\frac{2y - 4x}{3} = 5$.

2 Make u the subject of $t^2 = \frac{2(s - ut)}{a}$.

3 If s is directly proportional to t and $s = 1960$ when $t = 2$, find s when $t = 3$.

4 If y is inversely proportional to x and $y = 10$ when $x = 5$, find x when $y = 20$.

2.5(i) CHANGING THE SUBJECT OF AN EQUATION

In the last section we discussed how to solve equations with one unknown. We also introduced some equations with two unknowns. Given the value of one of the unknowns in such an equation it is

For an equation with two unknowns there is a whole range of solutions.

always possible to find the corresponding value of the other unknown. We start this section by considering how to manipulate equations with two unknowns in order to make this process easier.

Consider the equation

$$m = 2t + 10$$

m is the *subject* of this equation since it is isolated on one side and everything else is on the other side. In general, if y, say, is the subject of the equation, then the equation has the form

y = (Some expression not involving y)

y may be isolated on the left hand side, or the right hand side.

or

TRY SOME YOURSELF

1 For each of the following equations state whether or not it has a subject. If the equation has a subject, state what it is.
(i) $y = 3x + 2$ (ii) $2r = 5 - 3t$ (iii) $y = 4x - 2y + 1$
(iv) $2a + b = c$ (v) $s = ut + \frac{1}{2}at^2$.

In the equation $m = 2t + 10$ it is quite easy to find the corresponding value of m for a given value of t. For example, when $t = 3$, $m = (2 \times 3) + 10 = 16$.

m is found by substituting for t into the right hand side of the equation.

This was easy because m is the subject of the equation.

However, it is more difficult to find the corresponding value of t for a given value of m. For example, when $m = 20$, the equation becomes $20 = 2t + 10$.

$20 = 2t + 10$
$10 = 2t$
$5 = t$

Finding the solution requires some algebraic manipulation. It's all very well if only a few values of t are required; each equation can be solved separately. But it would be easier to calculate the value of t if t was the subject of the equation. Making an unknown the subject of an equation requires the techniques of manipulation introduced in Section 2.4. The process is called *changing the subject of the equation.* For example, in the equation $w = l + 10$, w is the subject. Subtracting 10 from both sides gives

*In the equation $20 = 2t + 10$, t is **not** the subject.*

$$w - 10 = l$$

In this equation l is the subject.

The balance of the equation must be maintained; whatever is done to one side must also be done to the other. The strategy which applies to solving equations also applies to changing the subject.

CHECK YOUR ANSWERS

1 (i) x is the subject of the equation. *Section 2.5(i)*

(ii) $2y - 4x = 15$

$$-4x = 15 - 2y$$

$$x = \frac{2y - 15}{4}$$

2 $at^2 = 2(s - ut)$ *Section 2.5(ii)*

$$= 2s - 2ut$$

$$at^2 - 2s = -2ut$$

$$u = \frac{2s - at^2}{2t}$$

3 $k = \frac{s}{t} = \frac{1960}{2}$; when $t = 3$, $s = 2940$. *Section 2.5(iii)*

4 $k = xy = 50$; when $y = 20$, $x = \frac{50}{20} = 2\frac{1}{2}$. *Section 2.5(iv)*

EXAMPLE

Make t the subject of the equation $m = 2t + 10$ and hence calculate t when $m = 20$.

SOLUTION

$$m = 2t + 10$$

$$m - 10 = 2t \quad \text{(Subtract 10 from both sides.)}$$

$$\frac{(m - 10)}{2} = t \quad \text{(Divide by 2.)}$$

Leave t on the right. Take everything else to the left. Make coefficient of t equal to 1.

We can rewrite this as $t = \frac{(m - 10)}{2}$.

t is now the subject of the equation.

When $m = 20$, $t = \frac{(20 - 10)}{2} = 5$.

TRY SOME YOURSELF

2 In each of the following equations make y the subject of the equation and evaluate y when $x = 2$:
(i) $4x = 2y + 1$ (ii) $2x = 1 - 3y$ (iii) $y = 4x - 2y + 1$
(iv) $x + y = 10$.

The following strategy is an adaptation of the strategy used to solve equations with one unknown. It outlines the steps for changing the subject of a more complicated equation with two unknowns.

See Section 2.4(ii).

1 Remove fractions

2 Remove brackets

3 Take the symbol which is to be the subject to one side of the equation

4 Take everything else to the other side

As before, you will find that it is sometimes better to carry out step (2) before step (1).

The 'everything else' includes numbers and the other symbol.

The formula $C = \frac{5}{9}(F - 32)$ is used to convert from degrees Fahrenheit to degrees Centigrade when measuring temperature. This formula is an equation with two unknowns, C and F. In order to convert from degrees Centigrade to degrees Fahrenheit it is easier to change the subject of the equation to F.

Try converting 212° F to °C and 32° F to ° C by substituting for F. It's harder to convert 37°C to ° F.

EXAMPLE

Make F the subject of the equation $C = \frac{5}{9}(F - 32)$.

SOLUTION

Strategy:

$C = \frac{5}{9}(F - 32)$ — *Leave F on the right.*

$9C = 5(F - 32)$ (Multiply by 9.) — *Remove fraction.*

$9C = 5F - 160$ (Multiply out the brackets.) — *Remove brackets.*

$9C + 160 = 5F$ (Add 160 to both sides.) — *Take everything else to the left.*

$\frac{9C + 160}{5} = F$ (Divide by 5.) — *Make the coefficient of F equal to 1.*

Check this by converting 100° C to ° F.

TRY SOME YOURSELF

3(i) Make q the subject of the equation $p = \frac{(3q + 2)}{4}$.

(ii) Make d the subject of the equation $E = \frac{(3 - d)}{3}$.

(iii) Make x the subject of the equation $\frac{x}{2} + y - 1 = \frac{1}{2}$.

2.5(ii) COPING WITH MORE THAN TWO SYMBOLS

We mentioned earlier that

$$\text{Speed} = \frac{\text{Distance}}{\text{Time}}$$

See Section 2.2(v).

and that this general result can be represented by the formula

$$v = \frac{d}{t}$$

v denotes speed.
d denotes distance.
t denotes time.

This formula involves three symbols, but it is still possible to change the subject using the techniques outlined above. For example, if we wanted to find the distance travelled in $2\frac{3}{4}$ hours at a speed of 25 miles per hour it would be easier to make d the subject:

$v = \frac{d}{t}$ — *Leave d on the right.*

$vt = \frac{d}{t}t$ — *Multiply both sides by t to get rid of the fraction. (Remember we can multiply or divide by symbols as well as numbers.)*

$vt = d$

Now, when $t = 2\frac{3}{4}$ hours and $v = 25$ miles per hour,

$$d = 2\tfrac{3}{4} \times 25 = 68{\cdot}75 \text{ miles}$$

Notice that the units of d are also given by the formula. Put the units of v and t into the expression vt to get

$$\frac{\text{miles}}{\text{hour}} \times \text{hour} = \text{miles}$$

TRY SOME YOURSELF

4(i) Make R the subject of $V = IR$.
(ii) Make m the subject of $F = ma$.
(iii) Make T the subject of $I = \dfrac{PRT}{100}$.

$V = IR$, $F = ma$ and $I = \dfrac{PRT}{100}$ are all well known formulas for some general results. You may recognise them but don't worry if they're not familiar.

If the equation or formula contains brackets or fractions these can be dealt with by the same methods used for equations with two unknowns.

EXAMPLE

Make a the subject of $s = ut + \frac{1}{2}at^2$.

SOLUTION

$s = ut + \frac{1}{2}at^2$ — *Leave a on the right.*

$2s = 2ut + at^2$ (Multiply by 2.) — *Remove fraction.*

$2s - 2ut = at^2$ (Subtract $2ut$ from both sides.) — *Take everything else to the left.*

$\dfrac{2s - 2ut}{t^2} = a$ (Divide by t^2.) — *Make coefficient of a equal to 1.*

TRY SOME YOURSELF

5(i) Make l the subject of $P = 2(l + w)$.
(ii) Make u the subject of $s = \frac{1}{2}(u + v)t$.
(iii) Make u the subject of $s = ut + \frac{1}{2}at^2$.

Again, these are all well known formulas. You should recognise that

$$P = 2l + 2w$$

gives the perimeter of a rectangle.

2.5(iii) DIRECT PROPORTION

Suppose that a car travels 6 miles in 10 minutes
12 miles in 20 minutes
18 miles in 30 minutes
24 miles in 40 minutes
and 30 miles in 50 minutes

This information can be represented in a table.

Time t (mins)	10	20	30	40	50
Distance d (miles)	6	12	18	24	30
d/t	3/5	3/5	3/5	3/5	3/5

Notice that in each case

$$\frac{\text{distance}}{\text{time}} = \frac{d}{t} = \frac{3}{5}$$

The longer the car travels the farther it goes. In fact we can be more precise about this. The car travels twice as far in 40 minutes (24 miles) as it does in 20 minutes (12 miles). It travels five times as far in 50 minutes (30 miles) as it does in 10 minutes (6 miles).

This information can be used to find out how far the car travels in any given time.

EXAMPLE

Suppose that the car continues in the same way. How far will it travel in 60 minutes?

SOLUTION

In 10 minutes the car travels 6 miles.
In 20 (= 2 x 10) minutes it travels 12 (= 2 x 6) miles.
In 30 (= 3 x 10) minutes it travels 18 (= 3 x 6) miles.
In 50 (= 5 x 10) minutes it travels 30 (= 5 x 6) miles.
So in 60 (= 6 x 10) minutes it travels (6 x 6) = 36 miles.

Notice too that the car travels twice as far in 60 minutes as it does in 30 minutes.

In such a situation, where the distance (d) changes in the same way as the time (t), d is said to be proportional to t. This is usually written as

$$d \propto t$$

$d \propto t$ is pronounced 'd is proportional to t'.

More specifically, d is *directly proportional* to t.

In the above example $\frac{d}{t}$ always has the value $\frac{3}{5}$, a constant, so that $d = \frac{3}{5}t$.
$\frac{3}{5}$ is called the *constant of proportionality.*

You may recognise the symbols used here. Remember that the formula

$$v = \frac{d}{t}$$

represents the relationship

$$\text{Velocity} = \frac{\text{Distance}}{\text{Time}}$$

The formula can be rearranged so that

$$d = vt$$

Compare this with $d = \frac{3}{5}t$, which we obtained above.

If the speed is constant then $d \propto t$, and v is the constant of proportionality.

We are interested in the way that d varies with t and vice versa. Because both the unknowns vary they are often known as *variables.* In this case the variable d is directly proportional to the variable t.

*Similarly in any equation with two unknowns, the unknowns are often called **variables** since they vary together. (Given the value of one variable, the equation gives the corresponding value of the other.)*

The next example shows how to determine whether two variables are directly proportional.

EXAMPLE

The table below indicates the distances travelled by a second car in various times.
Is the distance directly proportional to the time?

Time t (mins)	10	20	30	40	50
Distance d (miles)	6	10	20	25	30

SOLUTION

If $d \propto t$, then $\frac{d}{t}$ is a constant.

If $d \propto t$ then $d = vt$ and v is a constant. Since $v = d/t$, we need to check whether d/t is a constant.

We can check whether it is a constant by evaluating d/t for the different values given in the table.

Time t (mins)	10	20	30	40	50
Distance d (miles)	6	10	20	25	30
d/t	3/5	1/2	2/3	5/8	3/5

The values of d/t are different. So d/t is not a constant and therefore d is *not* directly proportional to t.

This illustrates that at different stages of the journey the car was travelling at different speeds.

TRY SOME YOURSELF

6(i) The table below gives the same information as the table on page 114, but here the time is given in hours.

Time t (hours)	1/6	1/3	1/2	2/3	5/6
Distance d (miles)	6	12	18	24	30

(a) Show that d is proportional to t.
(b) What is the constant of proportionality?
(c) How far would the car travel in 3 hours?

(ii) A hi-fi catalogue advertises prices of magnetic tape cassettes as follows:

Number of cassettes n	12	50	100
Cost in pounds c	20	75	140

Is the cost directly proportional to the number of cassettes bought?

More generally, the letter k is often used to stand for the *constant of proportionality*. Thus if $y \propto x$ then $y = kx$. In any particular instance the value of k can be found provided one pair of values of x and y is known. And if k is known, then it is easy to find the corresponding value of y for *any* given value of x, and vice versa.

Since the variables vary together we can talk about pairs of values.

EXAMPLE

The cost of a number of bottles of wine is directly proportional to the cost of one bottle.
A case of 12 bottles of wine costs £20·28. What is the cost of 18 bottles?

SOLUTION

Let x denote the number of bottles and y the cost in pounds. Then $y \propto x$, that is, $y = kx$.

k is the constant of proportionality.

When $x = 12$, $y = 20{\cdot}28$, so

$$k = \frac{y}{x} = \frac{20{\cdot}28}{12}$$

There is no need to calculate k at this stage.

Now when $x = 18$,

$$y = kx = \frac{20{\cdot}28 \times 18}{12}$$
$$= 30{\cdot}42$$

So the cost of 18 bottles of wine is £30·42.

This example indicates that the first thing to do is to find the constant of proportionality. Notice that unless you are specifically asked to find k, it can be left as a fraction.

TRY SOME YOURSELF

7(i) If $d \propto t$ and $d = 15$ when $t = 5$, what is the value of d when $t = 15$?
(ii) As a useful rule of thumb, the distance travelled by a car is directly proportional to the number of gallons of petrol used. A car travelled 177 miles on 4 gallons of petrol. How far would it go on 7 gallons?
(iii) A stack of 200 sheets of paper is ¾ inch high. How high would a stack of 350 sheets be?
(iv) In July 1980 one pound sterling was equivalent to 2·3 dollars. What was the sterling equivalent of 5 dollars?

The height of the stack is directly proportional to the number of sheets of paper.

2.5(iv) INVERSE PROPORTION

The table below indicates journey times from London to Edinburgh, travelling at different speeds.

Speed v (m.p.h.)	20	40	50	80	100
Time t (hours)	20	10	8	5	4

At 20 m.p.h. the journey takes 20 hours; at 80 m.p.h. it takes 5 hours.

As the speed increases the time taken decreases. In this case, doubling the speed from 20 m.p.h. to 40 m.p.h. halves the travelling time, from 20 hours to 10 hours. Increasing the speed 4 times from 20 m.p.h. to 80 m.p.h. reduces the journey time from 20 hours to one quarter of that, 5 hours. These two variables are said to be *inversely proportional.* In fact

$$t \propto \frac{1}{v}$$

$t \propto \frac{1}{v}$ is pronounced 't is inversely proportional to v'.

Again, this can be expressed in terms of the constant of proportionality. Thus

$$t = \frac{k}{v} \text{ or } k = vt$$

Notice also that $v \propto \frac{1}{t}$ since if $t = \frac{k}{v}$ then $v = \frac{k}{t}$.

You can check that k is a constant from the table.

v	20	40	50	80	100
t	20	10	8	5	4
vt	400	400	400	400	400

In each case $vt = 400$.

Now, since $t = \frac{400}{v}$, given any value of v it is easy to find the corresponding value of t and vice versa. For example, if $v = 70$, $t = \frac{400}{70} = 5{\cdot}71$.
So, travelling at 70 m.p.h. the journey takes about 5·71 hours.

(Rounded to two decimal places.)

The next example indicates how inverse proportion occurs in an everyday situation.

EXAMPLE

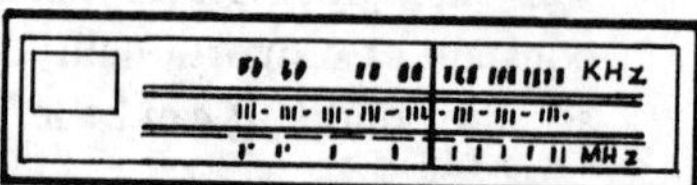

Radio frequency is inversely proportional to wavelength. Radio 1 is broadcast on a wavelength of 285 metres, which corresponds to a frequency of 1053 kilocycles per second. What is the frequency of Radio 3 broadcast on 247 metres?

Kilocycle means 1000 cycles per second.

SOLUTION

Let f be the frequency in kilocycles and l the wavelength in metres. Then f is inversely proportional to l and so

$$f = \frac{k}{l} \text{ or } k = fl$$

When $f = 1053$, $l = 285$, so

$$k = 1053 \times 285$$

Thus

$$f = \frac{1053 \times 285}{l}$$

When $l = 247$,

$$f = \frac{1053 \times 285}{247} = 1215$$

(Rounded to the nearest whole number.)

Thus Radio 3 is broadcast on 1215 kilocycles per second.

TRY SOME YOURSELF

8(i) y is inversely proportional to x. If $y = 2{\cdot}5$ when $x = 5$,
(a) What is the value of y when $x = 6{\cdot}25$?
(b) What is the value of x when $y = 50$?

(ii) This question is based on the information given in the example above.
(a) Radio 4 is broadcast on 1500 m. What is the equivalent frequency?
(b) What is the wavelength of a transmission at 1089 kilocycles per second?

(iii) The volume V occupied by a gas is known to be inversely proportional to the pressure P at which it is kept.
If $V = 24\text{ cm}^3$ when $P = 100$ cm, find V when $P = 64$ cm.

Volume is measured in cubic centimetres, pressure is measured in centimetres.

After you have worked through this section you should be able to

a Recognise whether or not an equation or formula has a subject
b Change the subject of an equation or formula involving two or more symbols
c Determine whether or not two variables are directly proportional
d Given that two variables are directly proportional, find the constant of proportionality, and given the value of one of the variables, find the corresponding value of the other
e Determine whether or not two variables are inversely proportional
f Given that two variables are inversely proportional, find the constant of proportionality, and given the value of one of the variables, find the corresponding value of the other

Finally, here are some exercises if you want more practice.

TRY SOME MORE YOURSELF

9(i) In each of the following equations make y the subject and evaluate y when $x = 4$:
(a) $2x + 3y = 6$ (b) $3x = 7 - 3y$ (c) $5y - 4x - 3 = 0$.

(ii) (a) Make y the subject of the equation

$$d = \frac{y - 5{\cdot}295}{0{\cdot}01}$$

(b) Make x the subject of the equation

$$y - 3 = \tfrac{2}{3}(x - 4)$$

(iii) (a) Make V the subject of $P = \dfrac{T}{V}$.
(b) Make v the subject of $\dfrac{t}{v} = \dfrac{1}{d^2}$.
(c) Make r the subject of $I = \dfrac{ne}{(R + nr)}$.
(d) Make s the subject of $p = \dfrac{100(s - c)}{c}$.
(e) Make h the subject of $A = 2\pi r(r + h)$.

(iv) The table below indicates times taken for a train to pass through tunnels of different lengths.

Length l (miles)	1	2	3	$4\frac{1}{2}$
Time t (secs)	30	60	90	135

Determine whether the distance is proportional to the time.

(v) The weekly cost of running a household is directly proportional to the number of people in the household. It costs £55 a week to run a household of 5.
(a) What is the weekly cost for a household of 2?
(b) If the weekly cost is £77, what is the household size?

(vi) 1 lb is equivalent to 2·2 kg. What is the imperial equivalent of 10 kg?

(vii) A road contractor estimated that it would take 400 men 27 months to build a motorway. He reckons that the time taken to complete the job is inversely proportional to the number of men employed.
(a) How long would it take to build the motorway using 200 men?
(b) How many men would it take to complete the job in 2 years?

Section 2.1 Solutions

1

(i) Think of a number: x
Add 2: $x + 2$
Multiply by 3: $(x + 2) \times 3$
Subtract 4: $[(x + 2) \times 3] - 4$

(ii) Think of a number: y
Multiply by 6: $y \times 6$
Add 2: $(y \times 6) + 2$
Divide by 3: $\dfrac{(y \times 6) + 2}{3}$
Add 1: $\dfrac{(y \times 6) + 2}{3} + 1$

(iii) Think of a number: A
Multiply by 3: $A \times 3$
Subtract 2: $(A \times 3) - 2$
Multiply by the original number: $[(A \times 3) - 2] \times A$

You probably chose different letters to represent the unknown number. That doesn't matter–just check that you get the same sequences.

2

(i) $r = 4$, $(4 + 4) - 6 = 8 - 6 = 2$
(ii) $[((-1) + 3) \times 2] + 7 = (2 \times 2) + 7 = 11$
(iii) $[(2 \times 0{\cdot}12 + 16] \div 8 = 2{\cdot}03$

3

(i) $4a$
(ii) $-2a$ (We usually omit the brackets around the negative number.)
(iii) $3y$
(iv) $(-3) \times (-2) \times b = 6b$
(Remember $(-) \times (-) = (+)$.)
(v) $3(x + 5)$
(vi) $-4(b + 2)$
In each case the sign comes first, then the number, then the symbol.

4

(i) $3(5 + 5) = 30$
(ii) $\dfrac{3(4 + 5) - (3 \times 4) + 1}{2} - 4$
$= \dfrac{27 - 11}{2} - 4 = 4$
(Remember that $3x$ means $3 \times x$.)
In each case all you have to do is rewrite the expression replacing the symbol with the numerical value; then you can calculate the answer.

5

(i) The term is $-6x$.
(ii) The term is $6pq$.
(iii) Each of the terms x, xy and z has coefficient 1. The term $-y$ has coefficient -1.

6

(i) (a) $y = 3x - 2$
(b) When $x = -2$, $y = [3 \times (-2)] - 2 = -8$.
(ii) (a) $y = 2(-2x + 1)$
(b) When $x = 3$, $y = 2[(-2 \times 3) + 1] = -10$.
(iii) (a) $y = x\left(\dfrac{3x - 2}{4} + 4\right)$
(b) When $x = 2$, $y = 2\left(\dfrac{(3 \times 2) - 2}{4} + 4\right) = 10$.

7

(i) Let the cost be c pounds. Then $c = 1{\cdot}7x$.
(ii) We don't know the number of employees or the number of guests. Let n stand for the number of employees and m stand for the number of guests.
(a) If W was the wages bill in pounds, $W = 50n$.
(b) If I was the income that week in pounds, $I = 80m$.
(c) The total profit in pounds that week was $80m - 50n$. But we are told that the total profit was £1100. Therefore $1100 = 80m - 50n$.

8

(i) (a) $6(4t - 2) + t$ (b) $[-(p - 7) + 14] - p$

(ii) (a) $7a$ (b) $-12b$ (c) $3z$ (d) $-0{\cdot}12r$

(iii) (a) 6·3 (b) 55·58

(iv) (a) −7

(b) Coefficient of z is 1, coefficient of w is −1.

(v) (a) $c = 1{\cdot}7x + 0{\cdot}4$ (b) $c = 1{\cdot}7x + 0{\cdot}4y$

(c) If $x = 9{\cdot}8$ and $y = 2$,

$$c = (1{\cdot}7 \times 9{\cdot}8) + (0{\cdot}4 \times 2) = 17{\cdot}46$$

The total cost is therefore £17·46.

(vi) If n = number of employees, m = number of guests, W = weekly wages bill in pounds and I = weekly income in pounds, then

(a) $W = 30n$

(b) $I = (7 \times 17) \times m = 119m$

(c) $P = I - W$, so $P = 119m - 30n$.

(d) In 4 weeks the profit would be £$4P$, that is £$4(119m - 30n)$.

Section 2.2 Solutions

1 (i) $2x$ (ii) y (iii) $29p$

2

(i) (a) $6a + 2b + 7a + 3b = (6a + 7a) + (2b + 3b) = 13a + 5b$

(b) $-2r + 4p + 2p + 6r = (-2r + 6r) + (4p + 2p) = 4r + 6p$

(c) $2e + f - 3e + 5e = (2e - 3e + 5e) + f = 4e + f$

(d) $3 + 4a - 2 - a = (3 - 2) + (4a - a) = 1 + 3a$

Don't worry if you've written, for example, $6p + 4r$ instead of $4r + 6p$; that is just as good. We've put brackets around like terms to clarify what is happening. With practice you will find that it's not necessary to use brackets. In part (d) you may have noticed that numbers are collected just like any other term. In fact, numbers are often called *constant terms* since they involve no other symbol.

(ii) (a)

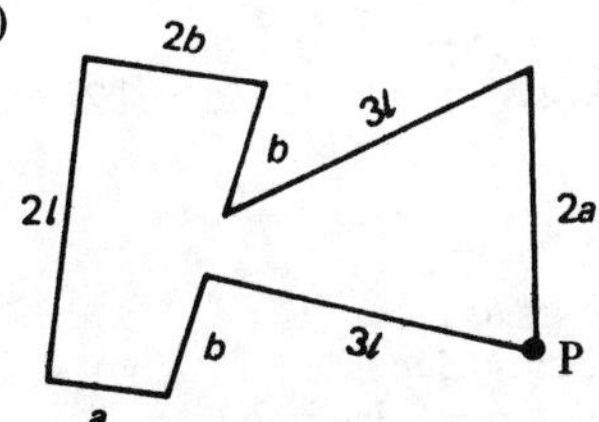

Starting at P and working anticlockwise around the diagram, we get the perimeter to be

$$2a + 3l + b + 2b + 2l + a + b + 3l$$

Collecting like terms, this becomes

$$(2a + a) + (3l + 2l + 3l) + (b + 2b + b)$$

or

$$3a + 8l + 4b$$

(b) When $a = 0{\cdot}25$ cm, $b = 1{\cdot}36$ cm and $l = 4{\cdot}27$ cm,

$$3a + 8l + 4b = 40{\cdot}35 \text{ cm}$$

3

(i) (a) $7cde$ (b) 172·2

(ii) (a) $-5xyz$

(b) 113·79 (Rounded to two decimal places.)

(iii) (a) $12acd$ (b) $12acd$ becomes $-34{\cdot}8d$.

(iv) (a) $6pqr$

(b) $6pqr$ becomes $3qr$. (No need for your calculator here!)

4

(i) $2 \times b \times 3b \times 4b = 24b^3$. When $b = 2$, $24b^3 = 192$.

(ii) $3a \times (-4a) = -12a^2$. When $a = -3$, $-12a^2 = -108$.

(iii) $6z \times 3 \times z \times (-4z) = -72z^3$. When $z = 2{\cdot}5$, $-72z^3 = -1125$.

5

(i) (a) $3 \times a \times c \times a \times c = 3 \times a \times a \times c \times c = 3a^2c^2$

(b) 108

(ii) (a) $6aba = 6aab = 6a^2b$

(b) When $a = 1$ and $b = 0$, $6a^2b = 0$. (Anything multiplied by zero is zero.)

(iii) (a) $-24a^2b^3$

(b) 2320·86 (Rounded to two decimal places.)

6

(i) (a) $(a + b) \div (c \times d) = (a + b) \times \dfrac{1}{(c \times d)} = \dfrac{a + b}{cd}$

(b) 2

(ii) (a) $(2x \times 3y) \div (5p \times 7q) = 6xy \times \dfrac{1}{(5p \times 7q)} = \dfrac{6xy}{35pq}$

(b) 1·59 (Rounded to two decimal places.)

(iii) (a) $\dfrac{a}{b} \times \dfrac{c}{d} \times \dfrac{a}{d} = \dfrac{a \times c \times a}{b \times d \times d} = \dfrac{a^2c}{bd^2}$

(b) 12

7

(i) $\dfrac{1}{z \times z \times z \times z} = z^{-4}$

(ii) $\dfrac{l}{p} = l \times \dfrac{1}{p} = l \times p^{-1} = lp^{-1}$

(iii) $\dfrac{s}{t^2} = s \times \dfrac{1}{t^2} = s \times t^{-2} = st^{-2}$

8

(i) ab and $5ab$ are like terms.

(ii) $3xy$, $-xy$ and $2xy$ are like terms.

(iii) c^3 and $4c^3$ are like terms.

$4c^2d$ and $-3c^2d$ are like terms.

$2d^2c$ and $-3cd^2$ are like terms because $2d^2c = 2cd^2$ when the symbols are written in alphabetical order. (Notice in particular that $4c^2d$ and $2cd^2$ are *unlike* terms.)

(iv) t and $-t$ are like terms.

9

(i) $8t^2$ (ii) $7y^2$

(iii) $4p^2 + pq - 3p^2 - 3pq + 5p^2$
$= 4p^2 - 3p^2 + 5p^2 + pq - 3pq$
$= 6p^2 - 2pq$

(iv) $2c^3 + 4c^2d - 3dc^2 + 2 = 2c^3 + 4c^2d - 3c^2d + 2$
$= 2c^3 + c^2d + 2$

10

(i) Letting l, w and h denote the dimensions of length, width and height, the volume V is given by

$$V = hlw$$

Putting $l = 5$, $w = 2$ and $h = 3$ gives $V = 30$. The units are (cm) × (cm) × (cm) = cm^3. Hence $V = 30\ \text{cm}^3$ (pronounced 30 cubic centimetres).

(ii) $V = d/t$. Putting $d = 5$, $t = 2$ gives $V = 2{\cdot}5$. The units are given by

$$\frac{\text{km}}{\text{hr}} = \text{km hr}^{-1}$$

(Again, hr is treated as one symbol–it doesn't mean h × r.)
Hence $V = 2{\cdot}5\ \text{km hr}^{-1}$ (pronounced 2·5 kilometres per hour).

(iii) $\text{Density} = \dfrac{\text{Mass}}{\text{Volume}}$

From (i) we know that cubic centimetres can be written symbolically as cm^3. The units of measurement for density are therefore

$$\frac{\text{g}}{\text{cm}^3} = \text{g cm}^{-3}$$

(pronounced grams per cubic centimetre).

11

(i) (a) $8y$ (b) $9r + 2d$ (c) $7 + s$

(ii) (a) $-8fq$ (b) $24r^2s$ (c) $24a^2b$ (d) r^4s^2

(iii) (a) $\dfrac{9xy}{3x + 4y}$ (b) $\dfrac{9 + x + y}{12xy}$ (c) $\dfrac{9x + y}{3(x + y)}$

(iv) (a) cs^{-1} (b) mv^{-1} (c) m^3t^{-2}

(v) (a) $3x^2y - 9xy^2 + x$ (b) $4 - 2r^2 + rs$
(c) $4st + 6s^2 - 5s^2t + st^2$

(vi) (a) $11x - xy$ (b) 8 (c) 30 (d) 13·9788

(vii) (a) Volume $= l \times l \times l = l^3$
$4^3 = 64$ and the units of measurement are ft^3. Therefore the volume is 64 ft^3 (64 cubic feet).

(b) $\text{Speed} = \dfrac{\text{Distance}}{\text{Time}}$

$\frac{14}{6} = 2{\cdot}33$ (rounded to two decimal places)
The units of measurement are km/s or km s^{-1}.
The speed was therefore $2{\cdot}33\ \text{km s}^{-1}$.

Section 2.3 Solutions

1

(i) $7 \times (8 - 3) = 7 \times 5 = 35$ (Evaluate the brackets first.)

(ii) $(3 - 5) \times 2 = (-2) \times 2 = -4$

(iii) $4 + [2 \times (6 + 3)] = 4 + [2 \times 9] = 4 + 18 = 22$ (In this case, evaluate the innermost brackets first.)

(iv) $-(3 + 4) = -3 - 4 = 7$

(v) $-(2 - 5 + 6) = -2 + 5 - 6 = -3$

2 (i) $y - (-z) = y + z$ (ii) $(-u) + (-v) = -u - v$
(iii) $(-t) \times r = -rt$ (iv) $x \times (-x) \times x = -x^3$
(v) $(-z) \times (-z) \times x \times (-x) = -x^2z^2$

3

(i) $3(x + y) = 3x + 3y$

(ii) $5(2x - y) = 10x - 5y$

(iii) $-4(a + 2b - c) = -4a - 8b + 4c$ (Each term inside the brackets is multiplied by (-4).)

(iv) $-5(a - b) = -5a + 5b$

4

(i) $x + (y + z) = x + y + z$

(ii) $(a + b) - (c + d) = a + b - c - d$

(iii) $(a + b) \times (-2) = -2(a + b) = -2a - 2b$

(iv) $(a - b + 2c) \times (-5) = -5(a - b + 2c)$
$= -5a + 5b - 10c$ (The final expression can be rewritten as $5b - 5a - 10c$ if you want to put the positive term first.)

5

(i) $r - (s - r) = r - s + r = 2r - s$

(ii) $3z + 2y - (y - 2z) = 3z + 2y - y + 2z$
$= 5z + y$

(iii) $3 + 2a - b - (3b - 4) = 3 + 2a - b - 3b + 4$
$= 7 + 2a - 4b$

6

(i) $x + y + 2(x - 2y) = x + y + 2x - 4y$
$= 3x - 3y$

(ii) $-(a + 2b) + 3(2c + b) = -a - 2b + 6c + 3b$
$= -a + b + 6c$

(iii) $4t - 2(4 - t) = 4t - 8 + 2t$
$= 6t - 8$

(iv) $2(r + 6s) - 3(4s - r) = 2r + 12s - 12s + 3r$
$= 5r$

7

(i) $a(b - c) = ab - ac$

(ii) $3a(b - c) = 3ab - 3ac$

(iii) $2r(3r + s - t) = 6r^2 + 2rs - 2rt$

(iv) $-3z(2y - z) = -6yz + 3z^2$

8

(i) $x(2x-3)-3x(5-2x) = 2x^2-3x-15x+6x^2$
$= 8x^2-18x$

(ii) $x(y-z)-y(x+z) = xy-xz-yx-yz$
$= -xz-yz$
(Remember that $-yx = -xy$.)

(iii) $2a(b-3a)-2b(3a-2b)$
$= 2ab-6a^2-6ba+4b^2$
$= 4b^2-4ab-6a^2$
(Remember that $-6ba = -6ab$.)

(iv) $-2p(p+q)+3p(2p-q)$
$= -2p^2-2pq+6p^2-3pq$
$= 4p^2-5pq$

9

(i) $6t-4s+3(s-4t) = 6t-4s+3s-12t$
$= -6t-s$
When $t=-1$ and $s=4$,
$-6t-s = ((-6)\times(-1))-4$
$= 6-4 = 2$

(ii) $2(r+6s)-3(4s-r) = 5r$
This is taken from Exercise 6(iv).
When $r=10$, $5r=50$.
(Notice that we could have managed without knowing a value for s.)

(iii) $x(2x-3)-3x(5-2x) = 8x^2-18x$
(This is taken from Exercise 8(i).)
When $x=1$, $8x^2-18x = 8-18 = -10$.

10

(i) $2(a+2b)-3(b-2a) = 8a+b$
(a) When $a=0{\cdot}1$ and $b=0{\cdot}7$, $8a+b = (8\times0{\cdot}1)+0{\cdot}7 = 1{\cdot}5$.
(b) When $a=143$ and $b=-91$, $8a+b = (8\times143)-91 = 1053$.

(ii) From Exercise 8 (i),
$x(2x-3)-3x(5-2x) = 8x^2-18x$
(a) When $x=0{\cdot}9$, $8x^2-18x = -9{\cdot}72$.
(b) When $x=-0{\cdot}31$, $8x^2-18x = 6{\cdot}35$
(Rounded to two decimal places.)

(iii) $6y^2-7x(3-2y)+y(2x-y)$
$= 6y^2-21x+14xy+2xy-y^2$
$= 5y^2+16xy-21x$
(a) When $x=0{\cdot}2$ and $y=0{\cdot}3$, $5y^2+16xy-21x = -2{\cdot}79$.
(b) When $x=-12$ and $y=-31$, $5y^2+16xy-21x = 11{,}009$.

In order to carry out these calculations on your calculator you must insert multiplication signs (remember $8a$ means $8\times a$ etc). You may also have to use brackets and the square key, [x2]. Sometimes it is just as quick to substitute into the original expression without simplifying it first—at other times it is not. You can always check your answer by substituting the numbers into both the original expression *and* the simplified result.

11

(i) $x(y-z) = xy-xz$

(ii) If $x=(y-z)$, $xy-xz = (y-z)y-(y-z)z$
$= y(y-z)-z(y-z)$
$= y^2-yz-zy+z^2$
$= y^2-2yz+z^2$

(iii) $(y-z)^2 = (y-z)(y-z) = y^2-2yz+z^2$

12

(i) $(x-1)(x+2) = x^2+2x-x-2$
$= x^2+x-2$

(ii) $(x+5)(4+3x) = 4x+3x^2+20+15x$
$= 19x+20+3x^2$
(This is usually written with the x^2 term first, that is $3x^2+19x+20$.)

(iii) $(a-3)(a+3) = a^2+3a-3a-9$
$= a^2-9$

(iv) $(2a-b)^2 = (2a-b)(2a-b)$
$= 4a^2-2ab-2ab+b^2$
$= 4a^2-4ab+b^2$

(v) $(2p+q)(p-3q) = 2p^2-6pq+pq-3q^2$
$= 2p^2-5pq-3q^2$

13

(i) (a) $3b+3c$ (b) $-2s+2t$ (c) $-3x+12y-6z$

(ii) (a) $a+b-2c$ (b) $4a-3t-5$
(c) $22-t$

(iii) (a) $2z-2y$ (b) $2x-6y$ (c) $15t-3$

(iv) (a) $8a-6b$ (b) $8t-5s$ (c) $14-26x$

(v) (a) $10x^2+11x$ (b) $6xy+10y-3$
(c) $10pq-12pr+2qr$

(vi) (a) $r^2+2rt+t^2$ (b) x^2-6x+9
(c) $3x^2-5x+2$

(vii) (a) $2x(3x+y)-y(x+2y) = 6x^2+xy-2y^2$
(b) 12 (c) -32 (d) -2003 (To the nearest whole number.)

Section 2.4 Solutions

1

(i) (a) $2x+7$ **+1** $= 13$ **+1**
$2x+8 = 14$
(b) $2x+7$ **+7** $= 13$ **+7**
$2x+14 = 20$
(c) $2x+7$ **−7** $= 13$ **−7**
$2x = 6$

(ii) (a) $2x-7$ **+1** $= 13$ **+1**
$2x-6 = 14$
(b) $2x-7$ **−7** $= 13$ **−7**
$2x-14 = 6$
(c) $2x-7$ **+7** $= 13$ **+7**
$2x = 20$

(iii) (a) $t-3 = 6$
Add 3 to both sides to get
$t-3$ **+3** $= 6$ **+3**
$t = 9$

(b) $z + 17 = 12$

Subtract 17 from both sides to get

$z + 17 - 17 = 12 - 17$

$z = -5$

(c) $x - 12 = -4$

Add 12 to both sides to get

$x - 12 + 12 = -4 + 12$

$x = 8$

2

(i) $3x + 2 = 5$

Subtract 2 from both sides to get

$3x = 3$

(ii) $4x - 7 = 3$

Add 7 to both sides to get

$4x = 10$

(iii) $5 - 2x = 2$

Subtract 5 from both sides to get

$-2x = -3$

(iv) $4x - 3 = 2x + 1$

(Take x to the left hand side.)

Subtract $2x$ from both sides to get

$2x - 3 = 1$

(Take the numbers to the right.)

Now add 3 to both sides to get

$2x = 4$

(These operations could equally well have been carried out in the opposite order.)

3

(i) $3x = 4$

Divide both sides by 3:

$\frac{3x}{3} = \frac{4}{3}$

$x = \frac{4}{3}$

(ii) $5x = -3$

Divide both sides by 5:

$\frac{5x}{5} = -\frac{3}{5}$

$x = -\frac{3}{5}$

(iii) $-2x = 11$

Divide both sides by -2:

$\frac{-2x}{-2} = \frac{11}{-2}$

$x = -\frac{11}{2}$

(iv) $-\frac{1}{4}y = 7$

Multiply both sides by -4 (which is the same as dividing by $-\frac{1}{4}$):

$\frac{-(-4)}{4}y = 7 \times (-4)$

$y = -28$

(v) $-\frac{4}{9}z = -2$

First multiply both sides by -9:

$\frac{(-4) \times (-9)}{9}z = (-2) \times (-9)$

$4z = 18$

Now divide both sides by 4:

$\frac{4z}{4} = \frac{18}{4}$

$z = \frac{18}{4} = \frac{9}{2}$

This could equally well have been done in one go by dividing both sides by $(-\frac{4}{9})$, which is the same as multiplying by $(-\frac{9}{4})$.

4

(i) $2x + 7 = -3681$

$2x = -3688$ (Subtract 7 from both sides.)

$x = -1844$ (Divide by 2.)

(ii) $1 - 3x = 19$

$-3x = 18$ (Subtract 1 from both sides.)

$x = -6$ (Divide by -3.)

(iii) $5y + 3 = 2 + y$

$4y + 3 = 2$ (Subtract y from both sides.)

$4y = -1$ (Subtract 3 from both sides.)

$y = -\frac{1}{4}$ (Divide by 4.)

(iv) $1 - 4t = 10 - 7t$

$1 + 3t = 10$ (Add $7t$ to both sides.)

$3t = 9$ (Subtract 1 from both sides.)

$t = 3$ (Divide by 3.)

In parts (iii) and (iv) the first two steps could have been carried out in the opposite order (that is, the numbers could equally well have been taken to the right hand side first).

5

(i) $2x + 1 = \frac{x}{2} + 4$

$2(2x + 1) = 2(\frac{x}{2} + 4)$ (Multiply by 2.)

$4x + 2 = x + 8$

$3x + 2 = 8$ (Subtract x from both sides.)

$3x = 6$ (Subtract 2 from both sides.)

$x = 2$ (Divide by 3.)

(ii) $6x + 6 = 1 - \frac{x}{4}$

$24x + 24 = 4 - x$ (Multiply by 4.)

$25x + 24 = 4$ (Add x to both sides.)

$25x = -20$ (Subtract 24 from both sides.)

$x = \frac{-20}{25} = \frac{-4}{5}$ (Divide by 25.)

(iii) $1 - \frac{x}{3} = 2 + x$

$3 - x = 6 + 3x$ (Multiply by 3.)

$3 - 4x = 6$ (Subtract $3x$ from both sides.)

$-4x = 3$ (Subtract 3 from both sides.)

$x = \frac{-3}{4}$ (Divide by -4.)

6

(i) $5(2x - 3) = 30$

$10x - 15 = 30$ (Multiply out brackets.)

$10x = 45$ (Add 15 to both sides.)

$x = \frac{45}{10} = 4{\cdot}5$ (Divide by 10.)

(ii) $\dfrac{x+4}{3} = x$

$x + 4 = 3x$ (Multiply by 3.)

$4 = 2x$ (Subtract x from both sides, leaving x on the right.)

$\frac{4}{2} = 2 = x$ (Divide by 2.)

(iii) $b - \dfrac{3(b+1)}{5} = 1$

$5b - 3(b + 1) = 5$ (Multiply by 5.)

$5b - 3b - 3 = 5$ (Multiply out brackets.)

$2b - 3 = 5$ (Simplify the left hand side.)

$2b = 8$ (Add 3 to both sides.)

$b = 4$ (Divide by 2.)

(iv) $a + 2 = \dfrac{5(1-a)}{4}$

$4a + 8 = 5(1 - a)$ (Multiply by 4.)

$4a + 8 = 5 - 5a$ (Multiply out brackets.)

$9a + 8 = 5$ (Add $5a$ to both sides.)

$9a = -3$ (Subtract 8 from both sides.)

$a = -\frac{3}{9} = -\frac{1}{3}$ (Divide by 9.)

You may have multiplied out the brackets before removing the fractions. You may also have carried out some of the other steps in a different order to the one we used. This doesn't matter–as long as you got the right solution!

7

(i) Let the original number be x. Then we get

$3(2x - \frac{1}{2}) + 2$

The final answer is $\frac{5}{2}$ so we get the equation

$3(2x - \frac{1}{2}) + 2 = \frac{5}{2}$

The equation can be solved as follows (in this case it is much easier to multiply out the brackets first):

$3(2x - \frac{1}{2}) + 2 = \frac{5}{2}$

$6x - \frac{3}{2} + 2 = \frac{5}{2}$ (Multiply out brackets.)

$12x - 3 + 4 = 5$ (Multiply by 2.)

$12x + 1 = 5$ (Simplify left hand side.)

$12x = 4$ (Subtract 1 from both sides.)

$x = \frac{4}{12} = \frac{1}{3}$ (Divide by 12.)

Thus the original number was $\frac{1}{3}$.

(ii) If x denotes the number of units consumed in the quarter, the overall cost in pounds under Tariff 1 is

$2{\cdot}86 + \frac{2{\cdot}51}{100}x$

whereas under Tariff 2 it is

$3{\cdot}77 + \frac{1{\cdot}65}{100}x$

If the overall cost is to be the same,

$2{\cdot}86 + \frac{2{\cdot}51}{100}x = 3{\cdot}77 + \frac{1{\cdot}65}{100}x$

$286 + 2{\cdot}51x = 377 + 1{\cdot}65x$

$286 + 0{\cdot}86x = 377$

$0{\cdot}86x = 91$

$x = \frac{91}{0{\cdot}86} = 106$

Thus 106 units would need to be consumed in a quarter for the costs to be the same.

8

(i) (a) When $w = 5$, $l = 5 + 10 = 15$.
(b) When $w = 35$, $l = 35 + 10 = 45$.
(c) When $l = 5$, $5 = w + 10$.
So $w = -5$.

(ii) (a) When $x = 3$, $y = 9 - 5 = 4$. 4.
(b) When $x = -2$, $y = -6 - 5 = -11$.
(c) When $y = 10$, $10 = 3x - 5$.
So $x = 5$.

(iii) (a) $P + N = 250$
(b) When $N = 150$, $P + 150 = 250$.
So $P = 100$.
The profit on the old car is £100.
(c) When $N = 300$, $P + 300 = 250$.
So $P = -50$
The profit is negative. This means that the dealer has actually made a loss of £50. In other words he paid more for the old car than he sold it for. This was probably an inducement to get the customer to buy the new car.

(iv) (a) $m = t + 10 + t$
$= 2t + 10$
(b) When $t = 3$, $m = 16$.
The round trip takes 16 seconds.
(c) When $t = 30$, $m = 70$.
The round trip takes 70 seconds.
(The minimum time for a signal to the moon and back, not allowing for response time, is about 6 seconds–quite a problem when dealing with a remote control TV camera from Earth and trying to follow the action through pictures.)
(d) When $m = 13$, $13 = 2t + 10$.
So $t = \frac{3}{2}$.
If the round trip takes 13 seconds then it takes 1·5 seconds for the radio signal to reach the space vehicle from Earth.
(e) When $m = 7$, $7 = 2t + 10$.
So $t = \frac{-3}{2}$.
This has no physical interpretation. We can conclude that it is not possible for the total time from transmission to reception to take 7 seconds.

9

(i) (a) $x = \frac{2}{3}$ (b) $b = 2$ (c) $z = -3$ (d) $d = 3$
(ii) (a) $x = -15$ (b) $a = -3$ (c) $y = 6$
(iii) (a) $x = 3/2$ (b) $d = 5/2$ (c) $r = -5$
(iv) 3
(v) 6
(vi) (a) $y = 33$ (b) $f = 2{\cdot}3$ (c) $f = 44$
(vii) Let the distance travelled be d tenths of a mile. Let the total cost be c pence. Then $c = 45 + 5d$ If $d = 4 \times 10 = 40$, then $c = 245$. The total cost was therefore £2·45.

Section 2.5 Solutions

1

(i) Yes. The subject is y.
(ii) No. The term on the left hand side is $2r$ rather than r.
(iii) No. Although y is isolated on the left hand side, there is also a term involving y $(-2y)$ on the right hand side.
(iv) Yes. The subject is c.
(v) Yes. The subject is s.

2

(i) We'll isolate y on the right hand side:

$4x = 2y + 1$
$4x - 1 = 2y$ (Subtract 1 from both sides.)
$\frac{(4x - 1)}{2} = y$ (Divide by 2.) (y is now the subject of the equation.)

When $x = 2$,

$\frac{8 - 1}{2} = y$ or $y = \frac{7}{2}$

(ii) $2x = 1 - 3y$
$2x - 1 = -3y$ (Subtract 1 from both sides.)
$-\frac{(2x - 1)}{3} = y$ (Divide by -3.) (y is now the subject of the equation.)

When $x = 2$,

$-\frac{(4 - 1)}{3} = y$ or $y = -1$

(iii) $y = 4x - 2y + 1$
$3y = 4x + 1$ (Add $2y$ to both sides.)
$y = \frac{(4x + 1)}{3}$ (Divide by 3.)

When $x = 2$,

$y = \frac{(8 + 1)}{3} = 3$

(iv) $x + y = 10$
$y = 10 - x$ (Subtract x from both sides.)

When $x = 2$,
$y = 10 - 2 = 8$

3

(i) $p = \frac{(3q + 2)}{4}$
$4p = 3q + 2$ (Multiply by 4.)
$4p - 2 = 3q$ (Subtract 2 from both sides, leaving q on the right.)
$\frac{(4p - 2)}{3} = q$ (Divide by 3.)

(ii) $E = \frac{(3 - d)}{3}$
$3E = 3 - d$ (Multiply by 3.)
$3E - 3 = -d$ (Subtract 3 from both sides, leaving d on the right.)
$3 - 3E = d$ (Divide by (-1).)

(iii) $\frac{x}{2} + y - 1 = \frac{1}{2}$
$x + 2y - 2 = 1$ (Multiply by 2.)
$x + 2y = 3$ (Add 2 to both sides.)
$x = 3 - 2y$ (Subtract $2y$ from both sides.)

4

(i) $V = IR$
$\frac{V}{I} = R$ (Divide by I.)

(ii) $F = ma$
$\frac{F}{a} = m$ (Divide by a.)

(iii) $I = \frac{PRT}{100}$
$100I = PRT$ (Multiply by 100.)
$\frac{100I}{P} = RT$ (Divide by P.)
$\frac{100I}{PR} = T$ (Divide by R.)

In each of these exercises it was easier to leave the subject on the right hand side. As a general rule you should always set out to isolate the subject on the side which seems to involve the minimum amount of manipulation.

5

(i) $P = 2(l + w)$
$P = 2l + 2w$ (Multiply out brackets.)
$P - 2w = 2l$ (Subtract $2w$ from both sides.)
$\frac{P - 2w}{2} = l$ (Divide by 2.)

(ii) $s = \frac{1}{2}(u + v)t$
$2s = (u + v)t$ (Multiply by 2.)
$2s = ut + vt$ (Multiply out brackets.)
$2s - vt = ut$ (Subtract vt from both sides.)
$\frac{(2s - vt)}{t} = u$ (Divide by t.)

(iii)

$$s = ut + \tfrac{1}{2}at^2$$

$$2s = 2ut + at^2 \quad \text{(Multiply by 2.)}$$

$$2s - at^2 = 2ut \quad \text{(Subtract } at^2 \text{ from both sides.)}$$

$$\frac{(2s - at^2)}{2t} = u \quad \text{(Divide by } 2t\text{.)}$$

Notice that in these last two examples it is customary to write the symbols ut and vt, rather than in alphabetical order.

6

(i) (a) We need to work out d/t for each of the values.

t	1/6	1/3	1/2	2/3	5/6
d	6	12	18	24	30
d/t	36	36	36	36	36

$\frac{d}{t}$ is equal to 36 in each case, so $d \propto t$.

(b) The constant of proportionality is 36. Since d is measured in miles and t in hours, the constant of proportionality represents the speed, 36 m.p.h.

Notice in the examples on page 114 the speed was

$\frac{3}{5}$ mile per minute.

Thus, if two quantities are proportional when measured in one set of units then they are also proportional when measured in a different set of units, although the units of the constant of proportionality will be different.

(c) When $t = 3$, $d = 36t = 108$. d is measured in miles, so the car travels 108 miles in 3 hours.

(ii) To determine whether c is directly proportional to n we need to calculate the various values of c/n.

n	12	50	100
c	20	75	140
c/n	1·67	1·5	1·4

In fact c/n is *not* constant. So the cost is not directly proportional to the number of cassettes bought.

7

(i) $d \propto t$. So $d = kt$ and $k = \frac{d}{t}$.

When $t = 5$, $d = 15$ and so $k = \frac{15}{5}$.

When $t = 15$, $d = \frac{15}{5} \times 15 = 45$.

(ii) Let m be the distance travelled in miles and n the number of gallons used.

Then $m \propto n$. So $m = kn$ and $k = \frac{m}{n}$.

When $m = 177$, $n = 4$ and so $k = \frac{177}{4}$.

When $n = 7$, $m = \frac{177}{4} \times 7 = 309{\cdot}75$.

The car therefore travelled 309·75 miles on 7 gallons. You may have used different symbols. That doesn't matter as long as you got this answer.

(iii) Let n be the number of sheets of paper and h their height in inches.

Then $h \propto n$. So $h = kn$ and $k = \frac{h}{n}$.

When $n = 200$, $h = \frac{3}{4}$ and so $k = \frac{3}{4 \times 200} = \frac{3}{800}$.

When $n = 350$, $h = \frac{3}{800} \times 350 = 1{\cdot}31$ (rounded to 2 decimal places).

The height of the stack is therefore 1·31 inches.

(iv) Let d be the number of dollars and l the number of pounds sterling.

Then $l \propto d$. So $l = kd$ and $k = \frac{l}{d}$.

When $d = 2{\cdot}3$, $l = 1$ and so $k = \frac{1}{2{\cdot}3}$.

When $d = 5$, $l = \frac{1}{2{\cdot}3} \times 5 = 2{\cdot}17$ (rounded to 2 decimal places).

Thus \$5 = £2·17.

8

(i) $y \propto \frac{1}{x}$. So $y = \frac{k}{x}$ and $k = xy$.

When $x = 5$, $y = 2{\cdot}5$ and so $k = 5 \times 2{\cdot}5$.

(a) When $x = 6{\cdot}25$, $y = \frac{5 \times 2{\cdot}5}{6{\cdot}25} = 2$.

(b) $y = \frac{k}{x}$ and so $x = \frac{k}{y}$.

When $y = 50$, $x = \frac{5 \times 2{\cdot}5}{50} = 0{\cdot}25$.

(ii) $f = \frac{k}{l}$ and $k = 1053 \times 285$.

(a) When $l = 1500$, $f = \frac{1053 \times 285}{1500} = 200$.

The equivalent frequency is therefore 200 kilocycles per second.

(b) $f = \frac{k}{l}$ and so $l = \frac{k}{f}$.

When $f = 1089$, $l = \frac{1053 \times 285}{1500} = 275$.

The equivalent wavelength is therefore about 275 metres.

(iii) $V \propto \frac{l}{P}$. So $V = \frac{k}{P}$ and $k = PV$.

When $P = 100$, $V = 24$ and so $k = 100 \times 24$.

When $P = 64$, $V = \frac{100 \times 24}{64} = 37{\cdot}5$.

The volume is therefore 37·5 cm^3.

9

(i) (a) $y = \frac{6 - 2x}{3}$; $-2/3$ (b) $y = \frac{7 - 3x}{3}$; $-5/3$

(c) $y = \frac{3 + 4x}{5}$; 3·8

(ii) (a) $y = 0{\cdot}01d + 5{\cdot}295$ (b) $x = \frac{(3y - 1)}{2}$

(iii) (a) $V = \frac{T}{P}$ (b) $v = d^2t$ (c) $r = \frac{(ne - IR)}{nI}$

(d) $s = \frac{(pc + 100c)}{100}$ (e) $h = \frac{(A - 2\pi r^2)}{2\pi r}$

(iv) In each case $\frac{l}{t} = \frac{1}{30}$, which is constant. Therefore $l \propto t$.

(v) (a) £22 (b) 7

(vi) 4·55 lbs (Rounded to two decimal places.)

(vii) (a) 54 months

(b) 2 years = 24 months. The job would therefore require 450 men.

MODULE 3

3.1 Co-ordinates

TRY THESE QUESTIONS FIRST

1 On the map opposite identify which feature is located at grid reference 583 574.

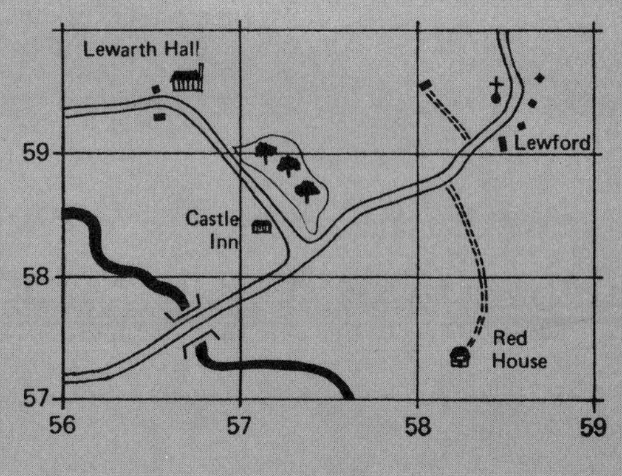

2

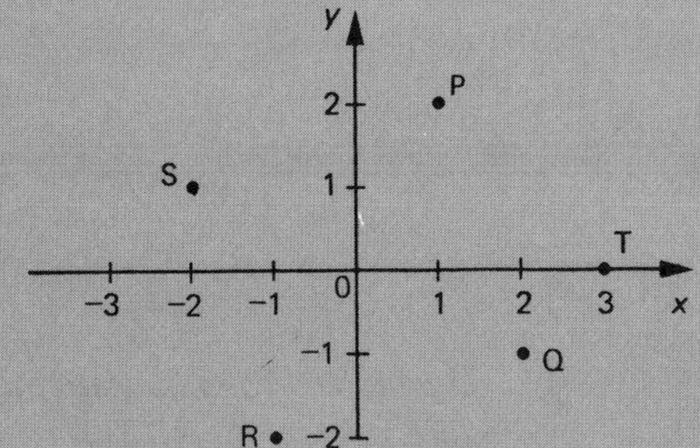

(i) Write down the co-ordinates of P, Q, R, S and T.
(ii) On the same diagram plot B (−2·5, −1·2) and C (−3·1, 1·3).

3 Write down the co-ordinates of K and interpret these co-ordinates in terms of the scales used for the axes.

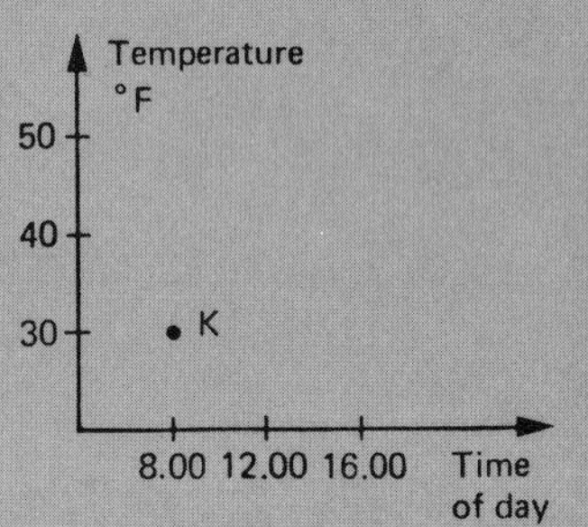

4 Plot the following points on a piece of graph paper: (3, 15) (−17, 23) (14, 62) (10, 3).

3.1(i) MAPS

The index in a London A-Z gives the reference for Rickland Place as

Rickland Place 17 3E

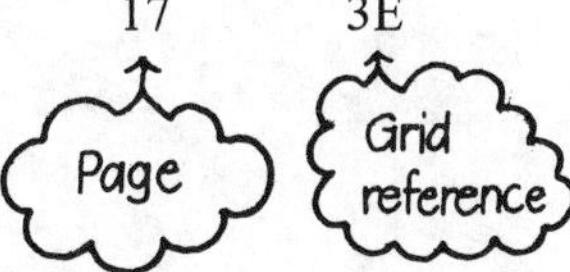

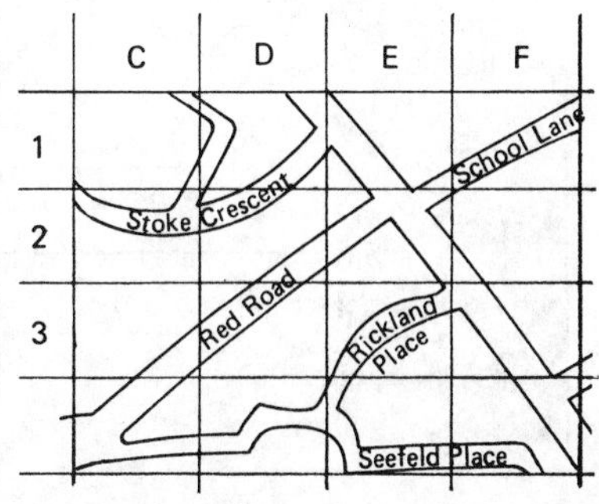

Having turned to page 17, the grid reference indicates in which square Rickland Place is located.

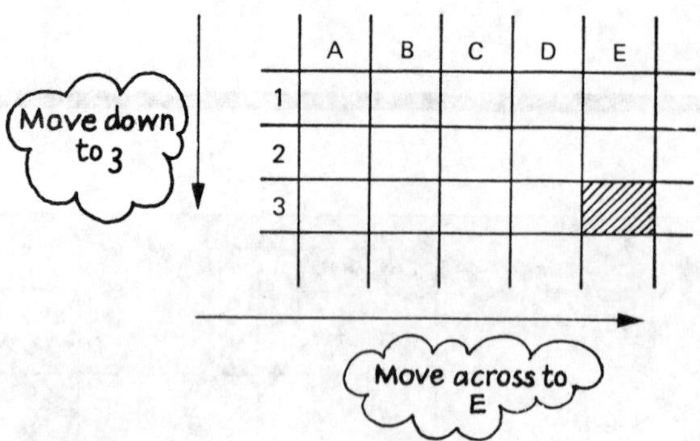

The arrows indicate the directions to move. Start at the top left hand corner and move **down** *and* **across.**

Grid references are often used to locate features on an Ordnance Survey map. In this case the system used is slightly different.

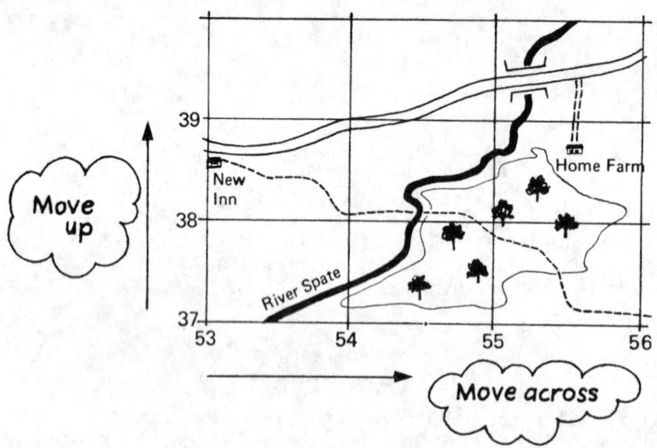

Again the arrows indicate the directions to move. This time, start in the bottom left hand corner and move **across** and **up**.

CHECK YOUR ANSWERS

1 Red House. *Section 3.1(i)*

2 (i) P(1, 2) Q(2, −1) R(−1, −2) S(−2, 1) T(3, 0) *Section 3.1(ii)*

(ii)

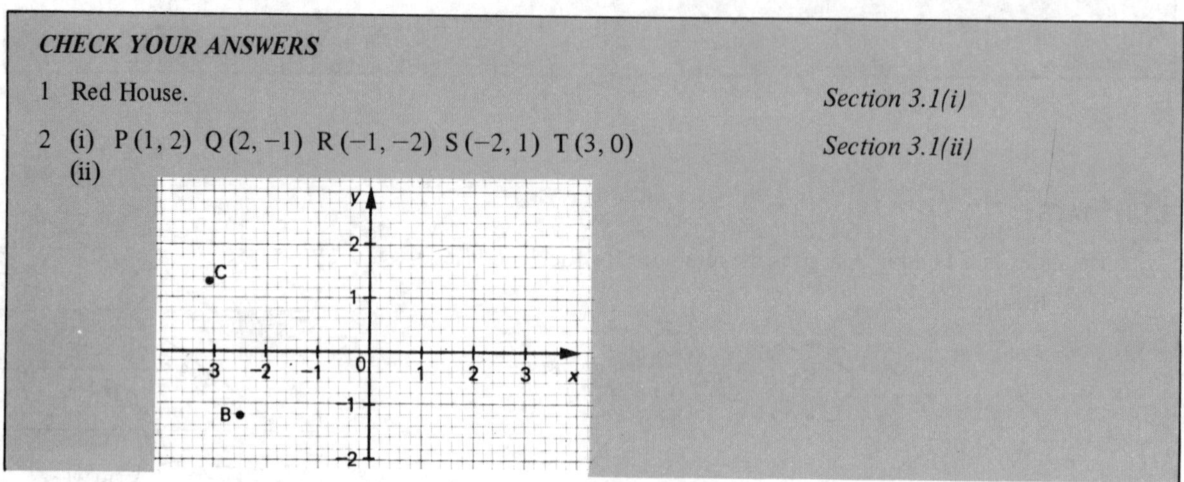

3 K has co-ordinates (8, 30). The temperature at 8 o'clock in the morning was 30° F. *Section 3.1(iii)*

4 This very much depends on the size of the graph paper you used. For an A4 sheet of paper suitable scales are *Section 3.1(iv)*

x-axis: 1 cm:2 units

y-axis: 2 cm:5 units.

The diagram opposite indicates the appearance—although this is considerably reduced in size.

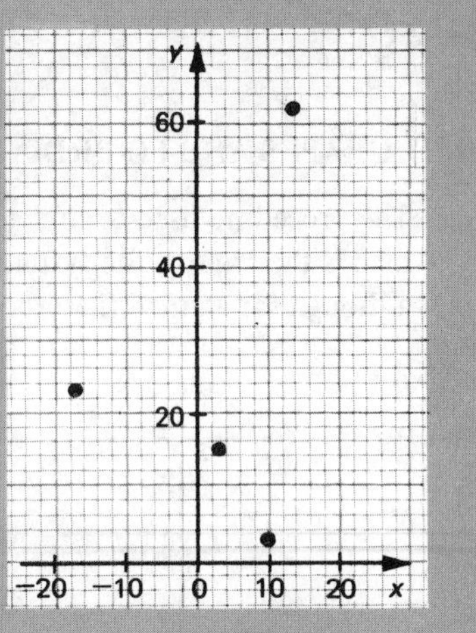

For example, the grid reference for Home Farm is

55 38

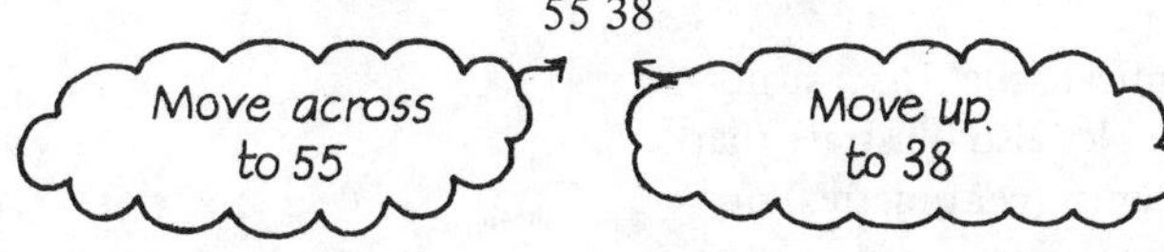

The distance across is given first. The distance up is given second. This grid reference identifies the bottom left hand corner of the square containing Home Farm.

Map references can be given more accurately than this by breaking down the squares into smaller units. For example, the precise grid reference for Home Farm is

556 387

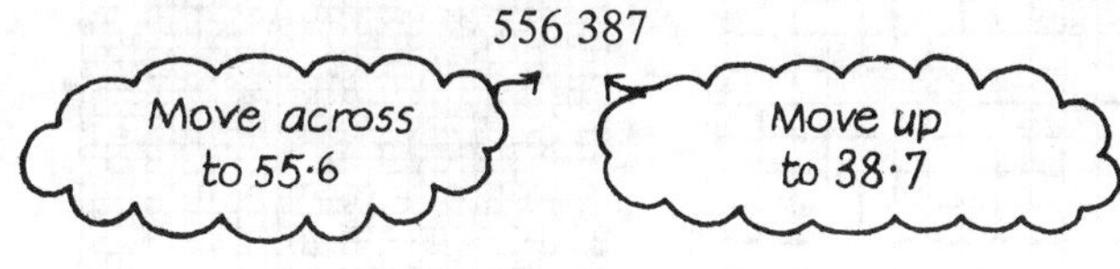

The first three digits give the distance across. The second three digits give the distance up.

EXAMPLE

Which feature on the map opposite is located at grid reference 481 382?

SOLUTION

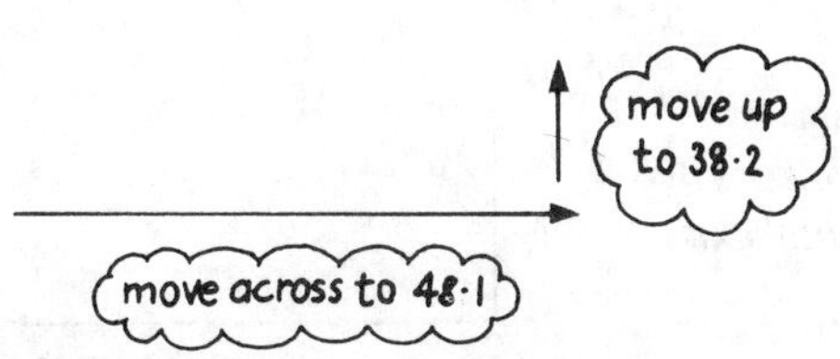

The map indicates a T at this point.

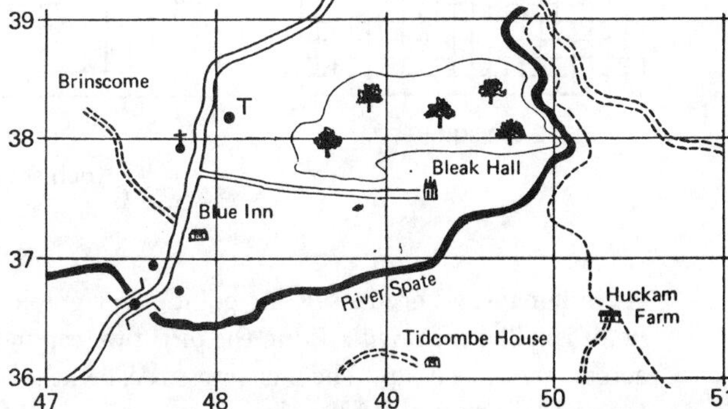

In fact, on an Ordnance Survey map T stands for Telephone Box.

TRY SOME YOURSELF

1 This question refers to the map given in the example above.
(i) Give the grid reference for the bridge over the river.
(ii) Which feature is located at grid reference 504 366?
(iii) Give the grid reference for Bleak Hall.

Remember that the distance across is given first. You will need to do some estimating.

3.1(ii) GRAPHS AND CO-ORDINATES

The procedure used to find grid references on an Ordnance Survey map is the same as the system used to identify points on a graph, although for a graph the process is a bit more formal.

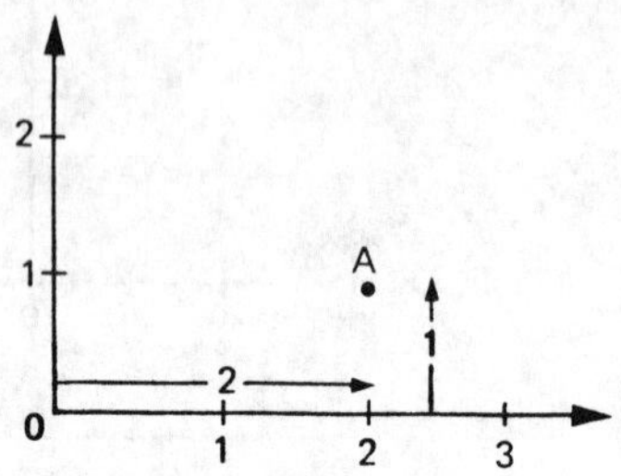

The point A is located by moving ***across*** *2 and moving* ***up*** *1. Again, the arrows indicate the directions to move.*

It's especially helpful if the points are plotted on graph paper, since this provides a useful 'grid system'.

The following samples of graph paper indicate that it is usually divided up into measures of 10. The samples also illustrate that graph paper may be measured in metric units (centimetres) or imperial units (inches).

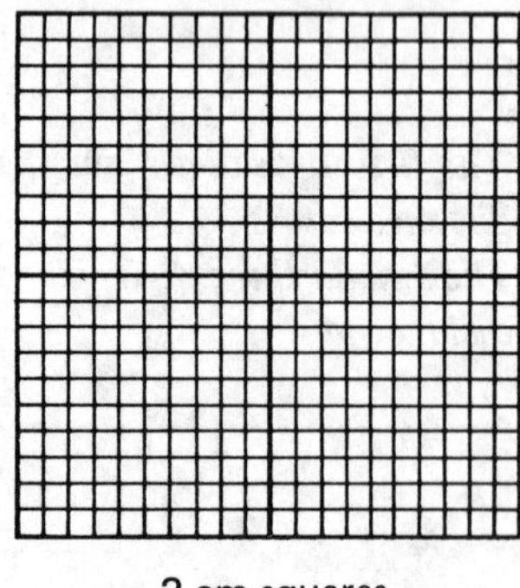

2 cm squares

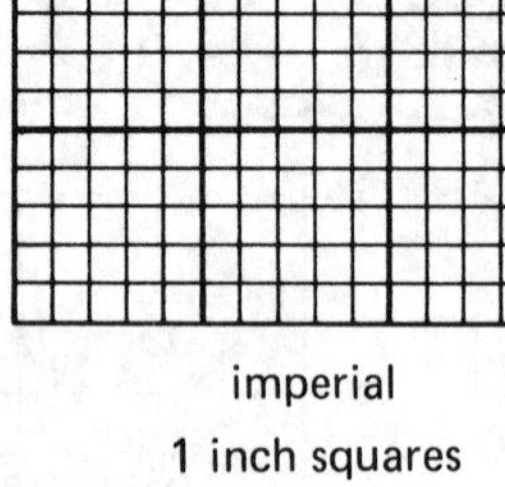

imperial
1 inch squares

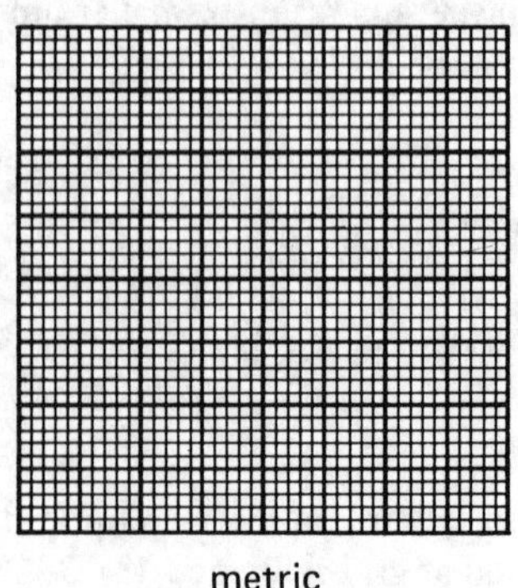

metric
1 cm squares

As with maps, we need a starting point. On a graph, the starting point is called the *origin*. From the origin we can move horizontally across or vertically up. The line across is called the *horizontal axis*; the vertical line is called the *vertical axis*.

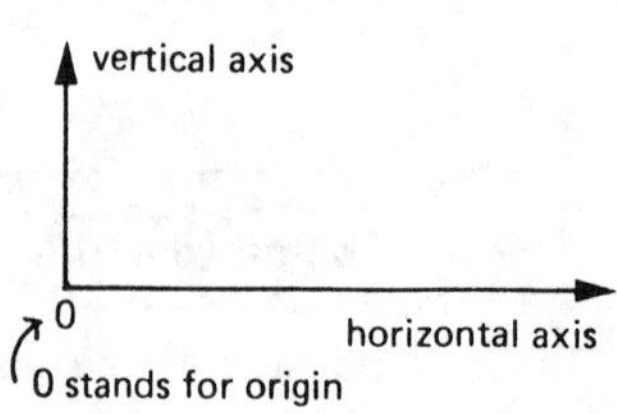

On mathematical graphs, the horizontal axis is usually labelled the *x-axis,* and the vertical axis is labelled the *y-axis*. Scales are indicated on the axes to locate points.

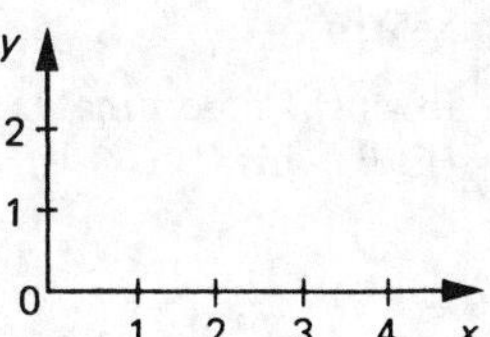

Reading co-ordinates

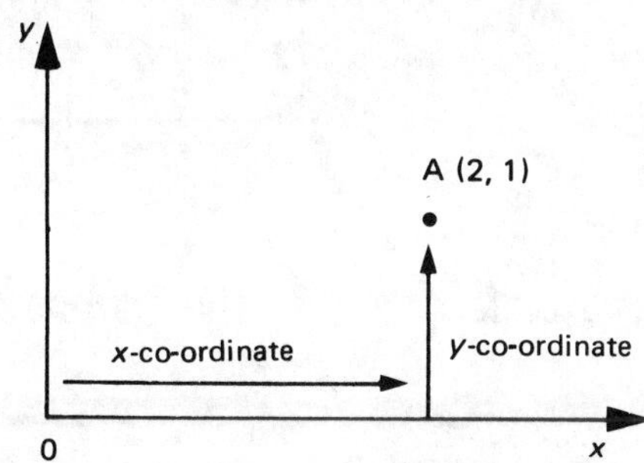

The distance *across* to A is called the *x-co-ordinate* of A and the distance *up* to A is called the *y-co-ordinate* of A. In this example, A is located at the point

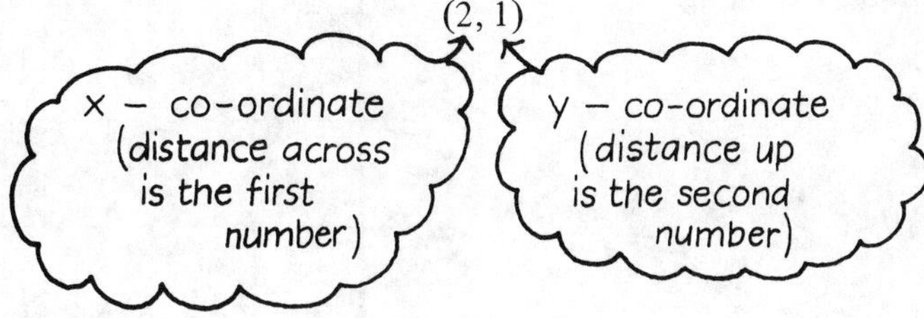

This can be written as

A is the point (2, 1).

The co-ordinates of A are written in brackets, separated by a comma.

(2, 1)

x-co-ordinate *y-co-ordinate*

EXAMPLE

Write down the co-ordinates of the points A and B.

SOLUTION

To locate A,

move *across* 3 units

move *up* 1 unit

The co-ordinates of A are (3, 1).

To locate B,

move *across* 1 unit

move *up* 0 units

The co-ordinates of B are (1, 0).

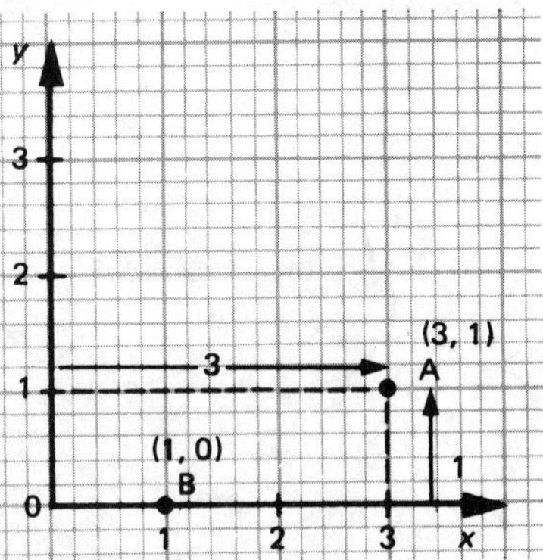

Remember the x-co-ordinate is given first. Since B is **on** *the x-axis, the y-co-ordinate is 0.*

TRY SOME YOURSELF

2 Write down the co-ordinates of
(i) A (ii) B (iii) C (iv) D.

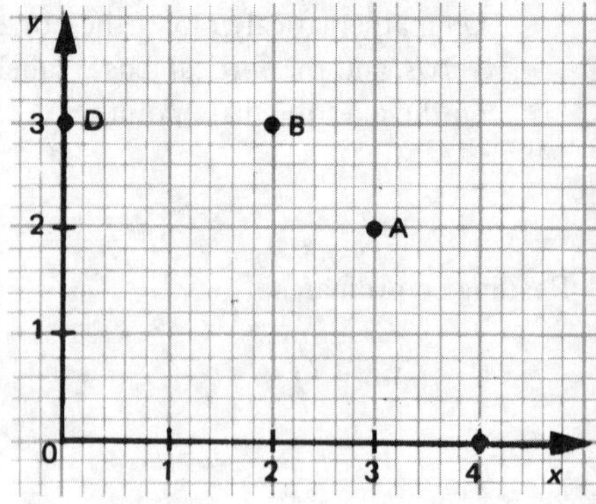

Plotting points

The same process is used to plot points. The x- and y-axes are marked off in scales to help measure across and up.

EXAMPLE

Plot the point P, with co-ordinates (1, 3) on the diagram at the right.

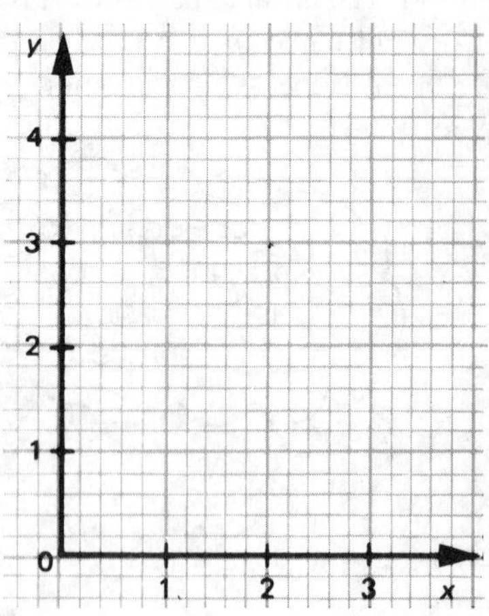

SOLUTION

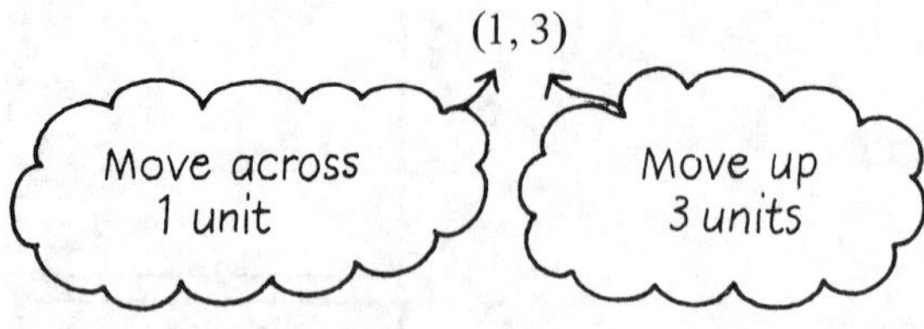

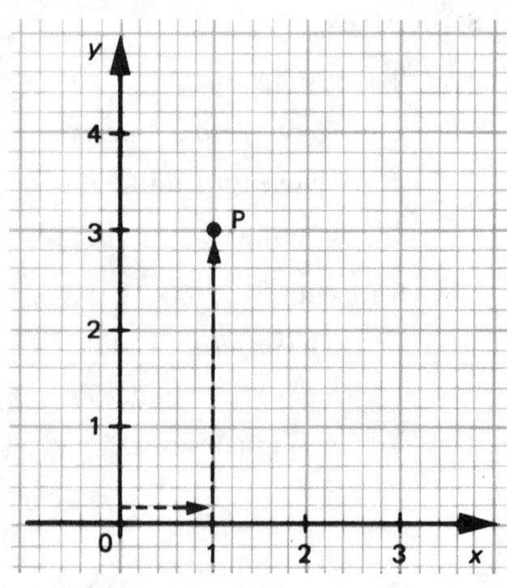

TRY SOME YOURSELF

3 On the diagram given in the example above plot the following points:

(i) A, with co-ordinates (2, 2)
(ii) B, with co-ordinates (3, 0).
(iii) C, with co-ordinates (0, 4).

The four quadrants

Up till now we have considered only those points with positive or zero co-ordinates, points like (3, 2) or (4, 7) or (0, 6). However, the system can be extended to cope with points involving negative co-ordinates like (−2, 3) or (−2, −3).

The diagram on the right shows how the x- and y-axes can be extended using the same principle we used to extend the number line to cope with negative numbers in Module 1.

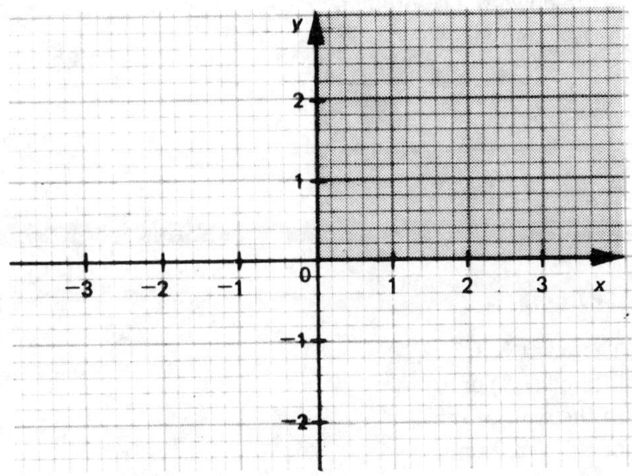

It may help to think of the co-ordinates for the point A as

(3, 2)

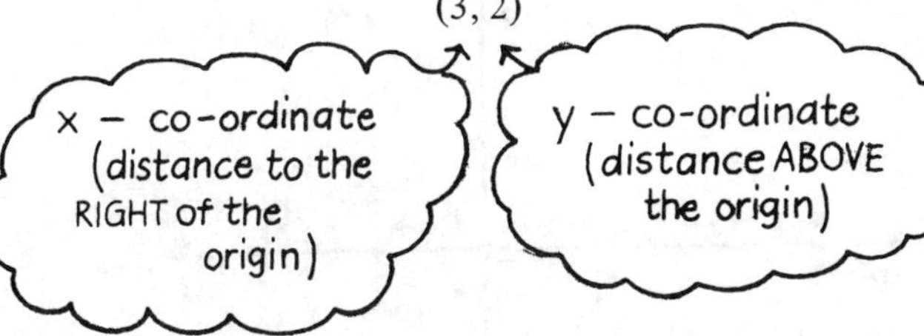

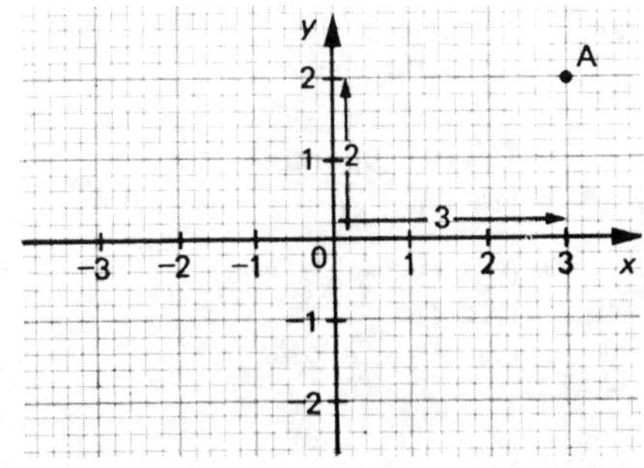

In this way if the point is to the left of the origin the x-co-ordinate is negative and if it is below the origin the y-co-ordinate is negative.

Just as negative numbers are located to the left of 0 on the number line:

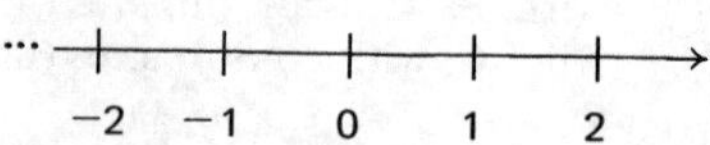

EXAMPLE

Write down the co-ordinates of B, C and D.

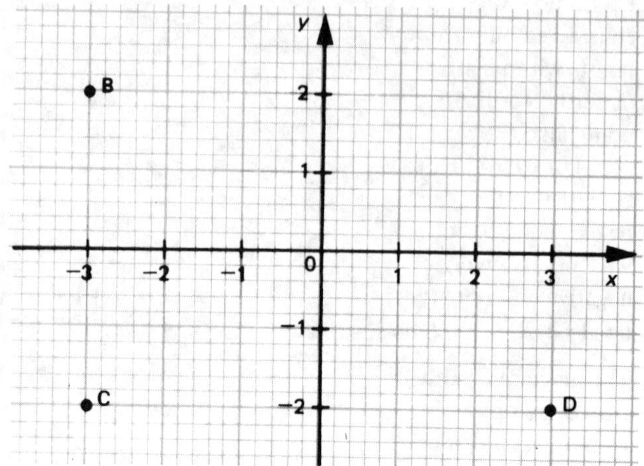

SOLUTION

The point B is located (−3) units to the right of the origin, and 2 units up from the origin.
So the co-ordinates of B are (−3, 2).

The point C is located (−3) units to the right of the origin, and (−2) units up.
So the co-ordinates of C are (−3, −2).

*Remember—the x-co-ordinate is given **first**, the y-co-ordinate is given second.*

The point D is located 3 units right of the origin and (−2) units above the origin.
So the co-ordinates of D are (3, −2).

TRY SOME YOURSELF

4 Write down the co-ordinates of each of the following points:
(i) A
(ii) B
(iii) C
(iv) D
(v) E

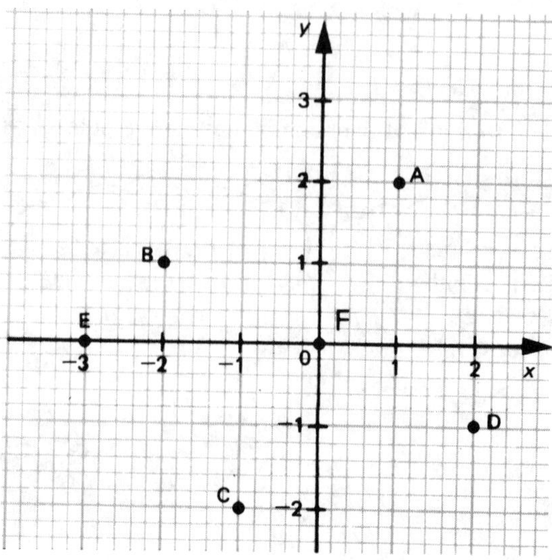

TRY SOME YOURSELF

5 Plot each of the following points on the diagram below:
(i) P (2, 3) (ii) Q (−2, 3) (iii) R (−3, 2)
(iv) S (−3, −2) (v) T (0, −2) (vi) V (2, −3).

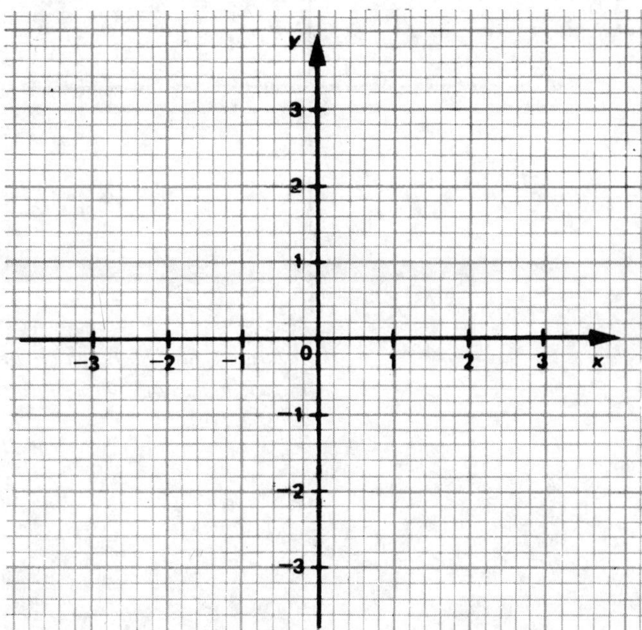

The co-ordinates of these points are all whole numbers. But just as Ordnance Survey map references can be given more precisely, co-ordinates of points can be given in decimals or fractions. Locating points whose co-ordinates are not whole numbers just requires more precision when measuring along the axes.

For example, here the co-ordinates of the point A are (2·5, 1·5).

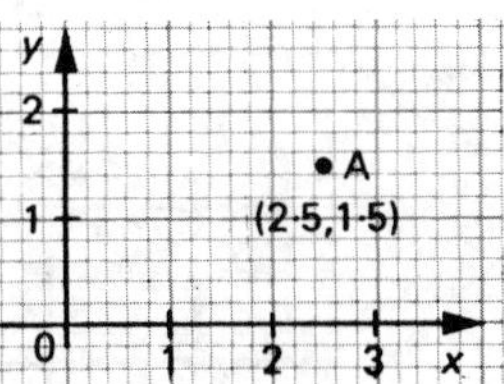

TRY SOME YOURSELF

6 Plot each of the following points on the diagram at the right.
(i) A $(1\frac{1}{2}, 0)$
(ii) B $(3{\cdot}2, -1{\cdot}8)$
(iii) C $(-1{\cdot}4, -2{\cdot}4)$
(iv) D $(-2\frac{1}{2}, 3\frac{1}{4})$

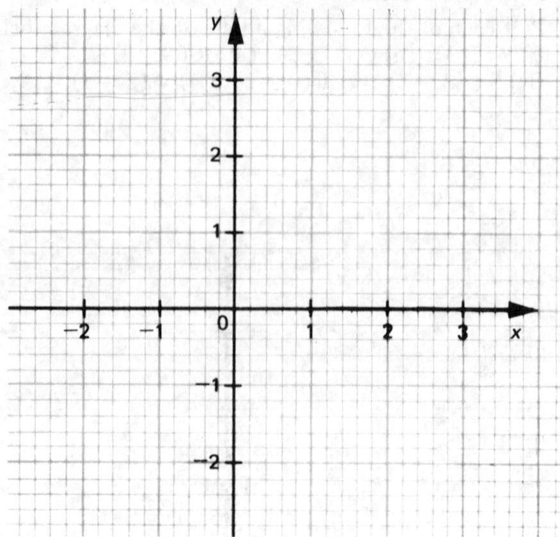

The accuracy with which you can plot points like these depends upon the *scales* of measurement indicated on the axes. A larger scale means that the points can be identified more accurately.

A larger scale is one which gives more squares per whole number on the axes.

3.1(iii) SCALES

The point P with co-ordinates $(0{\cdot}12, 0{\cdot}24)$ is difficult to plot on the axes given here because of the scales used. We can only plot the point approximately and the best we can do is to plot $(0{\cdot}1, 0{\cdot}2)$. However, the axes below are marked off in a larger scale and it is now quite easy to plot P reasonably accurately.

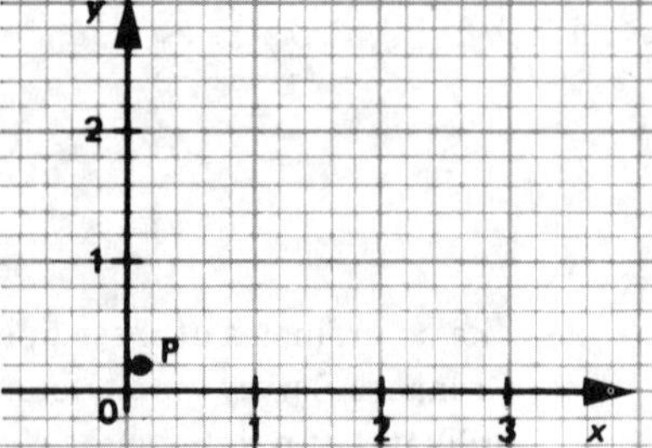

Here, the scale on both axes is 1 cm:1 unit or 1 cm per unit.

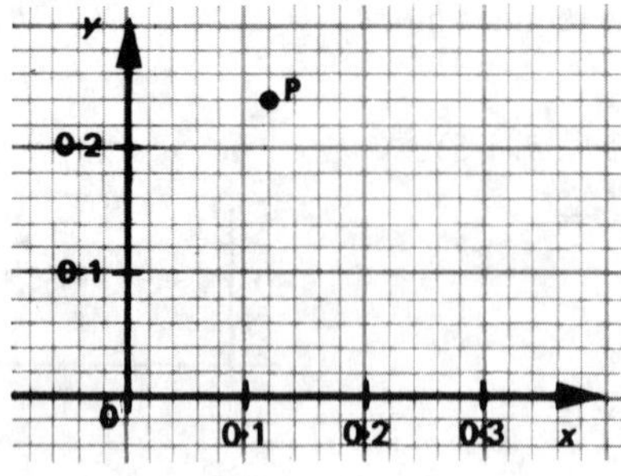

The scale on both axes is 1 cm:0·1 or 1 cm per 0·1 units (or 10 cm per unit).

Similarly, to plot very large co-ordinates it's necessary to use appropriate scales so that the points fit onto the paper!

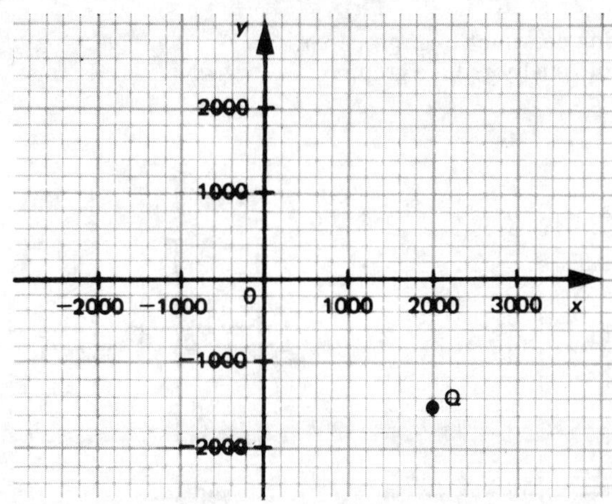

The scale on both axes is 1 cm:1000 units or 1 cm per 1000 units.

Using scales like these it's impossible to plot points exactly and the numbers need to be approximated. For example, the point Q with co-ordinates (2020, −1450) can probably only be plotted as (2000, −1500).

The scales on the x- and y-axes do not necessarily have to be the same. In this example, the scale on the x-axis is 1 cm:0·1 unit and the scale on the y-axis is 1 cm:10 units. Because scales play such an important part in the location of co-ordinates it's essential that they should be clearly indicated on the axes.

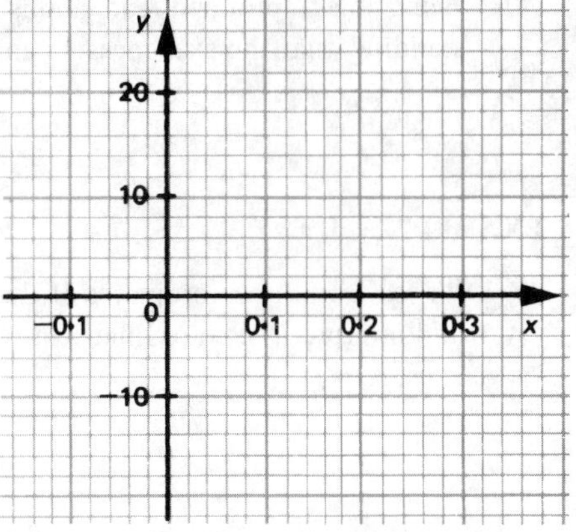

It's no good if you can't read the scale and even worse if no scale is indicated at all.

TRY SOME YOURSELF

7 What scales are used on the x- and y-axes in each of the following diagrams?

These graphs are drawn to actual size, using centimetre graph paper. However, because of lack of space, we will usually reduce graphs in size.

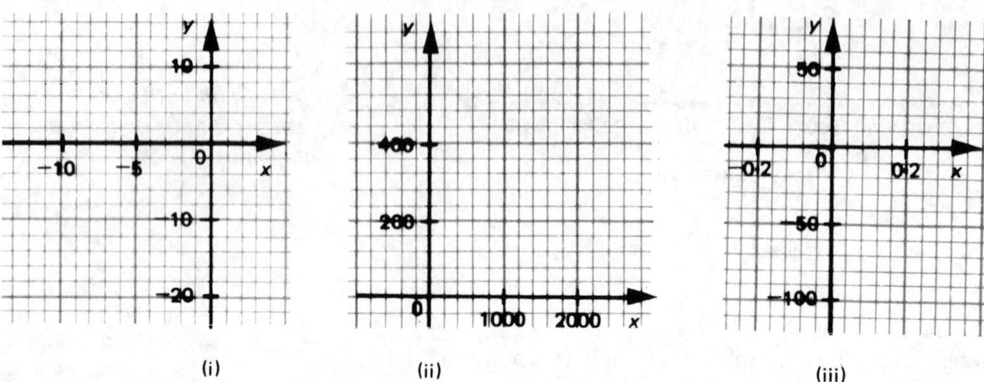

It's not always the case that the horizontal and vertical axes are labelled the x- and y-axes. More generally the axes indicate exactly what is being measured and the units of measurement used.

Being able to plot and read graphs very much depends on understanding what the axes measure and what scales are used.

EXAMPLE

Write down the co-ordinates of the point P on the diagram opposite and interpret the co-ordinates in terms of the scales used on the axes.

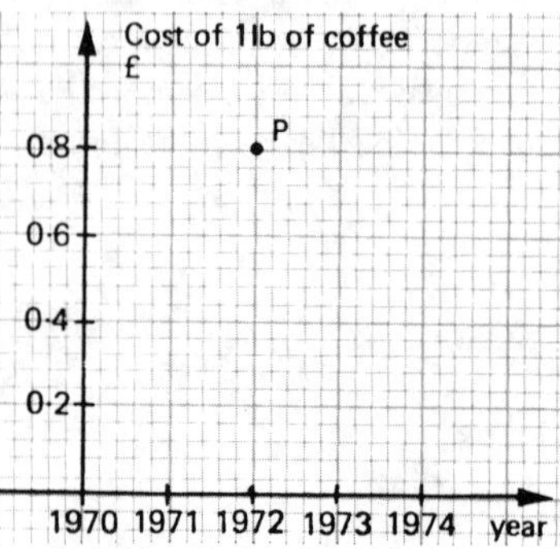

Notice that the horizontal axis starts at 1970, rather than 0.

SOLUTION

In this example the horizontal axis measures years and the vertical axis measures the cost of 1 lb of coffee in pounds (£). The co-ordinates of P are (1972, 0·8). This indicates that in 1972 the cost of 1 lb of coffee was £0·80 (or 80 pence).

TRY SOME YOURSELF

8 For each of the following diagrams write down the co-ordinates of the point P and interpret those co-ordinates in terms of the scales used on the axes.

Notice too that the units of measurement are also indicated on the axes. This means that the co-ordinates can be interpreted accurately.

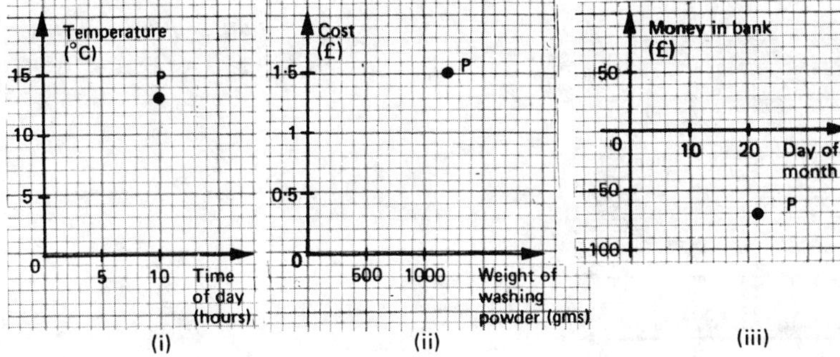

3.1(iv) CHOOSING YOUR OWN SCALES

In Section 3.1(iii) we introduced the use of scales to measure the axes. Usually if you've performed an experiment or gathered some information you will find that *you* have to decide what the axes should measure and what scales to use. This is probably the hardest part of plotting points and it's easy to go wrong at first. Ideally the points should be clear and should fill the available space sensibly. We're going to illustrate how to choose scales by plotting the set of points

(11·2, 21) (15, 35) (2, 43) (5·5, 52)

on the diagram below.

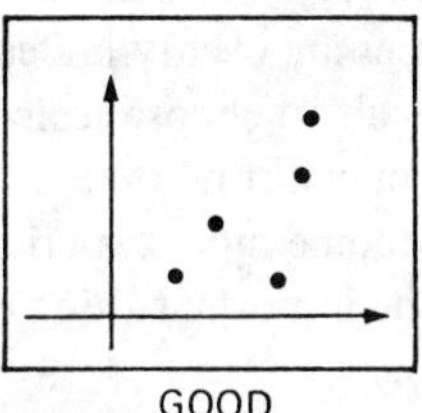

GOOD

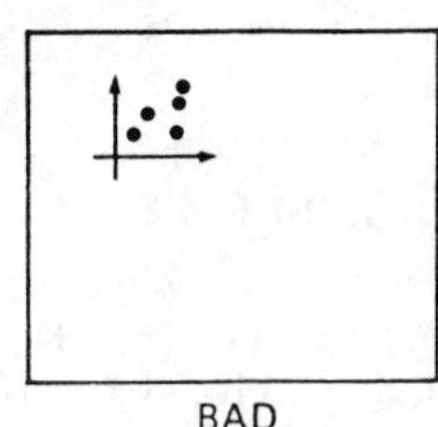

BAD

The second example is bad because it is squashed into one corner and the rest of the space is wasted as well as making it difficult to read the co-ordinates.

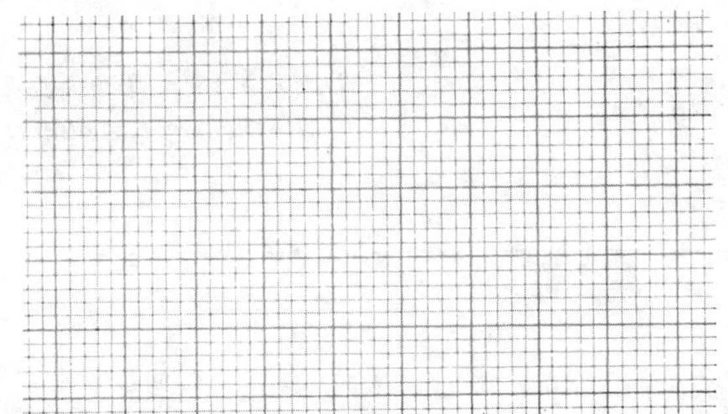

The first thing is to look at the range of the co-ordinates. The x-co-ordinates range from 2 to 15. This suggests that the x-axis should range from 0 to 15. Comparing the range of co-ordinates with the size of the graph paper, a suitable scale for the x-axis is 2 large squares:5 units.

The smallest is 2; the largest is 15.

But before starting to plot the points, it's necessary to work out a scale for the y-axis as well. The y co-ordinates range from 21 to 52. We could start at 0 and use a scale of 5 small squares:20 units.

You may have thought of another scale which would be suitable here but because graph paper is marked off in squares of 10 units or 5 units

However, since all the y-co-ordinates are greater than 20 it is probably more practical to start the scale at 20 and use a scale of 2 large squares: 20 units.

it makes more sense to use scales like

1 large square: 5 units
or 1 large square: 10 units
or 1 large square: 1 unit

Thus we get the following diagram.

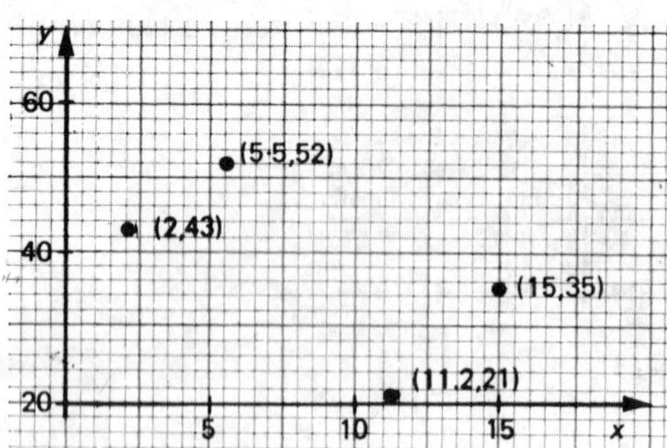

The origin (0, 0) does not necessarily have to be included. As long as the scales are clearly indicated there will be no confusion. It's very difficult to choose scales and to decide where the axes cross. At first you will find that it's mainly a question of trial and error. As you become more experienced you will find that it becomes easier. The important thing is to ensure that all the points fit onto your axes.

It's a good idea to use a pencil at first–then it's easy to correct any mistakes.
You should be able to judge for yourself whether the axes fit the paper well or whether they are squashed into one corner.

TRY SOME YOURSELF

9 Plot each of the following sets of points on suitable axes:
(i) (1, 21), (3, 14) (6, 43) (7, 32)
(ii) (100, 59) (340, 40) (410, 71) (220, 62)
(iii) (−3, 14) (14, 72) (−7, 9) (12, 61).

This is harder. Think carefully about the negative numbers.

After you have worked through this section you should be able to

a Locate grid references on a map
b Read and plot the co-ordinates of a point (involving both positive and negative co-ordinates)
c Read and interpret co-ordinates according to the scales of measurement used for the axes
d Given a set of points and a piece of graph paper, choose suitable scales and plot the points accordingly

Finally, here are some exercises if you want more practice.

TRY SOME YOURSELF

10(i) Write down the co-ordinates of all the points on the diagram opposite.

(ii) On the same axes plot the following points:
(a) A (2·6, 1·2) (b) B (−3·1, 0·7) (c) C (−2·2, 1·25)
(d) D (−2·2, −0·7).

(iii) Plot each of the following sets of points on some graph paper:
(a) (2, 510) (3, 540) (4, 570)
(b) (61, 0·2) (68, −0·1) (75, 0·7)
(c) (2, 15) (14, 82) (16, 110).

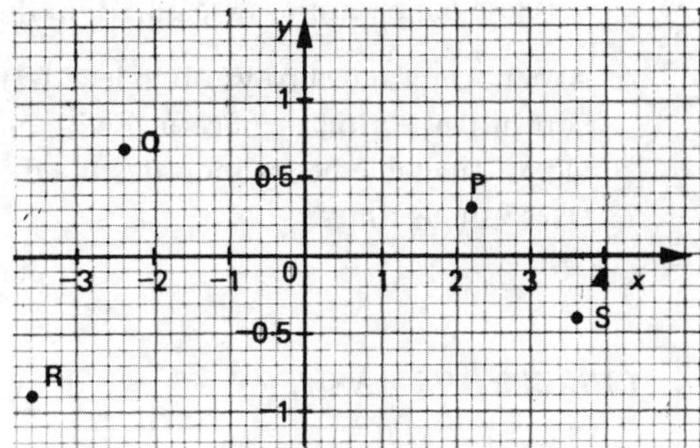

3.2 Plotting and Reading Graphs

TRY THESE QUESTIONS FIRST

1 The table below converts pounds sterling (£) to Deutschmarks (DM) based upon the exchange rate in July 1980.

£	1	2	3	4	5
DM	4·1	8·2	12·3	16·4	20·5

Plot the graph to illustrate this relationship.

2 From the graph find the sterling equivalent of 10DM.

3 Plot the straight line

$$d = 2 - 4f$$

3.2(i) GRAPHS FROM TABLES

In Section 3.1 we considered how to plot and read co-ordinates of points. This section goes one step further and we show how graphs are formed by joining up a set of points. A graph provides more information than the isolated points. It gives a better picture of a relationship and allows us to predict what happens in between the known points.

For example, this temperature chart indicates the hourly progress of a hospital patient. By joining up the points we can see clearly how the patient's temperature dropped. And although his temperature was not taken at 10·00, the graph indicates that it was about 98° F. The graph therefore illustrates the *relationship* between time and

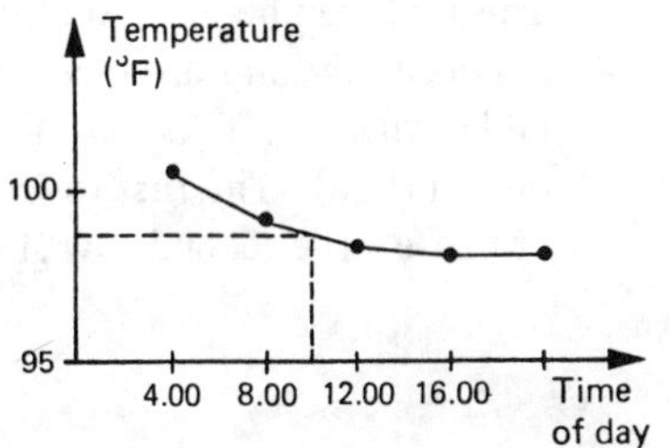

temperature. Notice that time is measured along the horizontal axis and temperature is measured up the vertical axis.

The scales again are clearly indicated on the axes.

Experiments and surveys often lead to tables of information. This part of the section is concerned with plotting a graph from a table. The table below indicates the distances travelled by a car after various times.

It is taken from Module 2, Section 2.5(iii).

CHECK YOUR ANSWERS

1 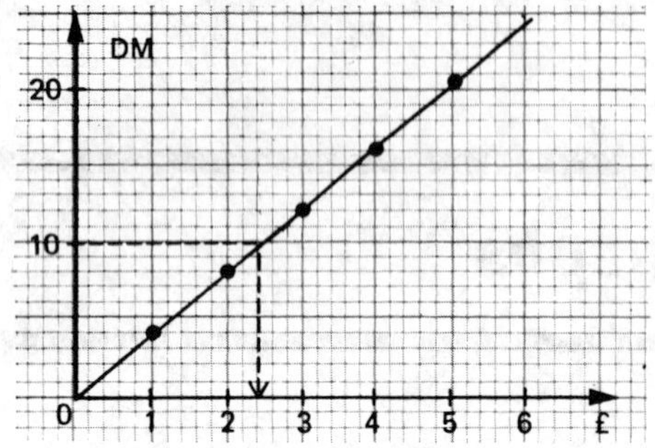 *Section 3.2(i)*

2 From the graph, 10DM ≏ £2·4. *Section 3.2(ii)*

3 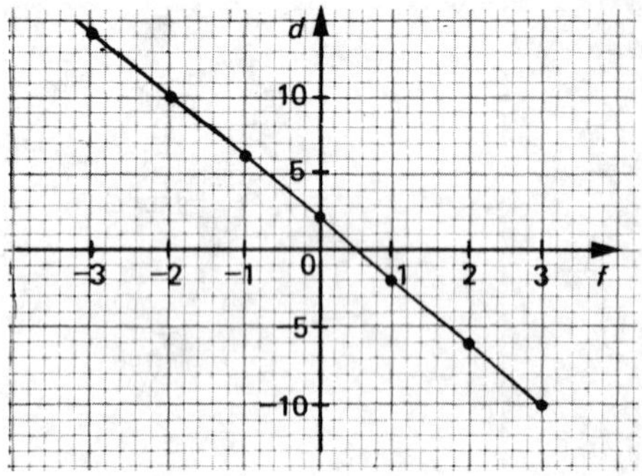 *Section 3.2(iii)*

Time t (mins)	10	20	30	40	50
Distance d (miles)	6	12	18	24	30

Time (t) is recorded in the first row.
The corresponding distance (d) is recorded in the second row.

After 10 minutes it had travelled 6 miles, after 20 minutes it had travelled 12 miles and so on.

This table can be used to plot a graph to illustrate the relationship between time and distance for this particular journey. From the table, when $t = 10$, $d = 6$. This suggests that we need to plot the point (10, 6). The first row of the table gives the horizontal co-ordinate. The second row gives the vertical co-ordinate.

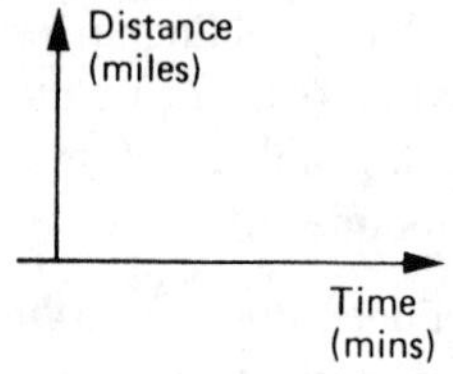

The axes could be labelled the other way round although it is conventional to measure time along the horizontal axis.

To determine appropriate scales for the axes it's necessary to look at the range of values for both co-ordinates.

The scale on the horizontal axis must go up to 50. The scale on the vertical axis must go up to 30.

t ranges from 10 to 50

d ranges from 6 to 30

This suggests the following choice of scales. We have already plotted the point (10, 6).

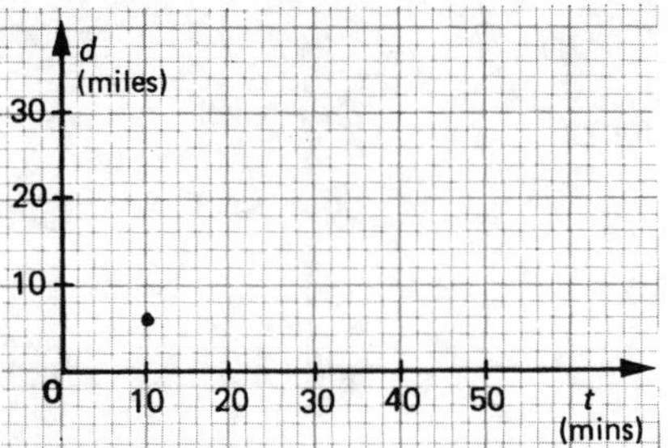

Since the vertical axis measures distance we have labelled it d. The units of measurement are miles. The horizontal axis measures t, and the units of measurement are minutes.

TRY SOME YOURSELF

1(i) Plot the remaining points on the axes above and join them up to complete the graph. You should get a straight line.

(ii) The table below converts pounds sterling (£) to dollars ($) based upon the exchange rate in July 1980.

Pounds (£)	5	10	15	20	25	30
Dollars ($)	11·5	23	34·5	46	57·5	69

The first row corresponds to the horizontal axis. The second row corresponds to the vertical axis.

(a) Plot the points on the paper below.

(b) Complete the graph by joining up the points in a straight line.

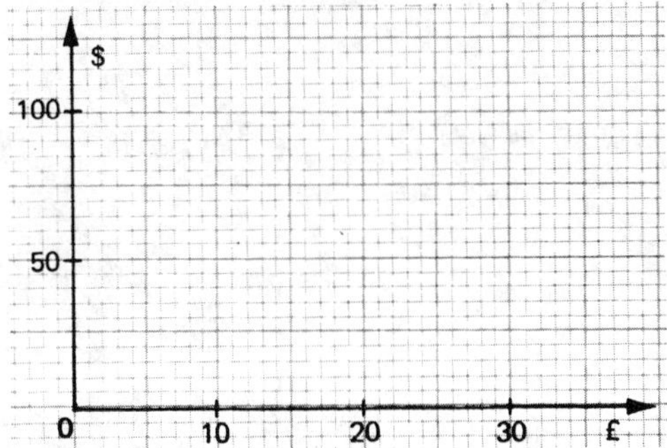

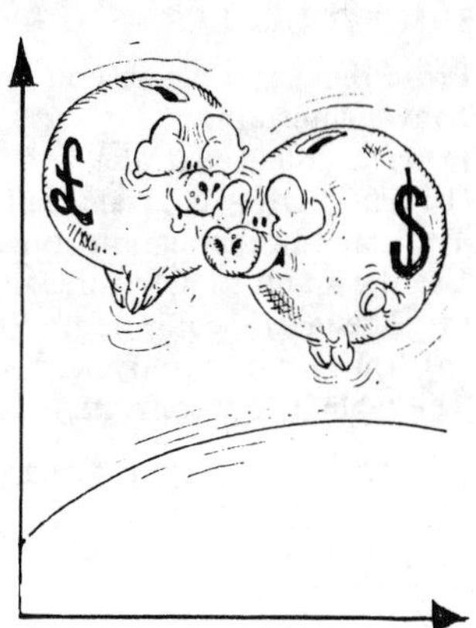

Again, the hardest part in drawing graphs is choosing appropriate scales. It is essential to mark these scales clearly on the axes and also to label the axes and indicate the units of measurement used.

3.2(ii) READING GRAPHS

The graph in the margin is taken from the solution to Exercise 1(i) above. Although the table of values only gives values of d corresponding to $t = 10, 20, 30, 40$ and 50, the graph enables us to find the corresponding value of d for *any* value of t and vice versa.

EXAMPLE

Use the graph to determine

(i) How far the car travelled in 35 minutes
(ii) How long it took for the car to travel 25 miles.

SOLUTION

(i) Here we are given $t = 35$.

First locate $t = 35$ on the horizontal axis. Then move vertically up to the graph and horizontally across to the d-axis. This gives $d \simeq 21$ miles. So the car travelled about 21 miles in 35 minutes. (The answer is only approximate. Its accuracy depends upon the scales used for the axes.)

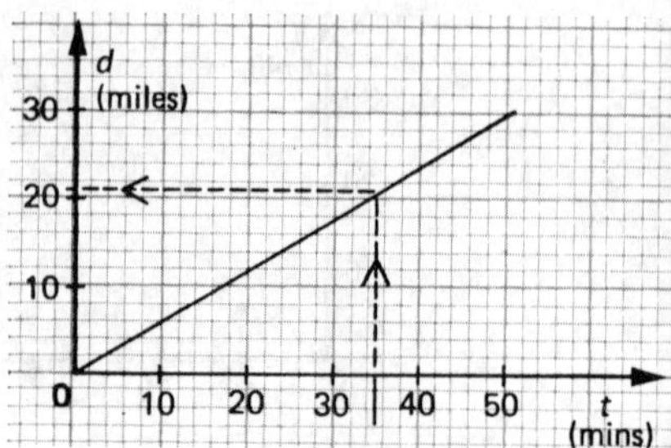

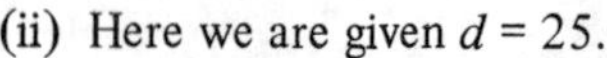

(ii) Here we are given $d = 25$.

First locate $d = 25$ on the d-axis. Then move horizontally across to the graph and vertically down to the t-axis. This gives $t \simeq 42$ minutes. So it took about 42 minutes to travel 25 miles. (The answer is only approximate. Its accuracy depends upon the scales used for the axes.)

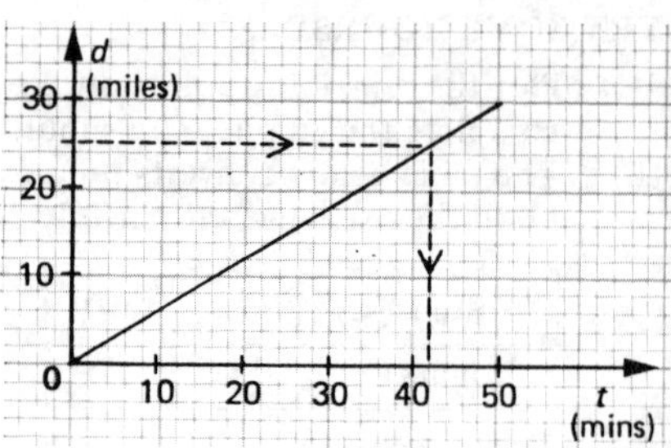

TRY SOME YOURSELF

2(i) From the graph which you obtained as the solution to Exercise 1(ii) above, determine
(a) The dollar equivalent of £12
(b) The sterling equivalent of $40.

(ii) The graph in the margin converts pounds (lbs) to kilograms (kg). Use the graph to determine:
(a) The metric equivalent of 4 lbs
(b) The imperial equivalent of 3 kg.

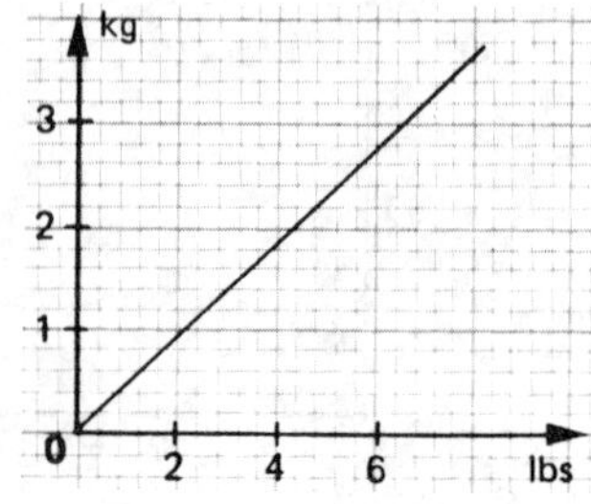

(iii) The table below converts miles to kilometres.

Miles	1	5	10	15	20
Kilometres	1·6	8	16	24	32

(a) Plot the graph to illustrate this.
(b) From your graph find the number of miles in 20 km.

Plotting a graph can therefore provide a lot more information than the table of values. The process of reading off values for intermediate points–in between the known points–is known as *interpolation.*

The shape of the graph–particularly if it is a straight line–enables us to draw further conclusions about the relationship. The interpretation of straight line relationships is discussed in Section 3.3.

Not all graphs are straight lines. We discuss some other common shapes in Section 3.4.

3.2(iii) PLOTTING THE GRAPH OF AN EQUATION

We've already considered how to plot a graph from a table. It is possible to plot the graph of an equation in two variables by constructing a table of values.

The equation

$$e = r + 7$$

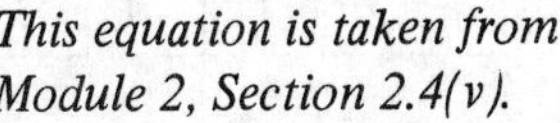

This equation is taken from Module 2, Section 2.4(v).

represents the relationship between the age of the Soviet cellist Rostropovich (r) and that of the comedian Jimmy Edwards (e). Given any value of r we can determine the corresponding value of e by evaluating $r + 7$. For example, when $r = 1, e = 1 + 7 = 8$.

Hence we can construct a table of values for selected values of r.

r	1	5	10	20
$e = r + 7$	8	12	17	27

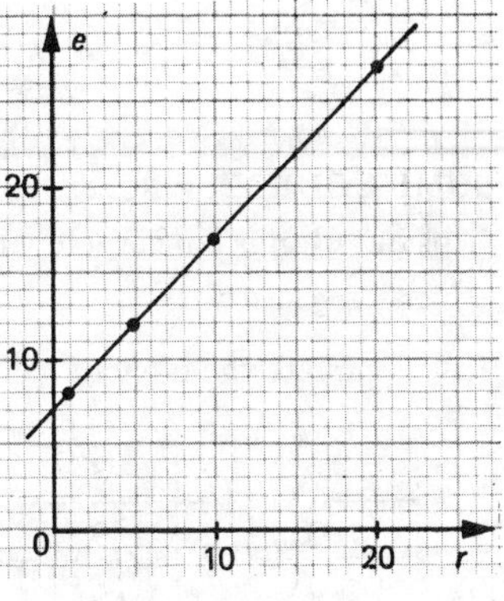

$e = r + 7$ *and* $r = e - 7$

Now the graph can be plotted using the procedure introduced in Section 3.2(i). It, too, is a straight line. The graph also gives additional information–given *any* value of r we can find the corresponding value of e, and vice versa.

For this particular equation you probably thought that it was easy enough to calculate corresponding values by algebraic manipulation. But equations are not always so straightforward and plotting the associated graph is often the quickest way of getting such information. The next example illustrates how to construct a table of values for a more complex equation.

EXAMPLE

(i) Construct a table of values for the equation

$$y = 2x + 3$$

by considering $x = -2, 0, 2, 4$ and 6.

(ii) Hence plot the graph of $y = 2x + 3$.

Mathematical equations are usually expressed in terms of x and y. In this case, as you might expect, x is recorded along the top row of the table and the corresponding values of y are evaluated and recorded in the bottom row.

SOLUTION

(i) The table of values is constructed row by row.

You can either work out the values of y step by step or you may want to use a calculator.

x	-2	0	2	4	6
$2x$	-4	0	4	8	12
$y = 2x + 3$	-1	3	7	11	15

(ii) The table indicates that x ranges from -2 to 6 and y ranges from -1 to 15. This information allows us to choose suitable scales for the axes and hence to plot the graph.

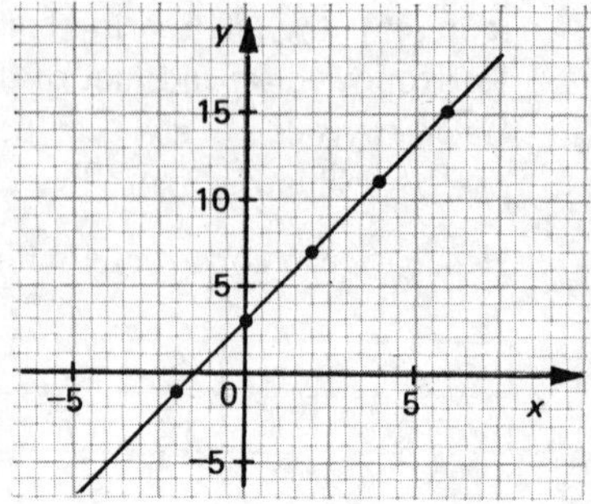

TRY SOME YOURSELF

3(i) This question is concerned with the equation

$$y = 3x - 7$$

(a) Complete the table of values below.

x	-3	-2	-1	0	1	2	3
$3x$	-9						
$y = 3x - 7$	-16						

(b) Use the table of values to plot the graph of $y = 3x - 7$.

Look at the range of values for each co-ordinate before you choose the scales. Remember to mark these scales clearly on the axes.

(ii) This question is concerned with the equation

$$t = -2s + 1$$

(a) Complete the table of values below.

s	-2	-1	0	1	2	3
$-2s$						
$t = -2s + 1$						

(b) Use this table to plot the graph of $t = -2s + 1$.

(iii) The equation

$$C = \tfrac{5}{9}(F - 32)$$

converts degrees Fahrenheit to degrees Centigrade and vice versa.

This is taken from Module 2, Section 2.5(iii).

(a) Complete the table of values below.

F (°F)	0	50	100	150	200	250
$F - 32$						
$C = \frac{5}{9}(F - 32)$						

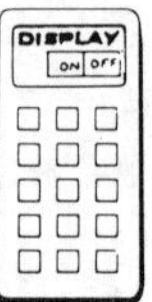

(b) Use the table of values to plot the graph of

$C = \frac{5}{9}(F - 32)$

(c) Use your graph to find the Fahrenheit equivalent of 100°C (boiling point of water).

(d) Use your graph to find the Centigrade equivalent of 98·4°F (blood temperature).

Check that the scales you choose allow you to plot each point from the table. Check, too, that your graph is sensible. Does it fill the space well? Is it cramped into one corner?

A quick method for straight lines

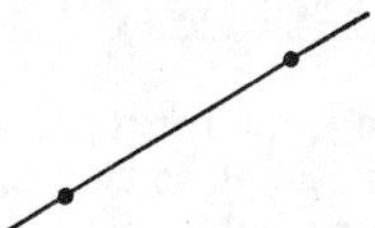

You may be wondering why we have asked you to plot so many points when in each case the graph was a straight line. After all, you really only need two points. Well, we've insisted on a lot of points in order to give you practice at plotting points and also to prepare you for those situations where the graphs are not straight lines. However, if you know the graph is going to be a straight line then we suggest you plot three points (just in case you make a mistake!)

Any equation such as $y = 2x + 6$, which involves a multiple of x but no higher powers (such as x^2), will give a straight line graph.

EXAMPLE

Plot the straight line given by

$s = 2t - 3$

SOLUTION

We pick easy values for t (say $t = 0, 1, 2$).

When $t = 0$, $s = (2 \times 0) - 3 = -3$ — *This gives the point (0, −3).*

When $t = 1$, $s = (2 \times 1) - 3 = -1$ — *This gives the point (1, −1).*

When $t = 2$, $s = (2 \times 2) - 3 = \ \ 1$ — *This gives the point (2, 1).*

The graph is illustrated below.

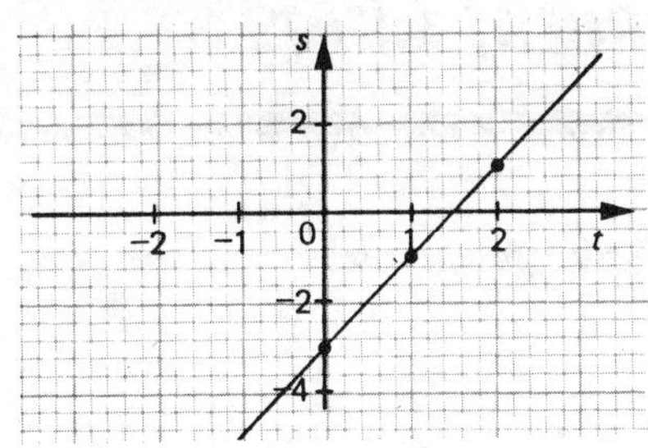

Although only 3 points are plotted, the graph can be extended indefinitely.

TRY SOME YOURSELF

4 Use this quick method to plot each of the following straight lines:
(i) $y = 4x + 2$ (ii) $P = 7t - 6$ (iii) $C = 7n + 16$.

After you have worked through this section you should be able to

a Plot a graph from a table of values, selecting your own scales and labelling the axes accordingly
b Given a value of x, read off the corresponding y-value from the graph and vice versa
c Complete a table of values for a given equation and so plot the graph
d Draw a straight line graph for a given equation by plotting three points

Finally, here are some exercises if you want more practice.

TRY SOME MORE YOURSELF

5(i) The table of values below converts pints to litres.

Pints	1	2	3	4	5
Litres	0·6	1·2	1·8	2·4	3·0

(a) Plot the graph for this conversion table.
(b) Use your graph to find the number of litres in 3·5 pints.
(c) Use your graph to find the number of pints in 4 litres.

(ii) (a) Complete the table of values for the equation $E = \frac{3-d}{3}$

d	-3	-2	-1	0	1	2	3
$3-d$							
$E = \frac{3-d}{3}$							

(b) Plot the graph of $E = \frac{3-d}{3}$.

(c) What value of E corresponds to $d = 1{\cdot}5$?
(d) What value of d corresponds to $E = 0{\cdot}1$?

(iii) Plot each of the following straight lines:
(a) $y = 4x - 3$ (b) $l = 2 - 3m$ (c) $q = 7p + 14$.

3.3 Characteristics of Straight Lines

TRY THESE QUESTIONS FIRST

1 Measure the gradient of this straight line graph.

2 What is the intercept of this straight line graph?

3 Hence write down the equation of the graph.

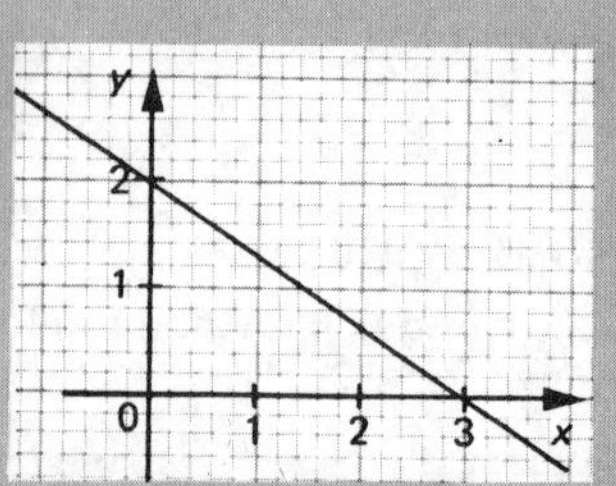

4 The graph in this diagram converts m.p.h. to km/h. Find the gradient of the line and hence find the conversion rate from m.p.h. to km/h.

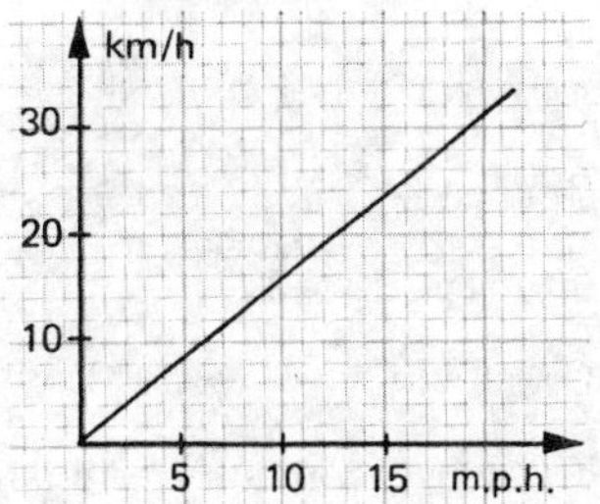

5 The graph below relates weekly wages to the number of hours worked overtime. From the graph find the overtime rate and the fixed weekly wage.

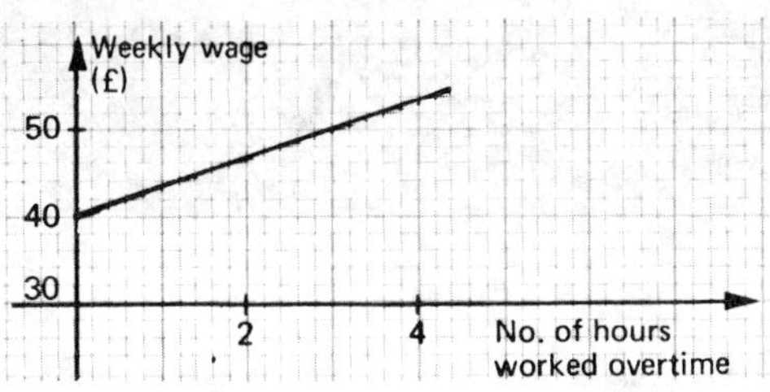

3.3(i) GRADIENTS

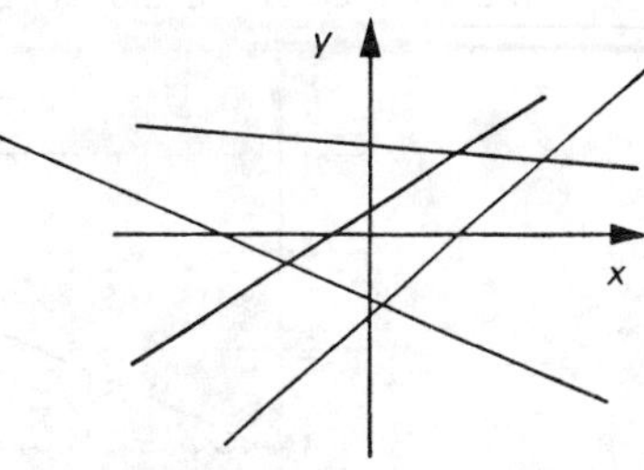

Straight lines are particularly useful graphs in that they can be analysed using relatively easy techniques. You will have plotted several straight line graphs by now, sufficient to notice that they slope in different directions. The steepness and direction of slope of a straight line are characteristics which can be measured.

The term *gradient* or *slope* is often used to describe how steeply a road rises. New road signs indicate the slope with a percentage. For example

10% means 1 in 10 or $\frac{1}{10}$

A 10% gradient means that the road rises by 1 unit for every 10 units travelled along the road. For example, it rises 1 *metre* for every 10 metres travelled.

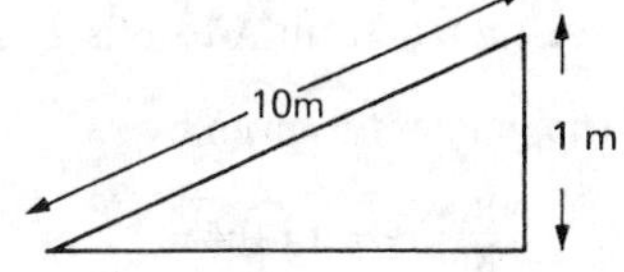

In mathematics the term gradient or slope is also used to describe how steeply a straight line rises. However, it is defined slightly differently.

A gradient of 1 in 10 or $\frac{1}{10}$ means a *rise in height* of one unit for every 10 units travelled *horizontally*.

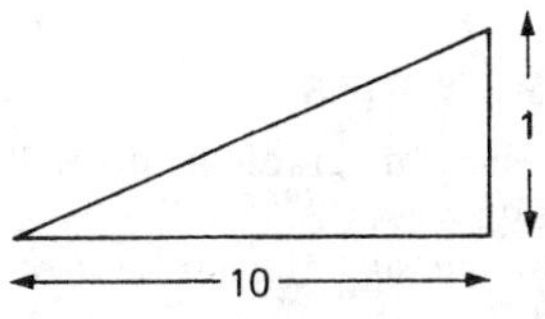

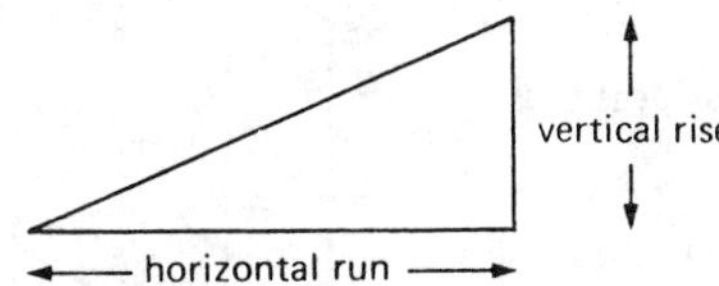

This gives a formula for the gradient or slope:

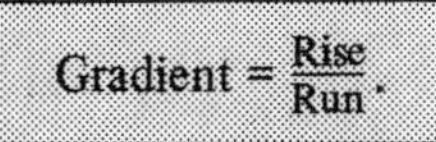

$$\text{Gradient} = \frac{\text{Rise}}{\text{Run}}.$$

The rise and run can be measured using a ruler. Of course, it is essential to measure the lengths using the appropriate scales.

CHECK YOUR ANSWERS

1 Gradient = $-\frac{2}{3}$.	*Section 3.3(i)*
2 Intercept is $y = 2$.	*Section 3.3(ii)*
3 Equation is $y = -\frac{2}{3}x + 2$.	*Section 3.3(iii)*
4 The conversion rate is 1 m.p.h. = 1·6 km/h.	*Section 3.3(iv)*
5 The weekly wage is £40 per week and the overtime rate is £3·50 per hour.	*Section 3.3(v)*

EXAMPLE

Find the gradient of the straight line AB in the figure below.

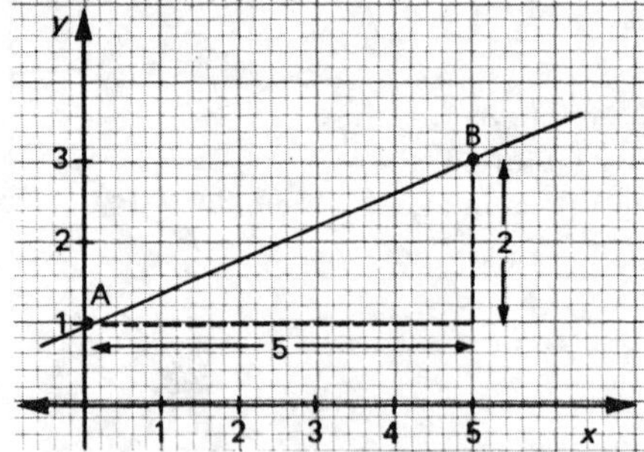

SOLUTION

From the graph the *horizontal run* from A to B is 5 units and the *vertical rise* from A to B is 2 units.

The gradient is given by

$$\frac{\text{Rise}}{\text{Run}} = \frac{2}{5} \text{ or } 0{\cdot}4$$

The gradient of a straight line is a constant. Whichever two points we pick on the line we will always find that the gradient is 0·4.

TRY SOME YOURSELF

1(i) Measure the gradient of the line in this diagram by considering the points A and B.

(ii) Now measure the gradient of the line by considering the points C and D.

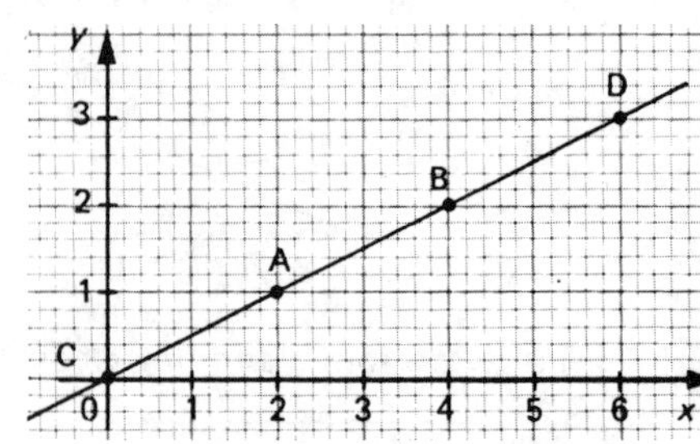

In fact we measure the horizontal run starting at *one point* and moving right to the other point. The vertical rise is obtained by starting *at the same point* and moving up. But what happens if the line *falls* rather than rises? The definition of gradient takes care of this eventuality. If the line falls, then the vertical rise is negative.

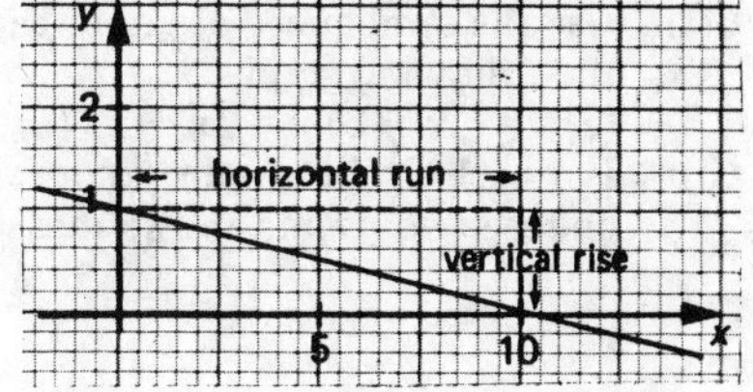

If the line slopes down to the right, then the vertical rise will be negative.

The gradient of the line in the diagram opposite is

$$\frac{\text{Rise}}{\text{Run}} = \frac{-1}{10}$$

Thus lines which slope up towards the right have positive gradients whereas lines which slope down to the right have negative gradients.

TRY SOME YOURSELF

2 Calculate the gradient for each of the straight lines below by considering the points P and Q.

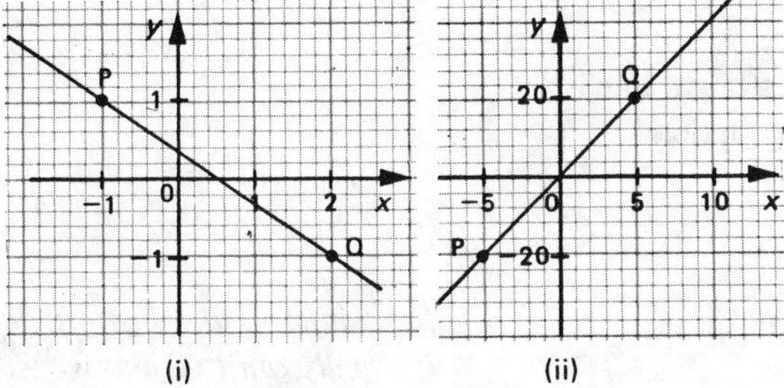

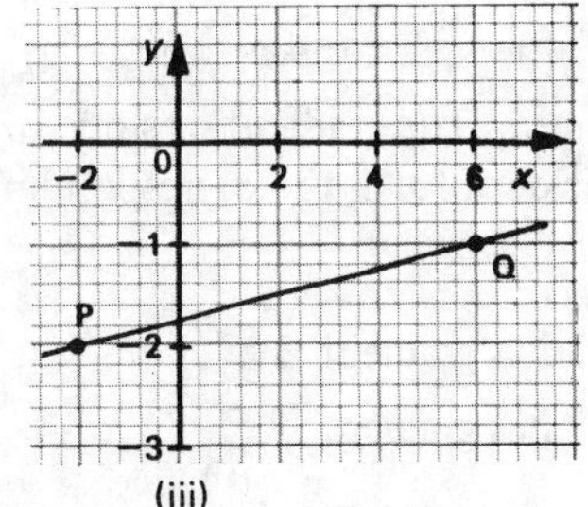

A line with zero gradient (0) is just a line which remains absolutely horizontal no matter how far it is extended.

$$\text{Gradient} = \frac{\text{Rise}}{\text{Run}} = \frac{0}{AB} = 0$$

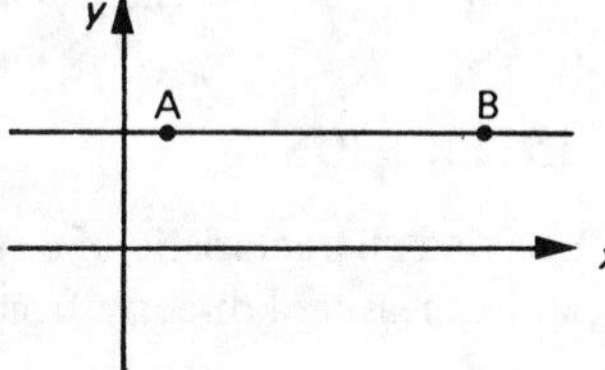

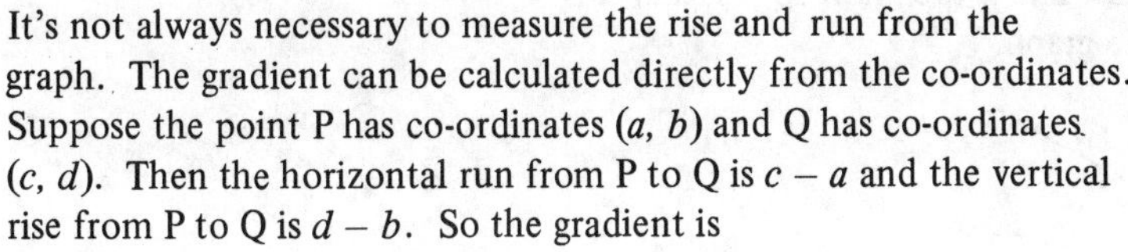

It's not always necessary to measure the rise and run from the graph. The gradient can be calculated directly from the co-ordinates. Suppose the point P has co-ordinates (a, b) and Q has co-ordinates (c, d). Then the horizontal run from P to Q is $c - a$ and the vertical rise from P to Q is $d - b$. So the gradient is

$$\frac{\text{Rise}}{\text{Run}} = \frac{d-b}{c-a}$$

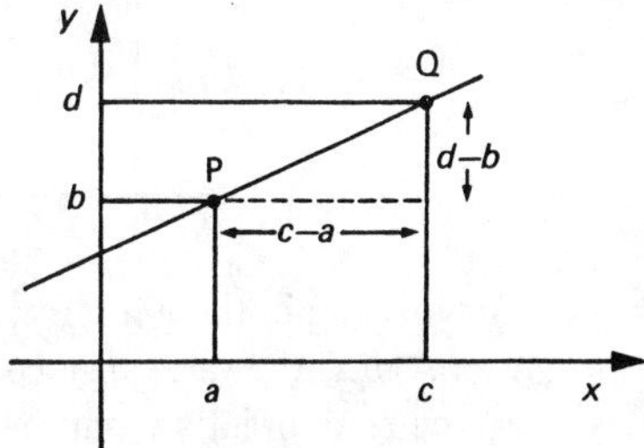

The following exercises form an investigation into gradients and straight lines.

TRY SOME YOURSELF

3(i) Use the same piece of graph paper and the same axes to plot each of the following graphs:
(a) $y = 2x$ (b) $y = 2x + 3$ (c) $y = -3x$ (d) $y = -3x - 4$.

(ii) For each of the graphs you drew for the first part of this question choose two points and so measure the gradients of all 4 lines.

Since all the graphs are to be plotted on the same axes, you will need to consider the range of values of ***all*** *the co-ordinates before deciding on suitable scales. Make sure that all four graphs fit onto the axes.*

Use your calculator if the numbers are awkward.

You should have discovered that

the gradient of $y = 2x$ is 2

the gradient of $y = 2x + 3$ is 2

the gradient of $y = -3x$ is -3

the gradient of $y = -3x - 4$ is -3

This suggests a pattern. The gradient or slope of a straight line is given by the coefficient of x in the corresponding equation.

So

$y = 198x + 1$ has gradient 198

$y = x - 7$ has gradient 1

and in general

$y = ax + b$ has gradient a

Remember that the coefficient 1 is usually omitted and we write x rather than 1x.

3.3(ii) INTERCEPTS

A second characteristic of a straight line is the *intercept.* Again this can be measured directly from the graph.

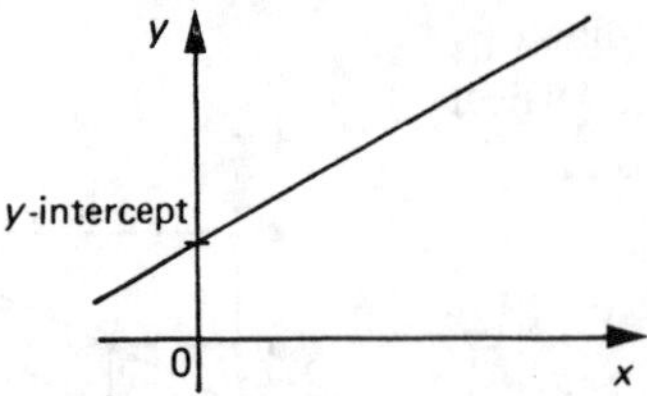

For this graph, the y-intercept is the point where the graph cuts the y-axis. It is the value of y corresponding to $x = 0$. This intercept can be read directly from the graph, providing the x-axis includes the point $x = 0$.

If the vertical axis measures another variable (t say) then the intercept may be known as the t-intercept.

If you look back at the graphs you obtained in Exercise 3(i) above you should find that

The y-intercept for $y = 2x$ is $y = 0$
The y-intercept for $y = 2x + 3$ is $y = 3$
The y-intercept for $y = -3x$ is $y = 0$
The y-intercept for $y = -3x - 4$ is $y = -4$

If the graph passes through the origin, the intercept is 0.

Like the gradient, the y-intercept can be identified straight from the equation.

So

$y = 198x + 1$ has y-intercept $+1$
$y = x - 7$ has y-intercept -7

and in general

$y = ax + b$ has y-intercept b

To summarise, any straight line corresponds to an equation of the form

y = (Gradient x x) + Intercept

Such an equation does not contain any other power of x (such as x^2). Nor does it contain any terms like $\frac{1}{x}$

And since the graph is a straight line, an equation such as $y = ax + b$ is called a *linear equation.*

The general linear equation is often written as

$$y = ax + b$$

or

$$y = mx + c$$

TRY SOME YOURSELF

4 Write down the gradient and intercept of each of the following equations:
(i) $y = 3x - 2$ (ii) $y = 3$ (iii) $y = 1 - 2x$ (iv) $y = -193x + 721$.

3.3(iii) DETERMINING THE EQUATION FROM THE GRAPH

In Section 3.3(ii) we discussed how to identify the slope and intercept from the equation. More often though, you will find that you start with the graph and will need to determine the characteristics (gradient and intercept) in order to write down the equation. The next example also illustrates that the variables do not necessarily have to be x and y.

EXAMPLE

Determine the equation of the straight line in this diagram.

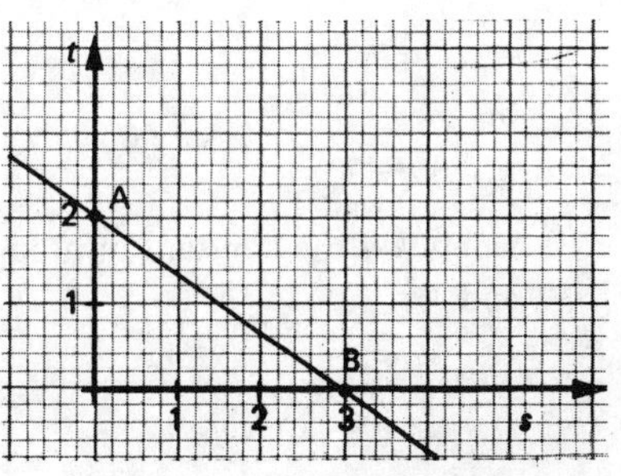

SOLUTION

We can measure the gradient by considering the two points A and B.

Horizontal run = 3

Vertical rise = −2

So the gradient = $-\frac{2}{3}$.

The intercept is obtained by considering the point where $s = 0$ (where the graph cuts the t-axis). From the graph, the intercept is $t = 2$. The equation is therefore

$$t = \frac{-2s}{3} + 2$$

The equation of the line is given by

t = (gradient x s) + intercept

Alternatively the equation may be written as

$$t = 2 - \tfrac{2}{3}s$$

putting the positive term first.

TRY SOME YOURSELF

5 Determine the equation of each of the straight lines in the diagram below.

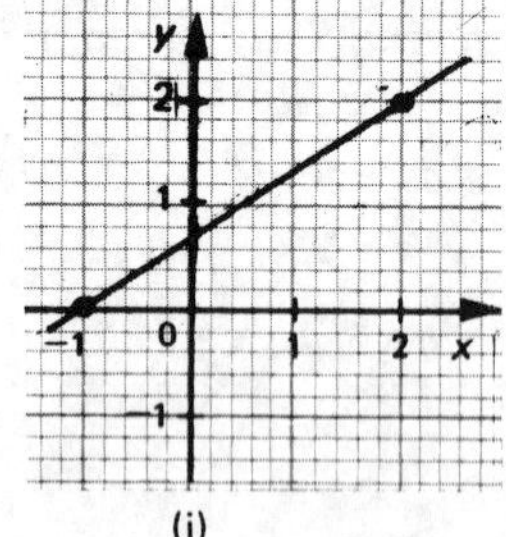

(i)

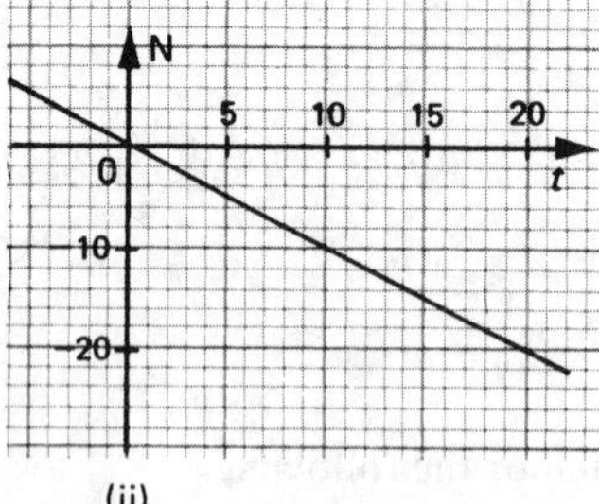

(ii)

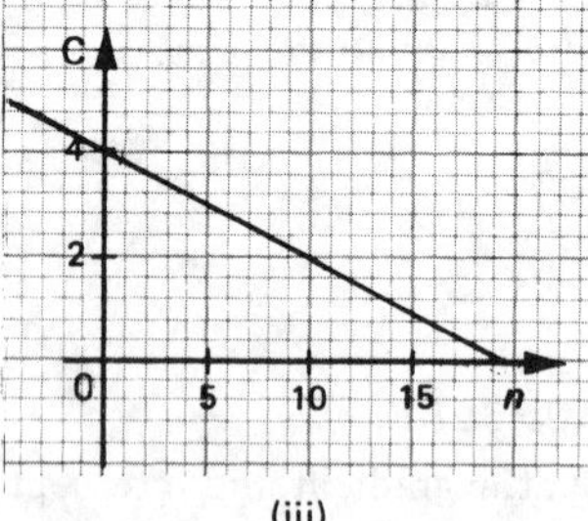

(iii)

Determining the equation is often the easy part. It is more difficult to interpret what the equation *means*. The next part of this section deals with the interpretation of the equation.

3.3(iv) PROPORTION

The graph in the margin illustrates the relationship between the distance travelled by a car and time. It is taken from page 147. You may recall from Module 2, Section 2.5(iii) that under these circumstances

$$d \propto t$$

or

$$d = kt$$

where k is the constant of proportionality. In this case $k = \frac{3}{5}$ and

$$d = \tfrac{3}{5}t$$

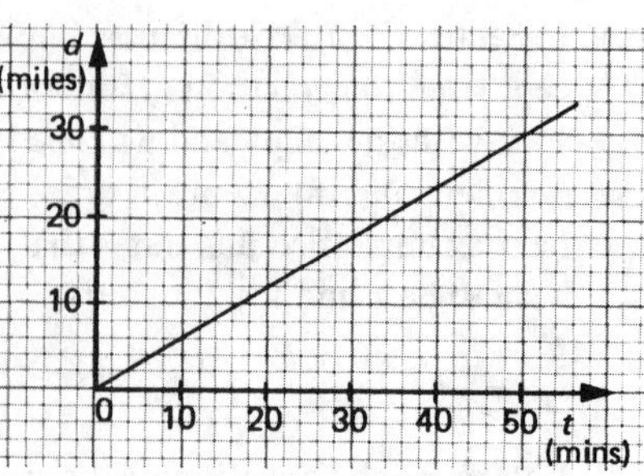

Measure the gradient of this straight line from the graph.

From Module 2, Section 2.5(iii).

We can now relate this to the characteristics of the graph.

The gradient of the line is $\frac{3}{5}$ and the intercept is 0.

Furthermore k (= $\frac{3}{5}$) represents the speed the car was travelling. Here the speed was $\frac{3}{5}$ miles per minute (or 36 m.p.h.).

So the equation of the graph is

$d = \frac{3}{5}t$

Since d is measured in miles and t is measured in minutes.

In fact any straight line which passes through the origin represents *direct proportion* and the *gradient* of the graph gives the *constant of proportionality.*

The next example illustrates how the gradient or constant of proportionality represents the *rate of increase.*

EXAMPLE

The graph in the margin illustrates the relationship between wages (w) and hours worked (t) for a certain factory. Use the graph to find the hourly rate of payment.

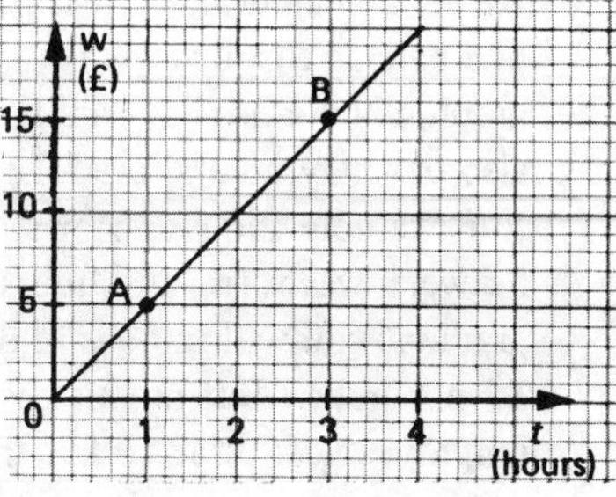

SOLUTION

Since the graph is a straight line which passes through the origin it must have the form

$w = kt$

where k is the gradient of the line. In fact k gives the hourly rate of pay.

From the graph, considering the points A and B,

Horizontal run = 2

Vertical rise = 10

$k = \text{gradient} = \frac{10}{2} = 5$

The units of measurement for k can be found by considering the units of measurement of w and t. Since $k = \frac{w}{t}$, k is measured in $\frac{\text{pounds}}{\text{hour}}$ (i.e. pounds per hour).

Therefore the hourly rate of pay is £5 per hour.

TRY SOME YOURSELF

6(i) Sue worked at a hairdressers where the hourly rate of pay was £2·50 per hour.
(a) Draw a graph to represent this.
(b) From your graph determine how much she earned in a 24-hour week.

(ii) The graph opposite converts pints to litres and vice versa.
(a) Use the graph to find the conversion rate from pints to litres.
(b) Use the answer to part (a) to determine the conversion rate from litres to pints.

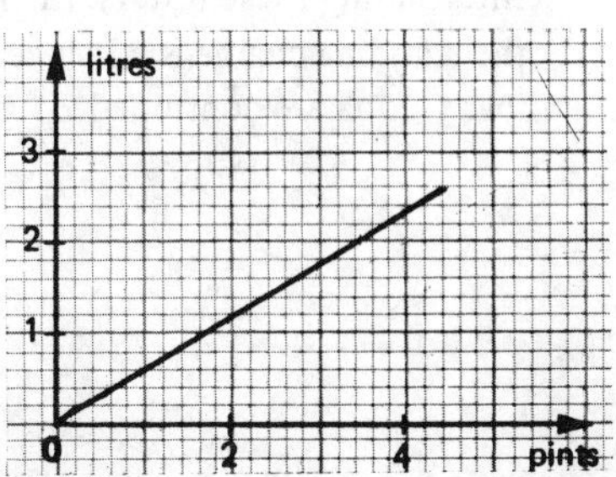

We hope that you now appreciate that a graph can be very useful in determining the type of relationship between two variables. If $y \propto x$ (say), then $y = kx$ and the corresponding graph is a straight line passing through the origin. Conversely, if the graph is not a straight line passing through the origin then the two variables cannot be directly proportional.

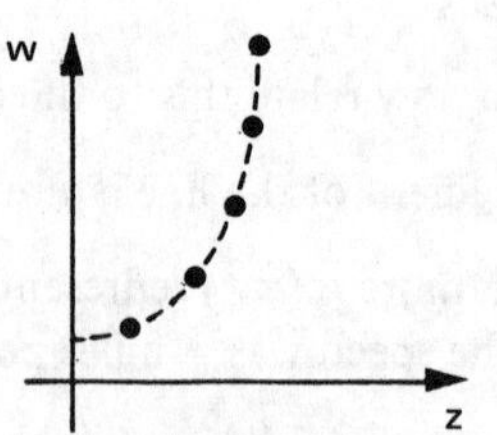

*This graph suggests that w is **not** proportional to z.*

3.3(v) INTERPRETING RATES AND INTERCEPTS

We've indicated that direct proportion corresponds to a straight line passing through the origin. But what about a straight line graph which does not pass through the origin? Such graphs can also be analysed and the relationship meaningfully interpreted. For example the graph opposite illustrates the relationship between wages (w) and hours worked (t). If you measure the gradient you will find, as in the previous example, that it represents £5 per hour. But in this case, even with *no* hours worked the employee is entitled to £5. This is a fixed sum of money payable no matter how many hours are worked. The intercept represents this fixed sum.

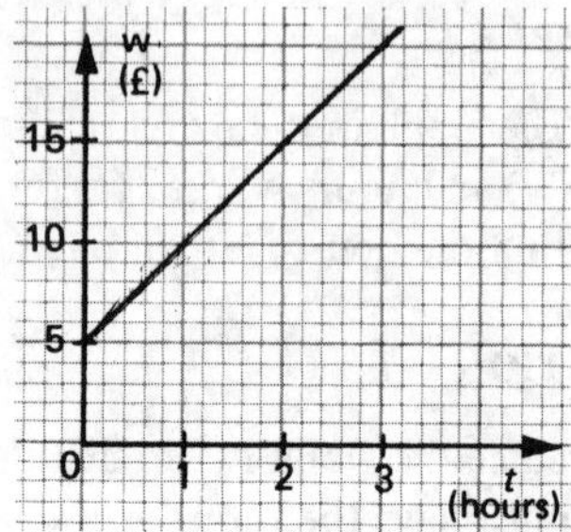

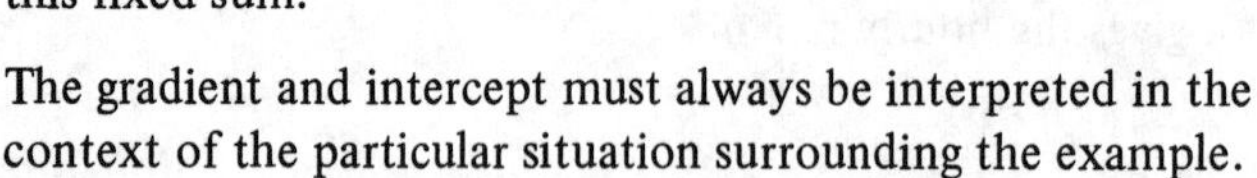
The gradient and intercept must always be interpreted in the context of the particular situation surrounding the example.

EXAMPLE

The graph opposite relates the cost of a quarterly electric bill to the number of units used. From the graph find the rate per unit consumed and the standing charge per quarter.

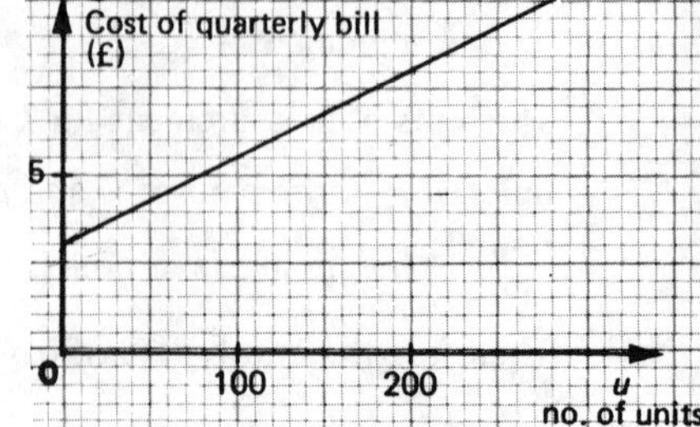

SOLUTION

From the graph the gradient = 0·025 and the intercept = 3. So

$$C = 0{\cdot}025\,u + 3$$

Interpreting these results, the rate per unit is £0·025 per unit and the standing charge is £3 since this must be paid no matter how many units are consumed.

TRY SOME YOURSELF

7 The table below gives the cost of a number of quarterly gas bills relative to the number of therms consumed.

Number of therms (n)	30	40	50	60	70
Cost (C) (£)	9	11	13	15	17

(i) Plot a graph to illustrate this information. Ensure that the axis measuring therms ranges from 0 to 70.
(ii) Measure the gradient and intercept of the graph.
(iii) Hence find the quarterly standing charge and the rate per therm.

After you have worked through this section you should be able to

a Find the gradient of a straight line graph
b Find the intercept of a straight line graph
c Relate the gradient and intercept to the equation of a straight line graph
d Given a straight line graph, find its equation
e Interpret a straight line graph passing through the origin in terms of direct proportionality
f Appreciate that the interpretation of the gradient and intercept depends upon the *context* of the example

Finally, here are some exercises if you want more practice.

TRY SOME MORE YOURSELF

8(i) Write down the gradient and intercept of each of the following linear equations:
(a) $t = 17y + 4$ (b) $s = -3$ (c) $t = r$ (d) $y = mx + c$.

(ii) (a) Find the gradient and intercept and hence the equation of each of the following straight lines:

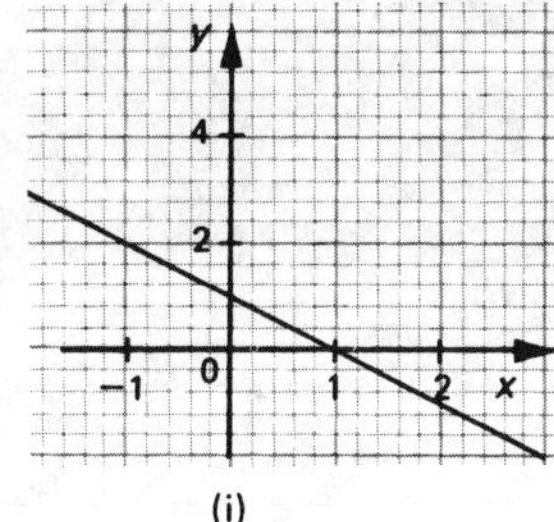

(i)

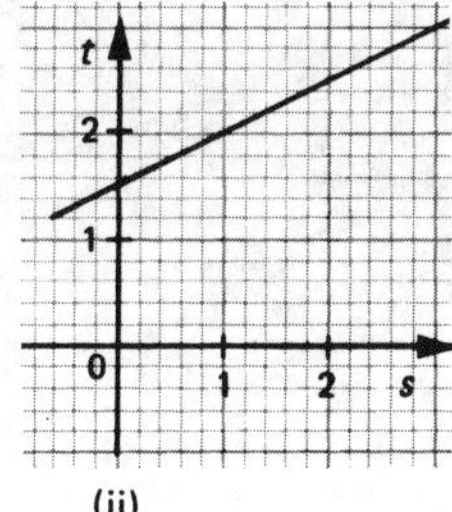

(ii)

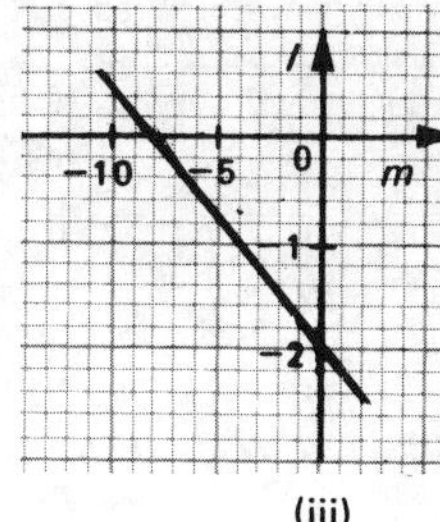

(iii)

(iii) The graph opposite converts pounds to kilograms.
(a) What is the conversion rate from pounds to kilograms?
(b) What is the conversion rate from kilograms to pounds?

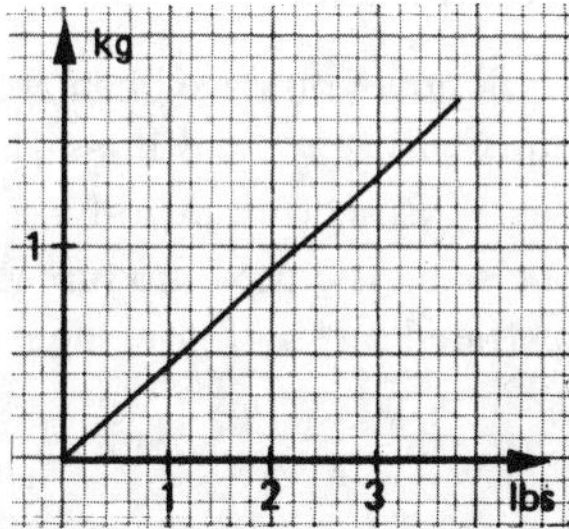

(iv) The table below gives the cost of a number of quarterly telephone bills.

Units	20	40	60	80	100
Costs (£)	12·8	13·6	14·4	15·2	16

Draw the graph and so find the quarterly rental cost and the charge per metered unit.

3.4 Curves

TRY THESE QUESTIONS FIRST

1 Complete the table of values below and so draw the graph of $s = 3t^2$.

t	-3	-2	-1	0	1	2	3
t^2							
$s = 3t^2$							

2 The kinetic energy of a body (E) is directly proportional to the square of the speed (v) at which it is moving. Describe the shape of the graph which illustrates this relationship.

3 Complete the table of values below and so draw the graph of $w = 2 + \frac{1}{z}$

z	1/3	1/2	1	2	3	5
$\frac{1}{z}$						
$w = 2 + \frac{1}{z}$						

4 The volume (V) occupied by a gas is known to be inversely proportional to the pressure (P). Describe the shape of the graph which illustrates this relationship.

3.4(i) PARABOLAS

Up till now you have been able to complete graphs by joining up the points with a straight line. However, graphs are not always straight lines. This section introduces graphs which consist of smooth curves and the next module looks at less regular graphs like those illustrated on the following page.

Module 4, Section 4.2.

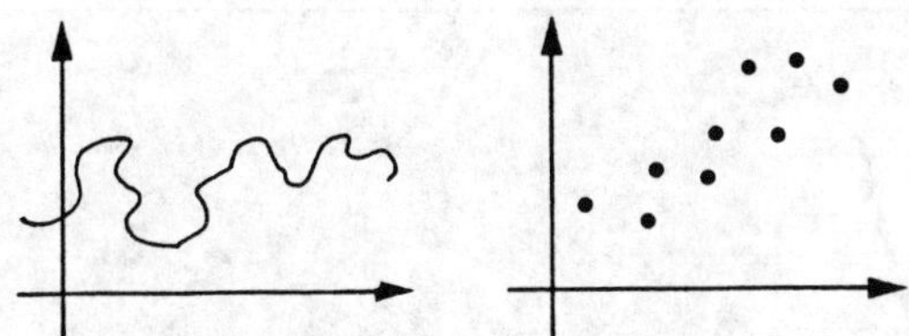

In Section 3.3 we indicated that a straight line graph corresponds to a linear equation, an equation of the form

$$y = ax + b$$

Such an equation contains no higher powers of x (such as x^2) and no reciprocals of x (such as $1/x$). This part of the section concerns equations which include a term involving x^2. We start by investigating the graph of

$$y = x^2$$

TRY SOME YOURSELF

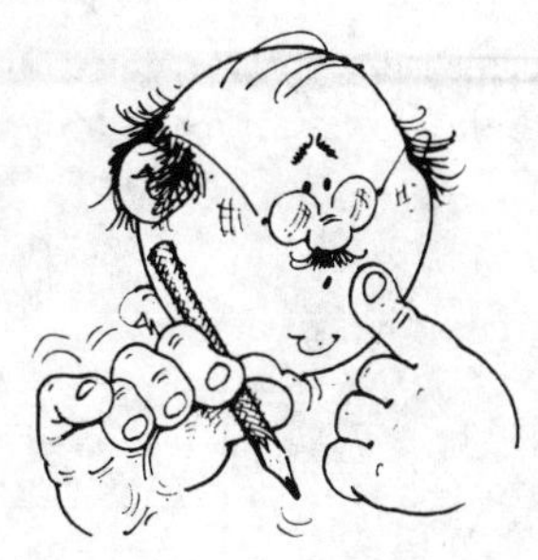

1(i) Complete the table of values below for $y = x^2$.

x	−3	−2	−1	0	1	2	3
$y = x^2$							

(ii) Plot the points from this table onto the diagram below. For the moment do not join up the points.

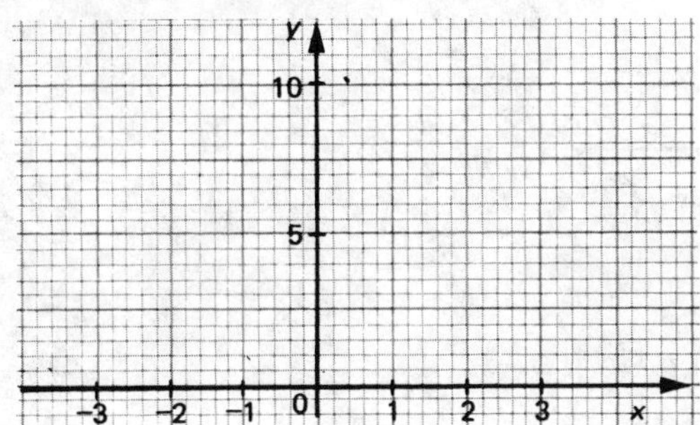

You may feel tempted to join up these points with straight lines–but wait a moment...

(iii) You should have noticed that these points do not lie on any one straight line. To get more idea of the shape of the graph complete the following table and add the points to your graph.

x	−2·5	−1·5	−0·5	0·5	1·5	2·5
$y = x^2$						

You will only be able to plot these points approximately–but this will still give a good idea of the shape of the graph.

(iv) You should by now have a reasonable picture of the graph, so join the points up to form a smooth curve.

The shape of this graph is quite distinctive. It is called a *parabola.*

You may have heard of a parabolic mirror.

CHECK YOUR ANSWERS

1 *Section 3.4(i)*

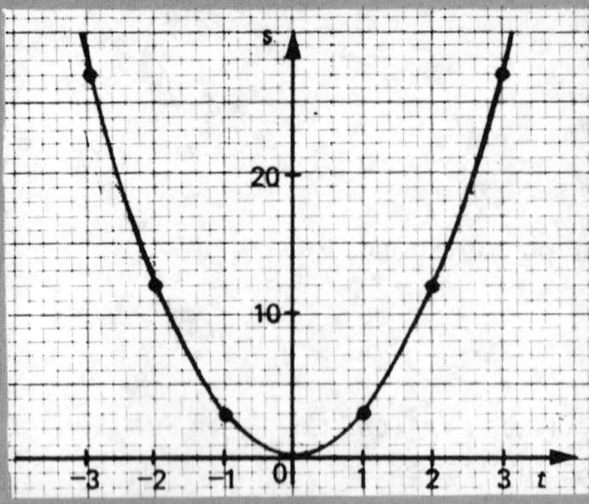

2 If E is directly proportional to v^2 then *Section 3.4(ii)*

$E = kv^2$ for some constant k

This is a quadratic equation so the corresponding graph is a parabola. In fact it looks something like

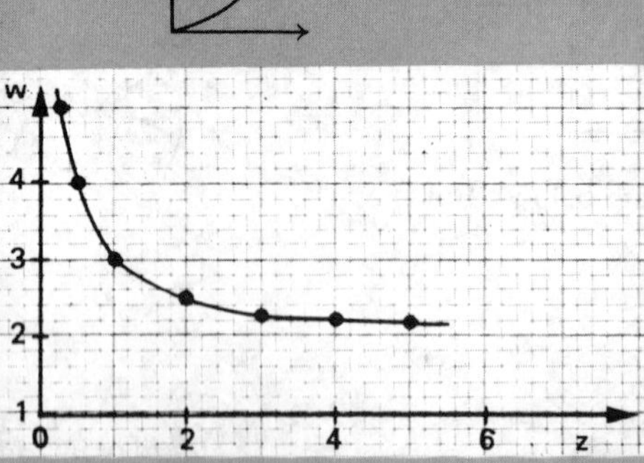

3 *Section 3.4(iii)*

4 If v is inversely proportional to p then *Section 3.4(iv)*

$v = \frac{k}{p}$ for some constant k

The corresponding graph will therefore be a hyperbola. In fact it looks something like

The following equations each contain a term involving x^2. The associated graph for each equation is a parabola.

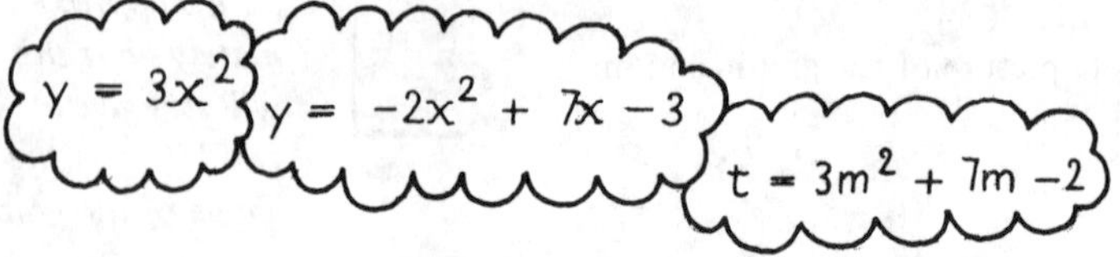

The variables do not necessarily have to be x and y.

Notice that these equations do not contain terms involving x^3 or higher powers of x. Nor do they include reciprocals of x. An equation which contains a term involving x^2, a term involving x and maybe a constant term, but no terms involving higher powers of x or reciprocals of x (such as $1/x$), is called a *quadratic equation.*

The expression may be written in terms of any symbol.

In general a parabola has equation

$$y = ax^2 + bx + c.$$

The following exercises give you more practice in plotting parabolas.

TRY SOME YOURSELF

2(i) (a) Complete the table of values below.

x	-3	-2	-1	0	1	2	3
x^2							
$y = 2x^2$							

Look at the range of values for both co-ordinates before choosing scales.

(b) Plot the points on some graph paper and so draw the graph of $y = 2x^2$.

You may want to include some extra points, such as $x = 0{\cdot}5$, $x = 1{\cdot}5$ *etc. to get a better idea of the shape.*

Make sure that your graphs are smooth. There should be no bumps or corners.

(ii) (a) Complete the table of values below.

x	-3	-2	-1	0	1	2	3
x^2							
$y = 0{\cdot}5x^2$							

(b) Plot the graph of $y = 0{\cdot}5x^2$ on the same axes which you used for Exercise 2(i) above.

This allows you to compare the shapes of the two graphs.

The graphs for $y = 2x^2$ and $y = 0{\cdot}5x^2$ indicate that, although the shapes are similar, the parabola, $y = 2x^2$, is much steeper than that of $y = 0{\cdot}5x^2$.

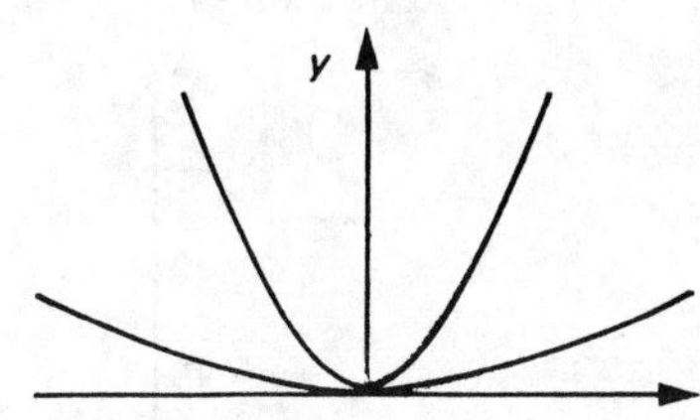

TRY SOME YOURSELF

3(i) Complete the table of values below.

x	-3	-2	-1	0	1	2	3
$y = -x^2$							

(ii) Hence draw the graph of $y = -x^2$.

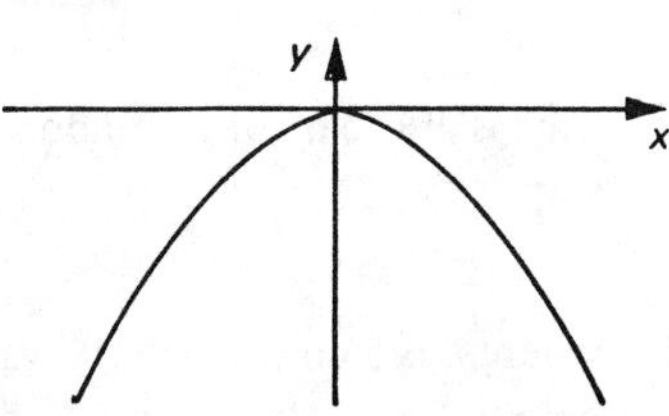

This graph has the same shape as $y = x^2$ but it is 'upside down'. What does this suggest about the parabolas for $y = -0{\cdot}5x^2$ and $y = -2x^2$? The *shapes* should be the same as $y = 0{\cdot}5x^2$ and $y = 2x^2$, but the parabolas will be 'upside down'. Thus parabolas have similar shapes although they may vary in steepness and direction.

Check this yourself if you want by plotting the graphs of $y = -2x^2$ *and* $y = -0{\cdot}5x^2$.

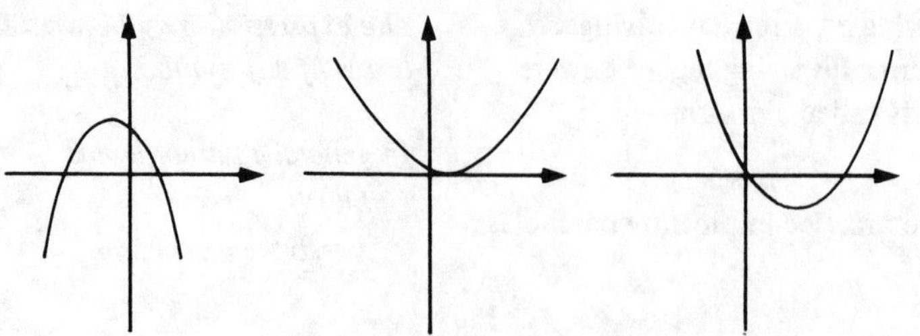

The parabola does not necessarily have to pass through the origin as these graphs illustrate. The exercises at the end of this section include examples of such graphs.

3.4(ii) INVESTIGATING QUADRATIC RELATIONSHIPS

It is extremely difficult to determine whether a given graph represents a quadratic equation since there are many other relationships which produce similar graphs (at least in part) and this is not the appropriate place for such an analysis as it requires a considerably more extensive mathematical background than we have assumed here. We do not expect you to be able to determine the *type* of relationship from the graph, we include only some well established examples.

This diagram illustrates a relationship which you have already met—the relationship between the area of a square (A) and the length of its sides (r). In fact $A = r^2$. Since the length of a side cannot be negative the resulting graph is just one half of the parabola.

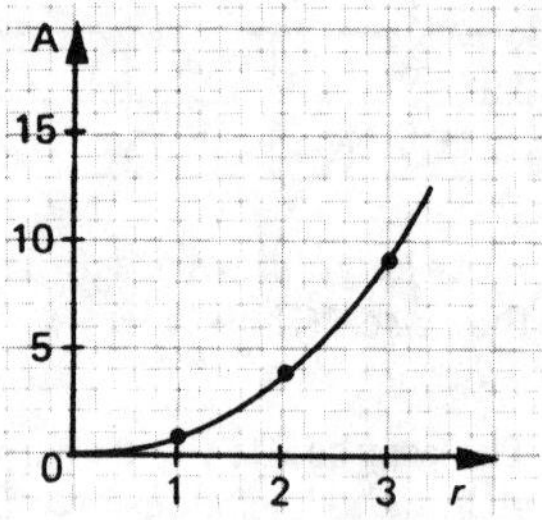

The diagram below illustrates the relationship between the area of a circle and its radius r.

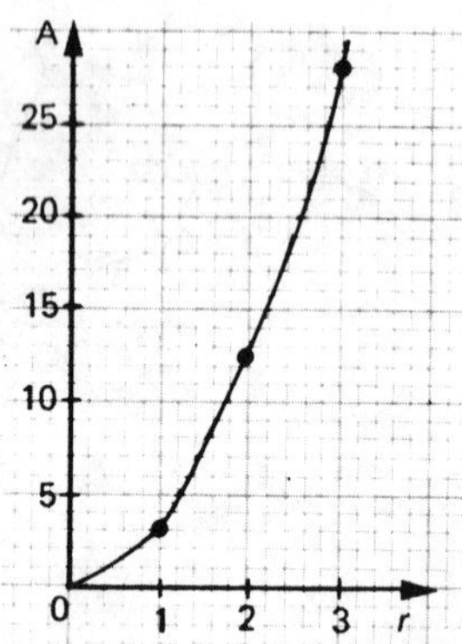

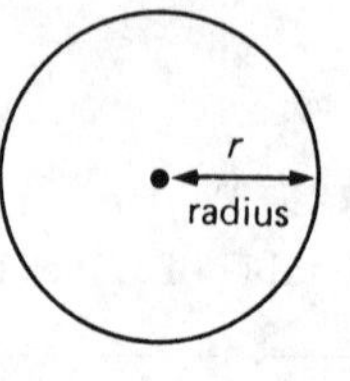

This is also part of a parabola and actually represents the relationship

$$A = kr^2$$

where k is a constant. The value of k is investigated in the exercise below.

You may already know that the area of a circle is equal to πr^2. The constant k is therefore equal to π.

TRY SOME YOURSELF

4(i) Determine the value of k in this relationship by considering $r = 1$ and $r = 2$ (this second value of r is included as a check).

From the graph you can determine the value of A for these given values of r. The value of k can then be determined from the equation $A = kr^2$.

(ii) This table of values is taken from the Highway Code and gives the braking distances for cars travelling at various speeds.

You may find this final example a bit unexpected.

Speed (m.p.h.)	20	30	40	50	60	70
Braking distance (ft)	20	45	80	125	180	245

(a) Plot the graph for this table of values.

The graph should be a parabola.

(b) Use your graph to find the braking distance for a car travelling at 55 m.p.h.

(c) What is the fastest a car can travel if it has to stop in less than 60 ft?

Again, we emphasise that graphs which look like parabolas often do not correspond to quadratic equations. If you find that an experiment produces a graph which seems to have this shape it is probably worth your while to be extremely cautious and to undertake some further investigation as to what the relationship might be.

3.4(iii) HYPERBOLAS

The graph of $y = \frac{1}{x}$ also has a distinctive shape which we investigate below.

TRY SOME YOURSELF

5(i) Complete the table of values below for $y = \frac{1}{x}$.

$\frac{1}{x}$ is the reciprocal of x.

x	$\frac{1}{5}$	$\frac{1}{3}$	$\frac{1}{2}$	1	2	3	5
$y = \frac{1}{x}$							

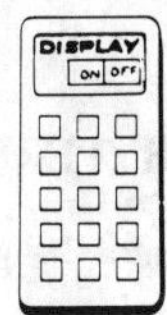

Use the reciprocal key, **1/x**.

(ii) Use your calculator to find
(a) The reciprocal of 1,000,000 (or 10^6)
(b) The reciprocal of 10^{-6}.

This exercise indicates that as x gets bigger, $\frac{1}{x}$ gets smaller and smaller, becoming closer and closer to zero. The mathematical expression for this is

$\frac{1}{x}$ *tends* to zero as x gets larger and larger

This can be written as $\frac{1}{x}$ tends to 0 as x tends to infinity. This is ***not*** *a number. 'x tends to infinity' just* ***means*** *'x gets larger and larger'.*

Similarly as x gets smaller, $\frac{1}{x}$ gets bigger and bigger. Use your calculator to try to find the reciprocal of zero. You should get an error display on your calculator. That is because $1 \div 0$ is undefined; no meaning can be attached to the expression. All we can say is that as x tends to zero, $\frac{1}{x}$ gets larger and larger. For this reason the point $x = 0$ is left out of the graph.

TRY SOME YOURSELF

6(i) Plot the graph of $y = \frac{1}{x}$ using the table of values you obtained in Exercise 5(i).

You will need to choose suitable scales for the axes.

(ii) (a) Complete the table of values below for $y = \frac{2}{x}$.

x	$\frac{1}{5}$	$\frac{1}{3}$	$\frac{1}{2}$	1	2	3	5
$y = \frac{2}{x}$							

Again, the point $x = 0$ must be excluded from the graph since $\frac{2}{0}$ is undefined.

(b) Use the table of values to plot the graph of $y = \frac{2}{x}$.

Curves such as the ones you obtained in Exercise 6 are called *hyperbolas*. Equations like

Each of these equations contains a reciprocal.

all correspond to hyperbolas.

Like parabolas, hyperbolas have varying degrees of steepness and different directions. The following graphs are all hyperbolas.

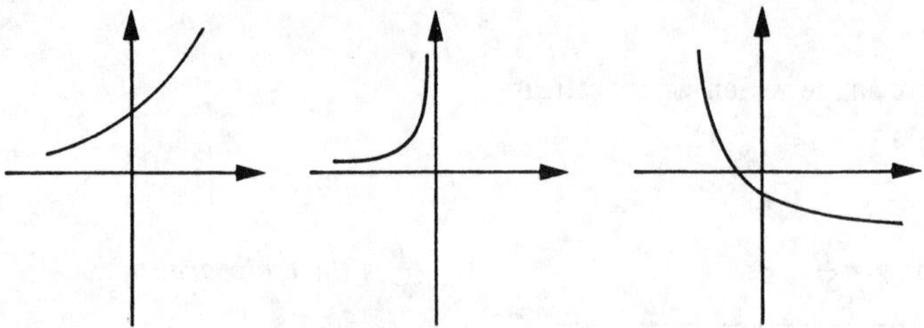

The hyperbola may be 'back to front' and it can also be in a different position relative to the axes. We have only considered positive values of x. You may like to investigate the graph of $y = 1/x$ when x is negative.

3.4(iv) INVESTIGATING HYPERBOLIC RELATIONSHIPS

Section 3.3(iv) discussed the relationship between direct proportion and straight line graphs. You may already have realised that a similar relationship exists between inverse proportion and hyperbolic graphs.

The table below relates journey times between Edinburgh and London to the travelling speed.

The table is taken from Module 2, Section 2.5(v).

Speed v (m.p.h.)	20	40	50	80	100
Time t (hrs)	20	10	8	5	4

Section 2.5(v) indicated that t is inversely proportional to v. vt is always equal to 400. So

$$t = \frac{400}{v}$$

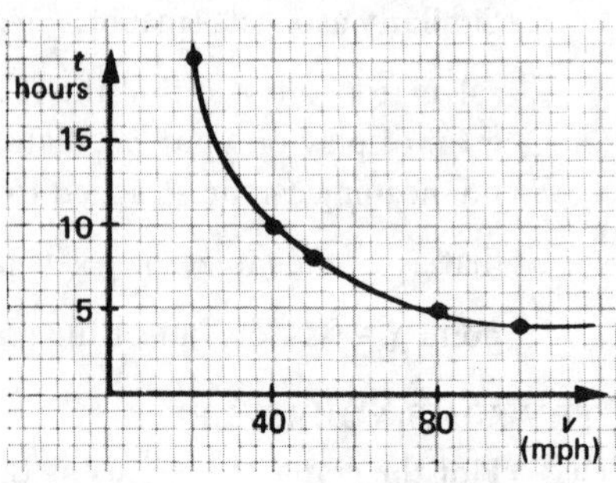

The graph corresponding to this equation is a hyperbola. The following exercise investigates more examples of inverse proportion.

TRY SOME YOURSELF

7(i) The table below is taken from a guide for the use of a photographic flash and relates the aperture setting to various distances from the subject.

Distance (m)	1·00	1·40	2·00	2·75	5·50
Aperture setting	22	16	11	8	4

You may recognise these aperture settings from your camera.

(a) Plot a graph to illustrate this information.
(b) Use your graph to determine the aperture setting when the distance to the subject is 4 metres.

(ii) The table below converts radio frequency to wavelength.

Frequency	500	1000	2000	3000	4000
Wavelength	600	300	150	100	75

Frequency is measured in kilo-cycles per second (kHz). Wavelength is measured in metres (m).

(a) Plot a graph to illustrate this information.
(b) Use your graph to find the wavelength equivalent of 1500 kHz.
(c) Use your graph to find the frequency equivalent of 433 m.

The graph should be a hyperbola.
(Radio 4)
(Radio 2)

Again, we urge you to proceed cautiously if you find that your results give a graph which seems to be a hyperbola. Such a graph may not necessarily represent inverse proportion since other relationships can produce similar graphs (at least in part).

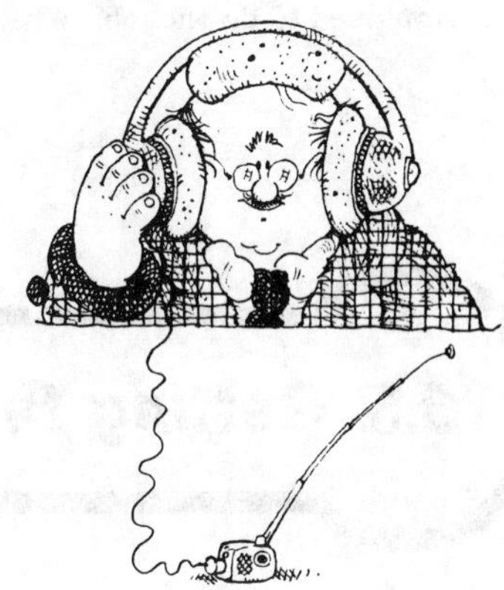

This section has illustrated the visual characteristics of some non-linear relationships. A graph provides a good picture of a relationship between two variables and, although this can only be approximate, plotting the graph is always a worthwhile first step.

After you have worked through this section you should be able to

a Plot a parabola by completing a table of values
b Recognise that the graph for a quadratic equation is always a parabola
c Plot a hyperbola by completing a table of values
d Recognise that the graph which represents inverse proportion is always a hyperbola

Finally, here are some exercises if you want more practice.

TRY SOME MORE YOURSELF

8(i) For each of the following equations draw up tables of values of the form

x	-3	-2	-1	0	1	2	3
$y =$							

and hence plot their graphs:
(a) $y = x^2 + 1$
(b) $w = t^2 - 2$
(c) $w = z^2 + 2z - 3$.

(ii) A stone is dropped from the top of a cliff. It is known that the distance it falls is proportional to the square of the time since it was dropped. What shape has the graph? If the cliff is 100 m high, draw a rough sketch to illustrate the flight of the stone.

(iii) For each of the following equations draw up tables of values of the form

x	1/5	1/3	1/2	1	2	3	5
$y =$							

and hence plot their graphs:
(a) $y = \frac{1}{x} + 2$
(b) $t = \frac{2}{s} - 3$
(c) $t = 2 - \frac{1}{m}$

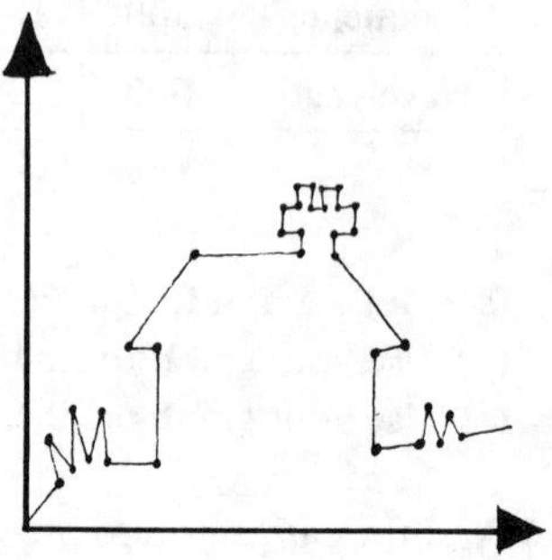

(iv) A building contractor reckons that the time taken to build a particular building is inversely proportional to the number of men employed to do the job. What shape has the graph?

3.5 Getting the Right Information from Graphs

TRY THESE QUESTIONS FIRST

1 This graph plots the cost of a taxi ride against distance travelled. From the graph it appears as though it costs a lot of money to travel a short distance. Redraw the graph so that it appears to cost very little to travel a long distance.

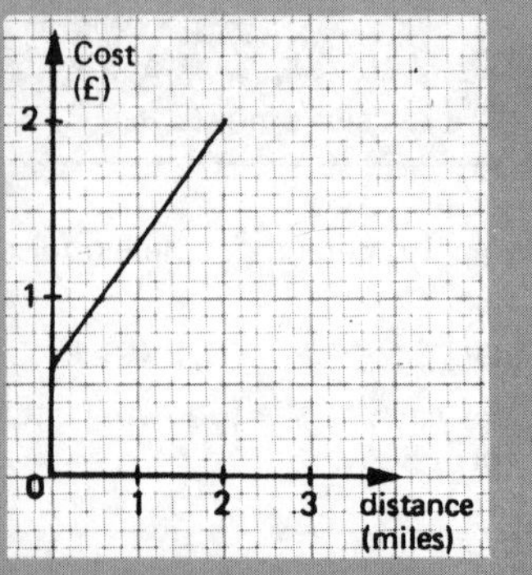

2 This graph shows the percentage increase in the cost of fresh fish. What does it indicate about the actual cost of fresh fish?

3 The graph only goes up to July 1977. If you were to extrapolate from this information what might you suggest would happen in 1978? What other factors may need to be considered?

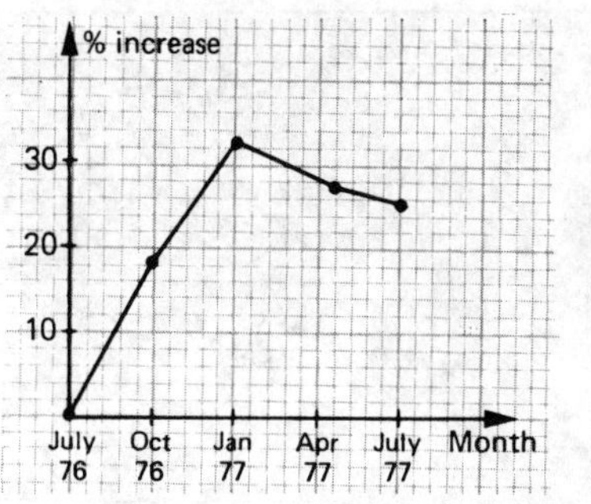

3.5(i) MISLEADING INFORMATION

This section is included to encourage you to take care when interpreting graphs, since first impressions are not always reliable. Graphs are often presented in forms which are deliberately intended to mislead the reader. We include some examples here which indicate that the actual information conveyed by a graph can be obscured by its appearance.

The graph in the margin represents the National Income in the USA for one particular year–from January to December. As you can see the National Income rises from $20 billion to $22 billion. This graph suggests that the National Income remains steady and high. The two graphs below give exactly the same information although they *look* very different.

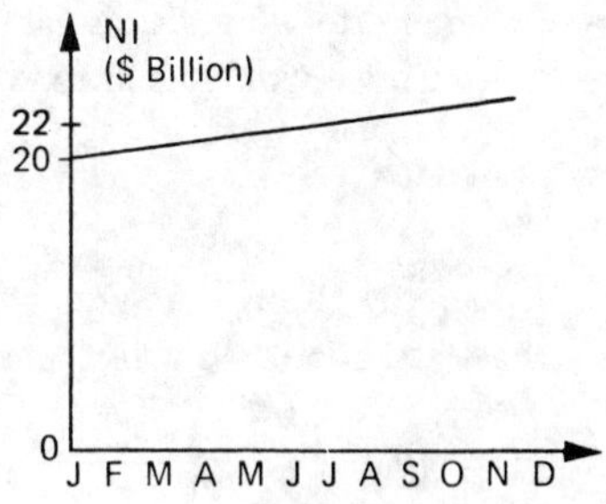

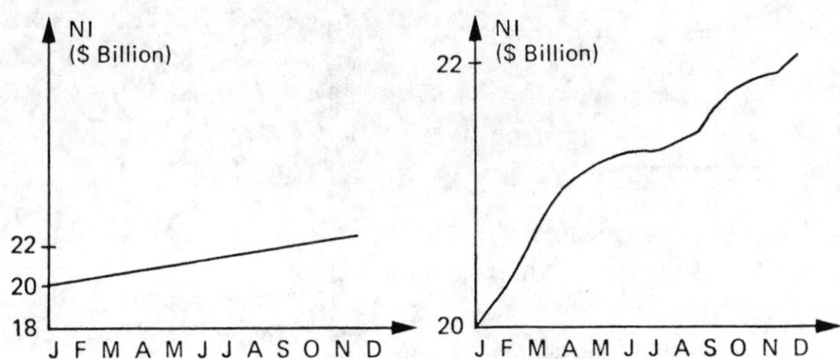

In all three graphs the scales used on the horizontal axis are the same –but the scales on the vertical axis are different. In the two graphs above the scales do not start at zero. The graph on the left suggests that although the National Income was steady it wasn't particularly high. The graph on the right uses a much larger scale and the National Income appears to rise dramatically.

The horizontal axis indicates the months of the year.

These examples emphasise that you should always read the scales carefully. In particular you should note where the scales start.

CHECK YOUR ANSWERS

1 The scale for the vertical axis must appear to squash down the cost and need only start at 50p or £0·5. The scale for the horizontal axis should appear to space out the distance. *Section 3.5(i)*

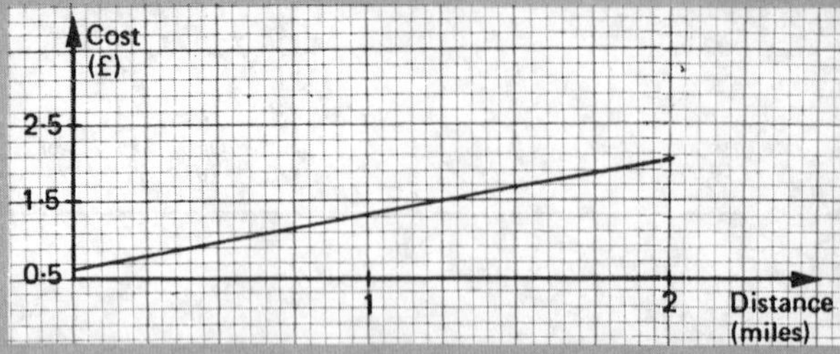

2 The vertical axis measures percentage increase—but are all the percentages based upon the price in July 1976, or are they calculated relative to the price in the previous quarter? It looks as though the prices are based on those in July 1976 in which case the graph showing the actual price of the fish will look the same, although we would need to know the base price in order to plot it. *Section 3.5(ii)*

3 It seems as though the graph will continue to decrease, suggesting that the actual cost will also continue to decrease. *Section 3.5(iii)*

% increase
30
20
10
0
July Oct Jan Apr July Oct Month

But if other factors are taken into account—such as availability of fresh fish and economical pricing policies—this may not be the case. In fact the price of fish increased considerably in 1978 owing to Common Market regulations and foreign competition in fishing waters.

The graphs overleaf both illustrate unemployment figures for one region in 1979. They give exactly the same information.

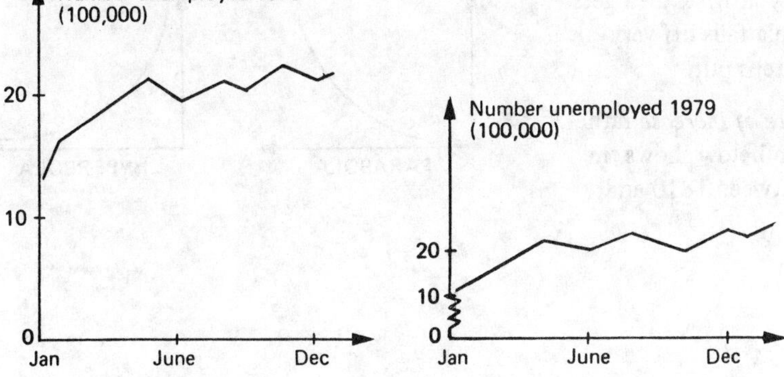

In the graph on the right, the vertical axis is broken. Although the scale **appears** *to start at 0 the jagged line indicates a sudden jump to 1,000,000.*

Thus graphs are often deliberately drawn to give a particular impression. The next exercise invites you to alter the appearance of a graph by drawing it to a different scale.

TRY SOME YOURSELF

1(i) This graph illustrates a train journey from London to Newcastle. The train appears to take a long time to cover a fairly short distance as the scales have been carefully chosen to give just that impression. Try redrawing the same graph to give the impression that the train covers a large distance in a short time. This involves choosing different scales for the axes.

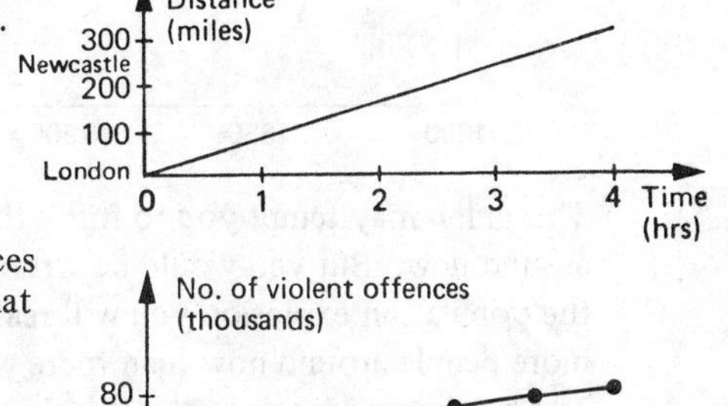

(ii) This graph indicates the rising number of violent criminal offences in England and Wales during the 1970s. Redraw the graph so that the figures appear to rise dramatically.

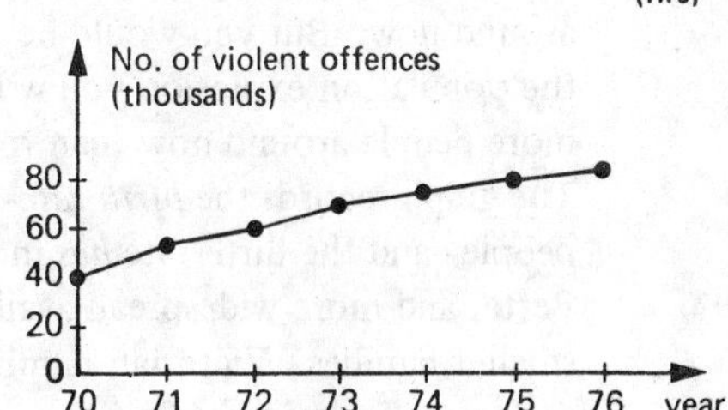

3.5(ii) RATE OF INCREASE

In Section 3.3 we looked at the interpretation of the gradient of a straight line graph and noted that the gradient represents the rate of increase. This graph illustrates the relationship between a quarterly electricity bill and the number of units used. The gradient gives the rate per unit and the intercept the quarterly standing charge.

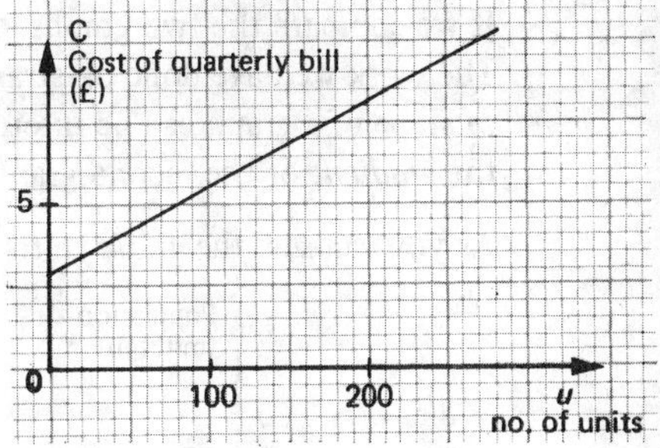

Rate: £0·025 per unit.
Standing charge: £3.

More generally, the gradient gives the rate at which the graph increases and for a straight line the gradient (or rate of increase) is constant. Having determined the gradient the information must be interpreted in terms of the particular example. This interpretation may also depend upon other factors not necessarily included on the graph.

This parabola increases very slowly at first, then gets steeper and steeper. This hyperbola falls off very steeply at first then gradually flattens out.

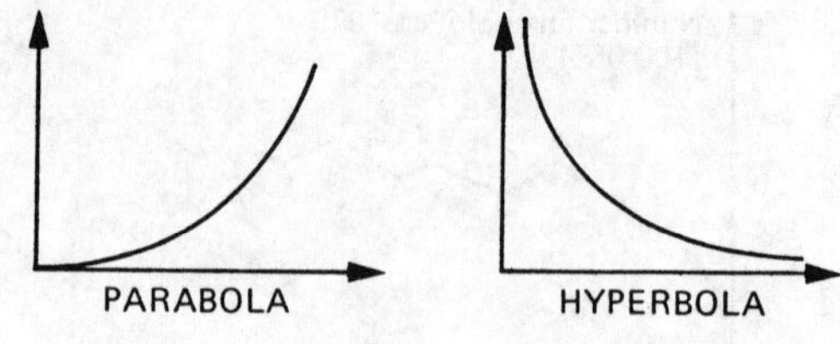

Sometimes a graph records the *rate of increase* rather than the increase itself. The graph below shows the birthrate in England and Wales between 1810 and 1970.

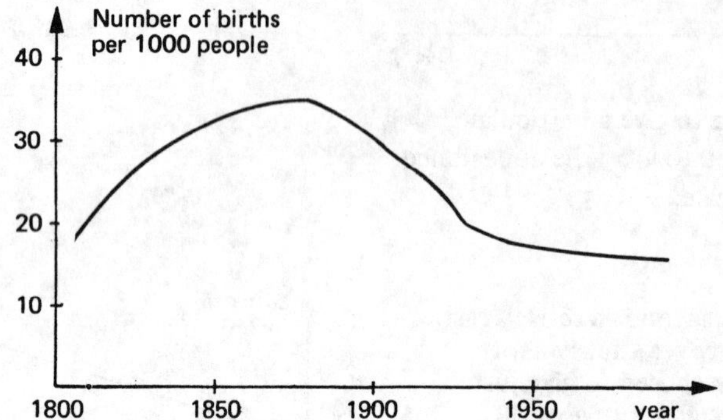

The graph indicates that the birth-rate increased steadily between 1810 and 1880. The graph then flattens off and the birth rate drops again.

This graph may tempt you to think that there are fewer people around now. But you would be wrong and if you reflect upon the population explosion you will realise that there are millions more people around now than there were in the 19th Century. The graph records the *birth rate*–the number of births per 1000 people–and the birth rate *has* indeed gone down this century. Better and more widespread availability of contraception has meant smaller families. Victorian families often had as many as ten children. Today most families are limited to two or three. But the population is still increasing, as the graph in the margin indicates.

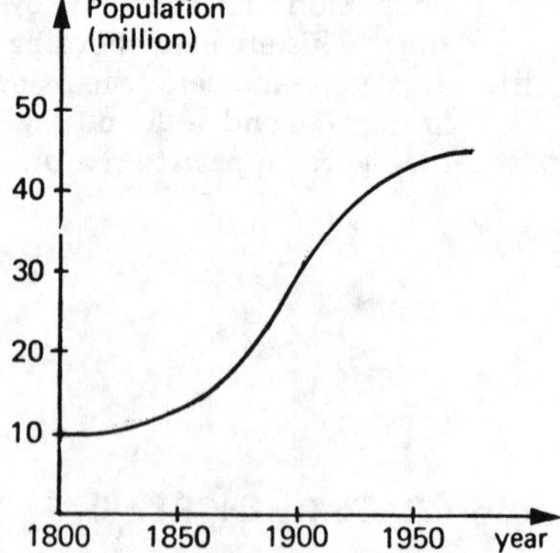

The birth rate is the rate of increase of the population.

If we compare the two graphs you may be able to see the link. The work we have done on straight lines suggests that the gradient or slope of a graph can be interpreted as the rate of increase. So the *gradient* of the second graph represents the birth rate.

As you can see, the graph starts off almost flat.

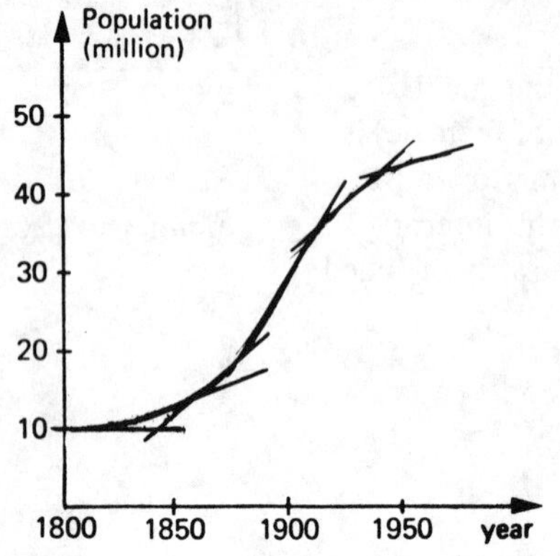

The straight lines approximate the gradient at various points. The steeper the straight line the greater the gradient.

Moving along the curve the gradient increases and the curve becomes steeper. Gradually the curve flattens off. This means that the gradient or birth rate decreases again. But although the curve flattens off it is still increasing.

The graph illustrating the birth rate also indicates that the birth rate increases then decreases again.

We can conclude, therefore, that the two graphs actually give the same information. However, the graph plotting the birth rate may be misunderstood if the reader does not appreciate the meaning of *rate*.

In the next example the vertical axis measures percentage increase.

EXAMPLE

This graph illustrates the percentage increase in the cost of 1 dozen eggs between July 1976 and September 1977 based upon the cost of 1 dozen eggs in July 1976 which was 50p. Draw a graph to illustrate the actual increase in cost.

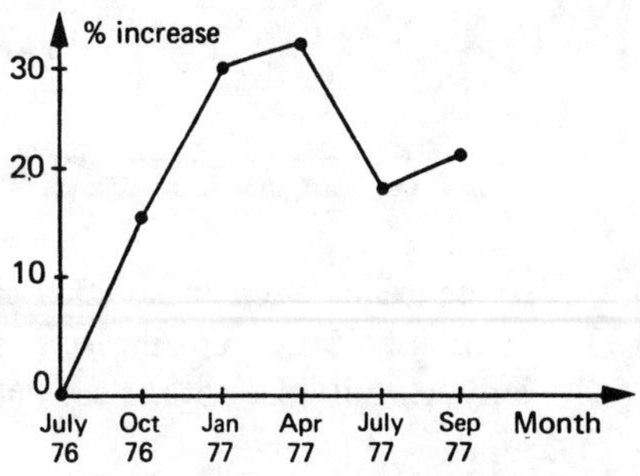

SOLUTION

We first need to interpret the meaning of percentage increase. For example, in October 1976 the percentage increase was about 15%. This means that the actual cost was 115% of 50p = 57·5p. In January 1977 the % increase was 30%. The actual cost was 130% of 50p = 65p.

Percentage increase is discussed in Module 1, Section 1.5(iii).

The remaining costs can be evaluated in the same way. In April 1977 the actual cost was about 132% of 50p = 66p. In July 1977 it was about 118% of 50p = 59p. And in September it was about 122% of 50p = 61p.

The graph of the actual costs can now be plotted. The graph *looks* the same but the vertical axis now indicates the actual *cost*. A graph which measures the percentage increase without giving a base cost can be very confusing. The examples below also illustrate the rising costs of food.

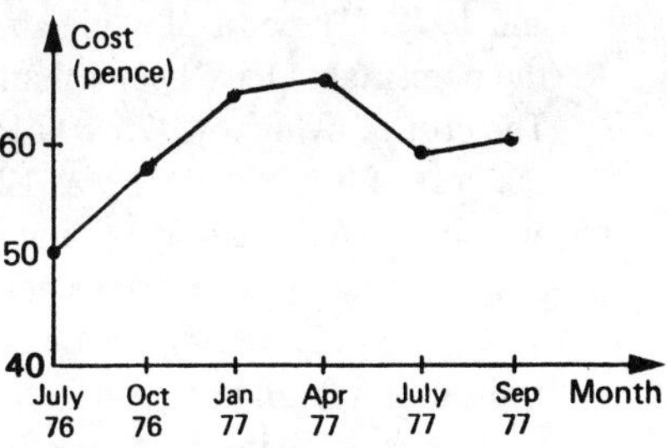

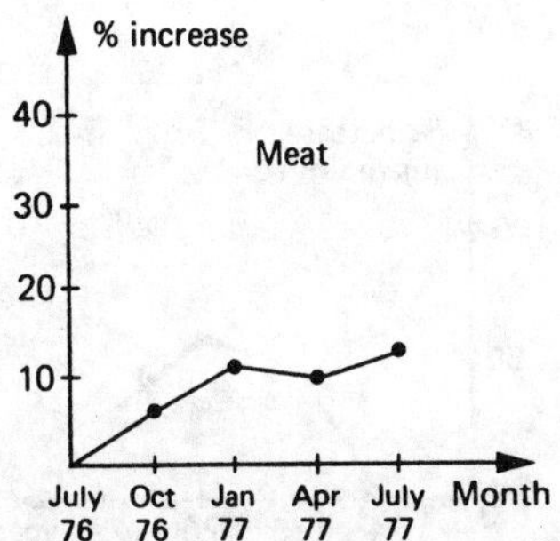

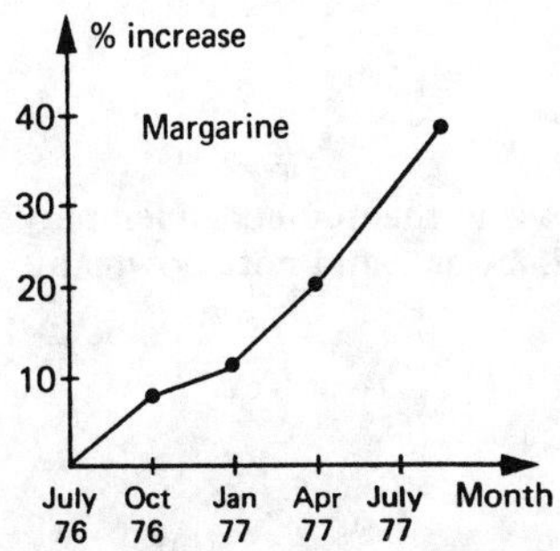

At first sight, the price of margarine has escalated whereas the cost of meat has hardly risen at all.

However, if the graphs are redrawn to show the actual increase in cost we can see that the cost of fresh meat (steak) is always considerably more than the cost of margarine, and has in fact gone up by more.

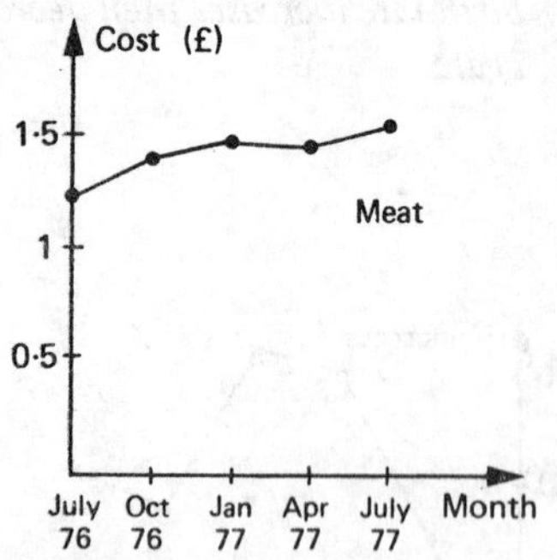

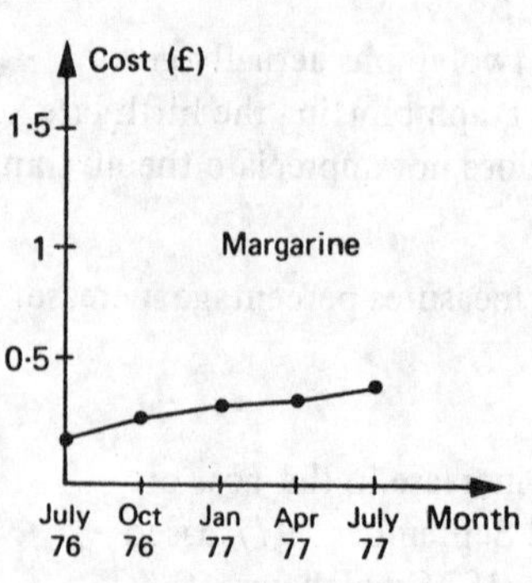

These examples emphasise that you must proceed extremely cautiously if the graph measures the percentage increase, particularly when it does not indicate any actual costs at all.

TRY SOME YOURSELF

2 This graph indicates the percentage increase in the numbers of unemployed people. There were about half a million unemployed in 1974. What does the graph indicating the *actual* numbers unemployed look like?

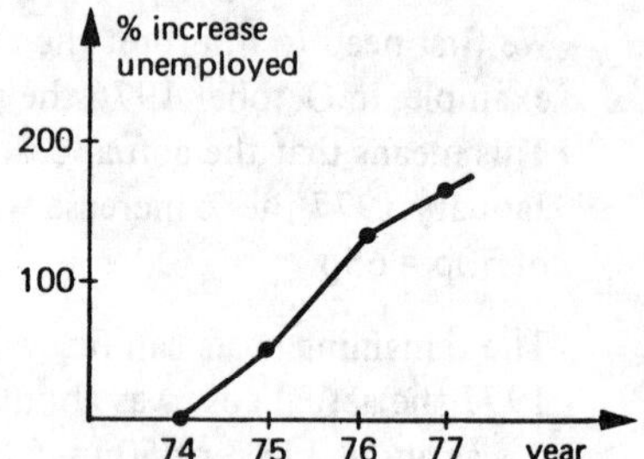

Occasionally the graph can be even more misleading. This graph illustrates the percentage increase in the cost of living between 1975 and 1979. The cost of living appears to be going down, but here the percentages have been calculated from one year to the next. The cost of living in 1975 was 15% above the cost of living in 1974, the cost of living in 1976 was 12% above the cost of living in 1975 and so on. At each stage the percentage increase is based upon the cost of living in the previous year. In fact this graph is really measuring the rate of increase of the cost of living. Like the previous example on birth rates, even though the *rate* decreases, the actual cost of living continues to rise.

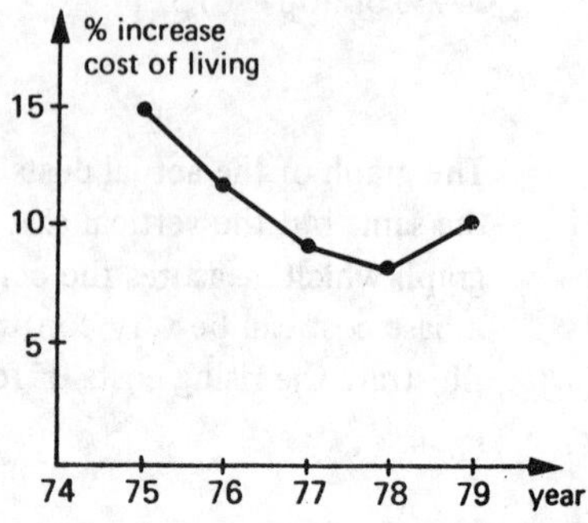

TRY SOME YOURSELF

3 This graph gives the percentage increase in the numbers unemployed. Compare it with the graph given in Exercise 2 and note down any observations.

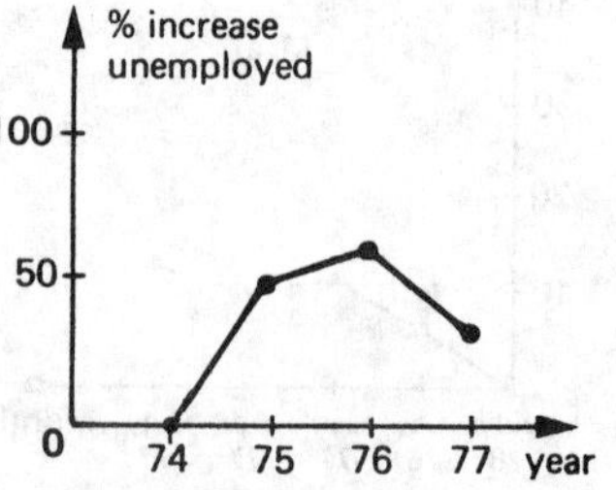

We hope therefore that you will always carefully consider what the axes measure. Check whether the vertical axis really does indicate the actual increase–or whether it gives the percentage increase or the rate of increase. The vertical axis is often used to measure the rate of increase to give a deliberately misleading picture.

3.5(iii) INTERPOLATION AND EXTRAPOLATION

In Section 3.2(ii) we introduced the concept of *interpolation.* It is easy enough to determine intermediate values if the graph is known to be a straight line.

This is the process of reading off values in between known points.

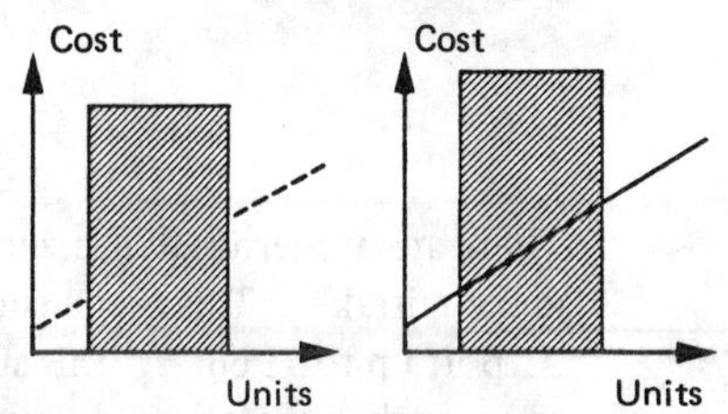

For example, suppose we only have some incomplete information taken from a number of quarterly electricity bills, leaving a gap in the graph. We know that this graph *should* be a straight line so we can complete the graph and read off intermediate values with confidence.

However, if we do not know the exact shape of the graph we must be very cautious.

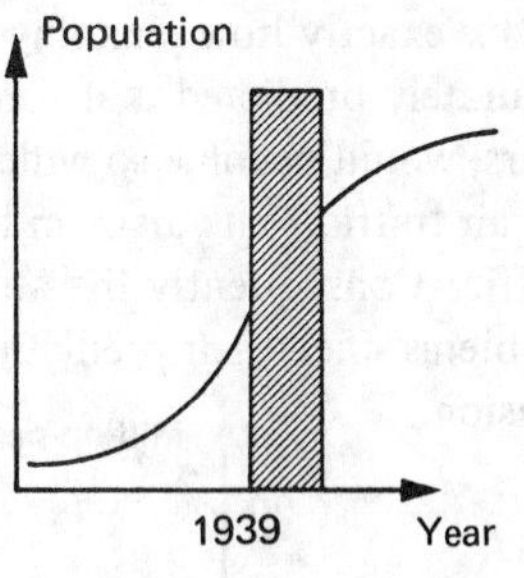

This graph records the population of England and Wales. However, there is a gap in the records between 1939 and 1945 owing to the Second World War. It is very tempting to join up the graph with a smooth line. But this may not be appropriate–for all we know there may have been a sudden drop during the war, or a great expansion.

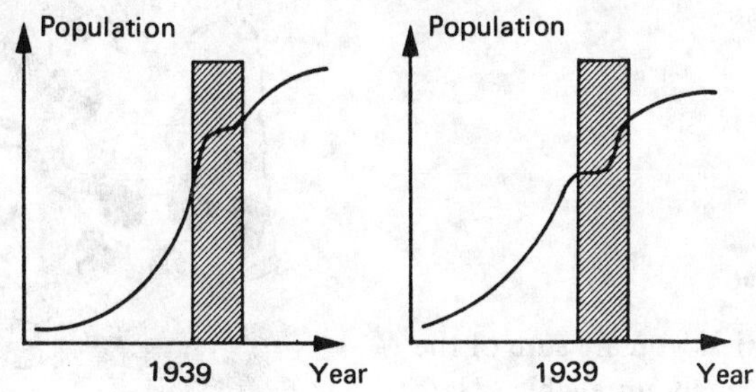

We have included this example to illustrate that although interpolation can be very useful in filling in gaps in our knoweldge, we must always acknowledge an element of uncertainty unless there is some other source of information which suggests otherwise.

Extrapolation is the process of making estimates *beyond* the available range of information. It is the process of predicting what might happen. As with interpolation we must take great care in making such predictions.

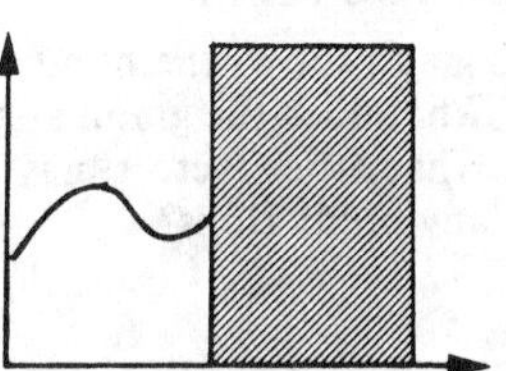

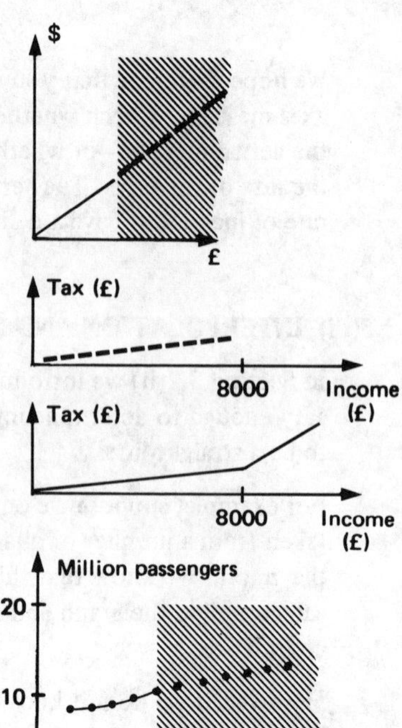

It is reasonable to suppose that a known straight line can safely be extrapolated. For example, a graph showing conversion from pounds to dollars can be extended as far as we please. However, there are instances where a graph produces a straight line at first, but then it suddenly changes direction.

This is illustrated by this graph which plots income tax against yearly income. At first, the points suggest that the graph is a straight line, but when the annual income reaches £8000 the tax rate increases from 30% in the pound to 45% in the pound. So the line changes direction.

There are numerous dangers in prediction and even the experts can make mistakes. This graph shows air passenger traffic at Heathrow Airport up to 1968. At this stage it would seem reasonable to suppose that there would only be a marginal increase in following years and so plans were made accordingly.

This is exactly how planning disasters originate. Nobody could have accurately predicted that foreign travel, and in particular package tours, would become so widespread and thus significantly increase the air traffic. The diagram below indicates the actual increase of traffic. Consequently the authorities are faced with overcrowding problems since their predictions did not allow for this level of expansion.

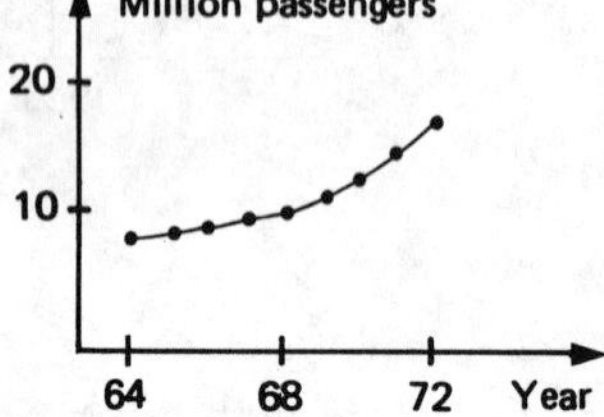

There lies the danger in extrapolation. Unless you are sure of the form of the graph, predictions can be extremely unreliable. Unfortunately, as this last example shows, predictions often have to be made at some stage and the consequences suffered at a later date.

TRY SOME YOURSELF

4 This graph shows the number of foreign visitors to a British city.
(i) What does the graph appear to suggest about future visitors?
(ii) What other factors may need to be considered in order to make any predictions?

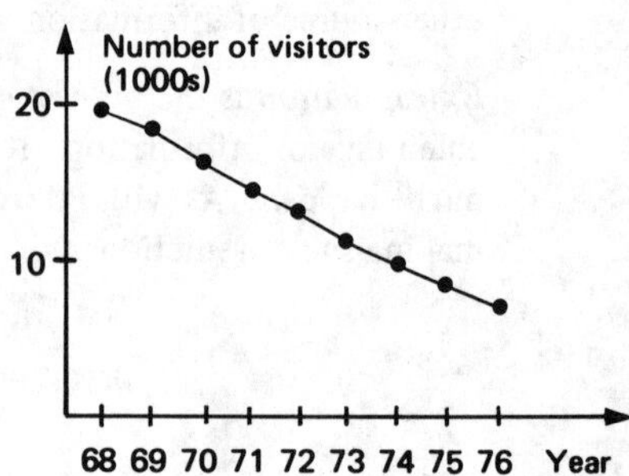

This section has really been a chance to reflect upon what graphs can and can't show; we don't expect you to be expert in interpreting graphs since this can only come with practice.

However, after reading this section you should

a Always read the scales on the axes carefully
b Always check where the axes intersect and whether one axis is 'broken' or not in an attempt to mislead
c Always check carefully what the axes are measuring and in particular whether the graph gives the actual increase, percentage increase or rate of increase
d Realise that interpolation and extrapolation can be very unreliable even if they seem reasonable in the context of the existing information

Since this section has been mainly explorative we do not include any further exercises.

Section 3.1 Solutions

1
(i) The bridge is located at 476 366.
(ii) Huckam Farm.
(iii) 493 376.

2
(i) A has co-ordinates (3, 2).
(ii) B has co-ordinates (2, 3).
(iii) C has co-ordinates (4, 0).
(iv) D has co-ordinates (0, 3).

3 The points are indicated on the diagram below.

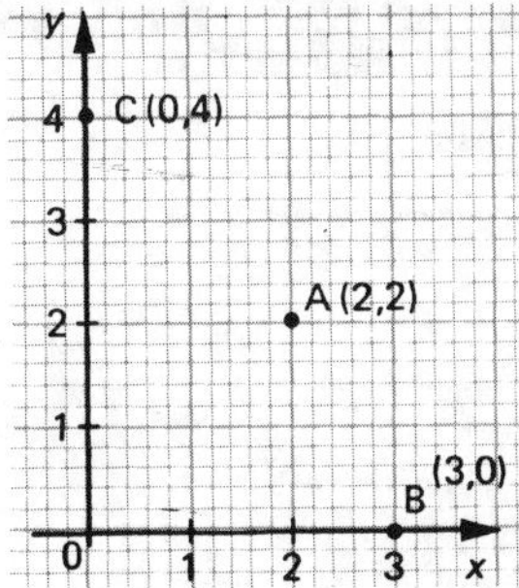

4
(i) A has co-ordinates (1, 2).
(ii) B has co-ordinates (−2, 1).
(iii) C has co-ordinates (−1, −2).
(iv) D has co-ordinates (2, −1).
(v) E has co-ordinates (−3, 0).

5 The points are indicated on the diagram below.

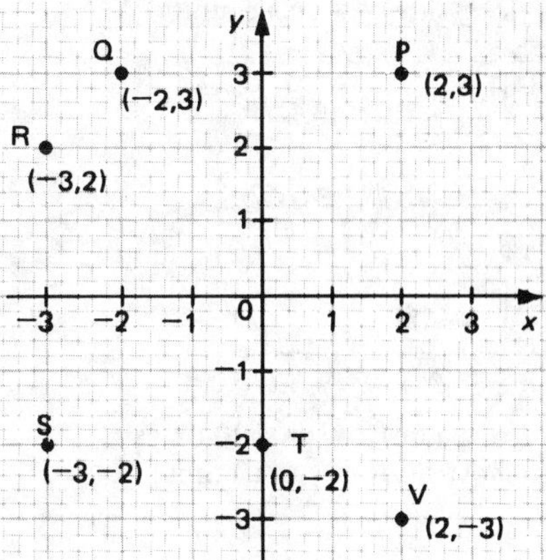

6 The points are indicated on the diagram below.

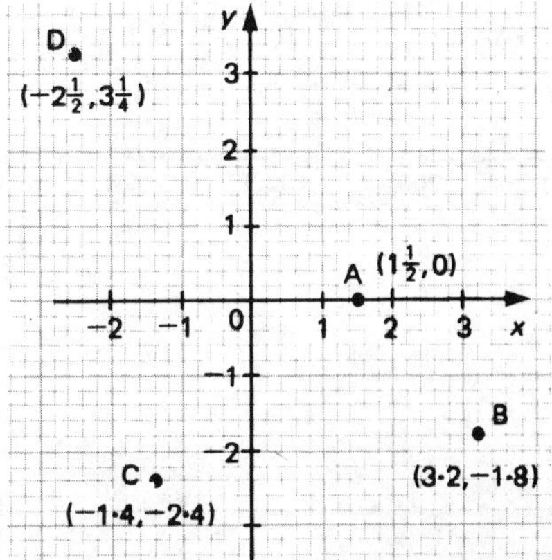

7

(i) x-axis: 1 cm: 5 units
y-axis: 1 cm: 10 units

(ii) x-axis: 1 cm: 1000 units
y-axis: 1 cm: 200 units

(iii) x-axis: 1 cm: 0·2 units
y-axis: 1 cm: 50 units

8

(i) P (10, 13). At 10 o'clock the temperature was 13° C.

(ii) P (1200, 1·5). The cost of 1200 gms of washing powder (1·2 kg) was £1·50.

(iii) P (22, −70). On the 22nd of the month there was £−70 in the bank (i.e. on the 22nd of the month the account was overdrawn by £70).

9 The scales you choose depend very much on the size of the paper used. The scales we indicate are for a sheet of A4 paper. We have chosen scales which are easy to measure. You may have selected scales which fill the paper better but which are harder to measure. In each case the diagram indicates where the axes intersect and the general shape of the points.

(i) x-axis (short side): 2 cm:1 unit
y-axis (long side): 4 cms:5 units

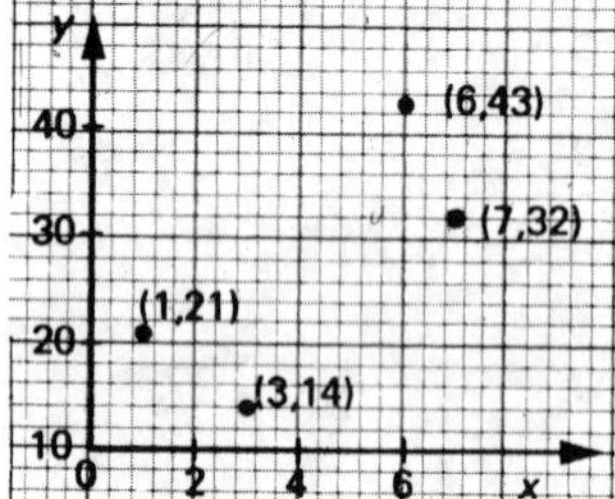

(ii) x-axis (long side): 4 cm:100 units
y-axis (short side): 4 cm:10 units

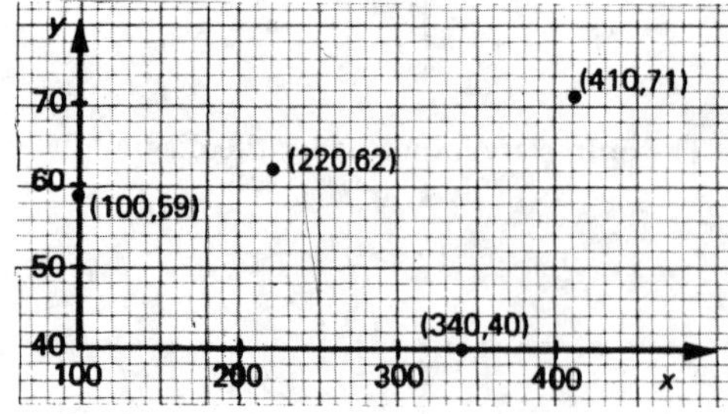

(iii) x-axis (short side): 2 cm:5 units
y-axis (long side): 2 cm:10 units

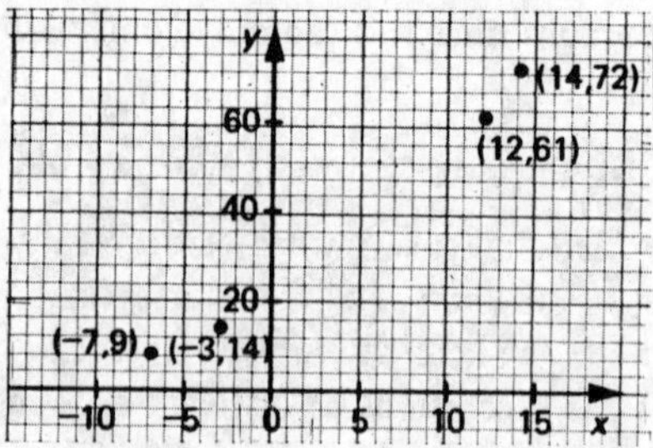

10

(i) P (2·2, 0·3); Q (−2·4, 0·7); R (−3·6, −0·9); S (3·6, −0·4)

(ii) The points are indicated on the diagram below.

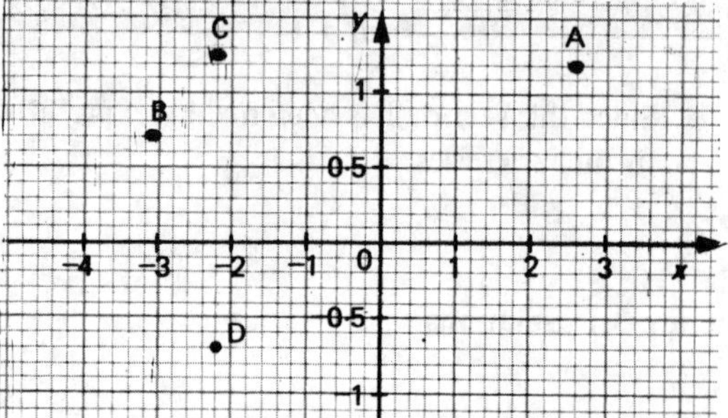

(iii) The scales you choose depend on the size of paper used. The diagrams below indicate the general shape of the points. You can check for yourself whether your diagram is sensible. Does it fill the space? Is it cramped? Is it clear?

(a)

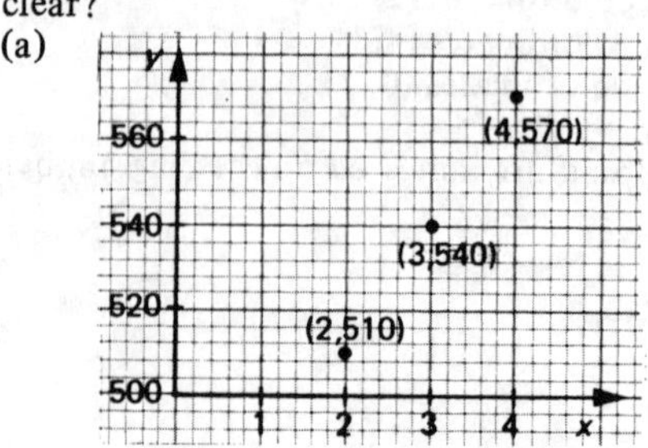

(b)

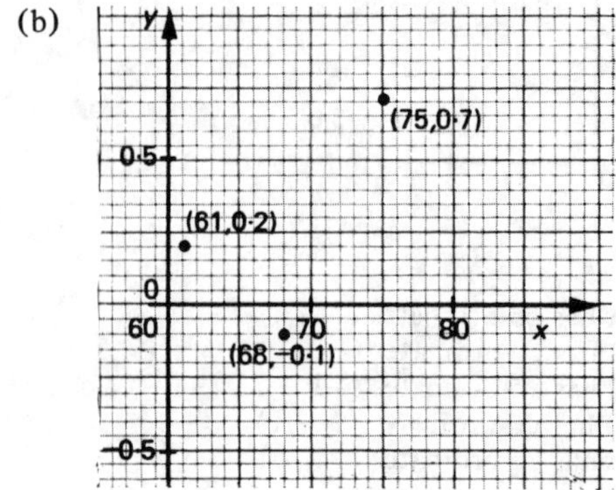

(c)

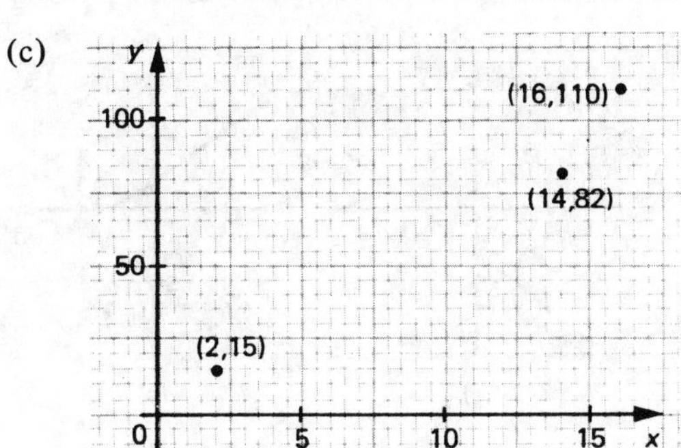

Section 3.2 Solutions

1

(i)

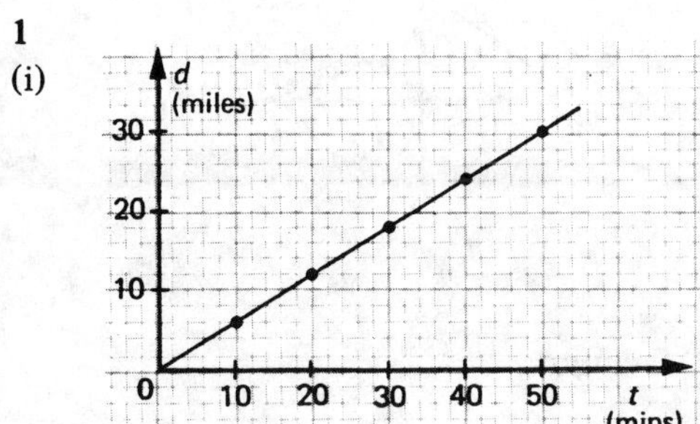

(ii)

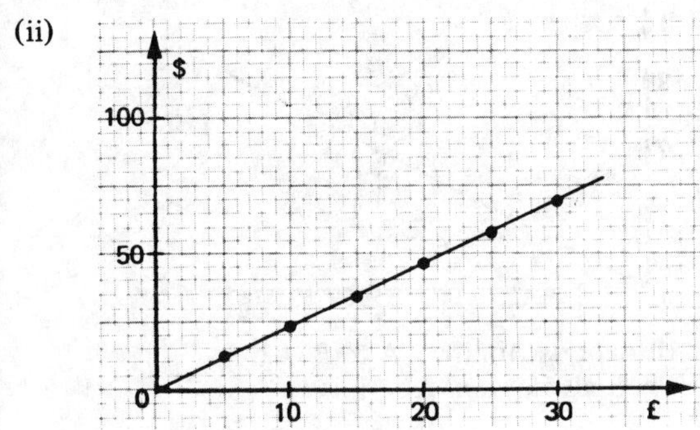

2

(i) (a) £12 ≏ \$28 (b) \$40 ≏ £17

(ii) (a) 4 lbs ≏ 1·8 kg (b) 3 kg ≏ 6·5 lbs

(iii) (a)

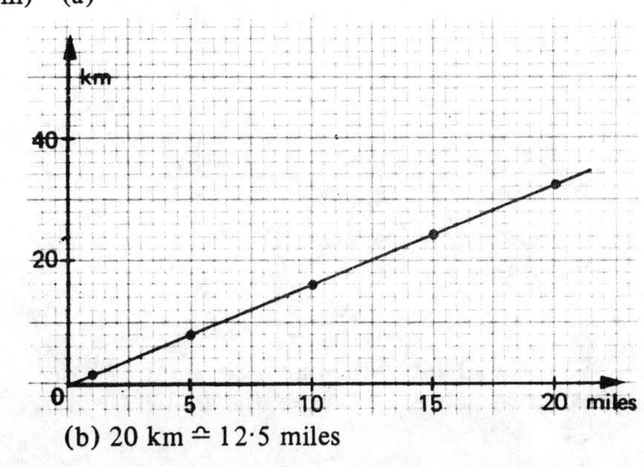

(b) 20 km ≏ 12·5 miles

3

(i) (a)

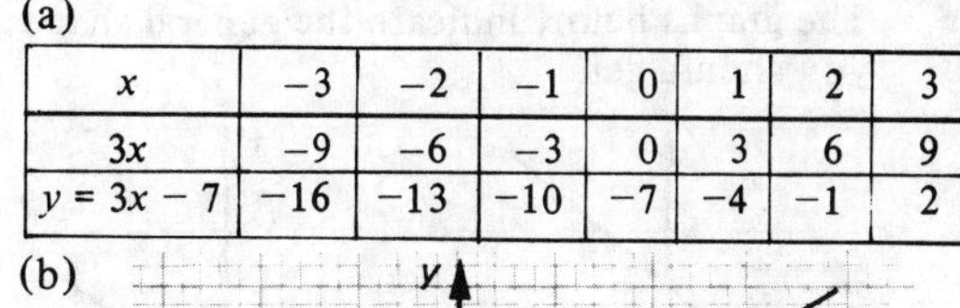

x	−3	−2	−1	0	1	2	3
$3x$	−9	−6	−3	0	3	6	9
$y = 3x - 7$	−16	−13	−10	−7	−4	−1	2

(b)

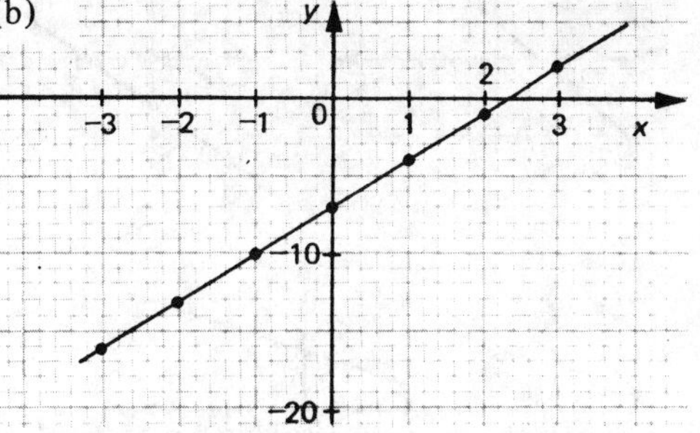

This diagram indicates where the axes cross. Check that your graph uses sensible scales.

(ii) (a)

s	−2	−1	0	1	2	3
$-2s$	4	2	0	−2	−4	−6
$t = -2s + 1$	5	3	1	−1	−3	−5

(b)

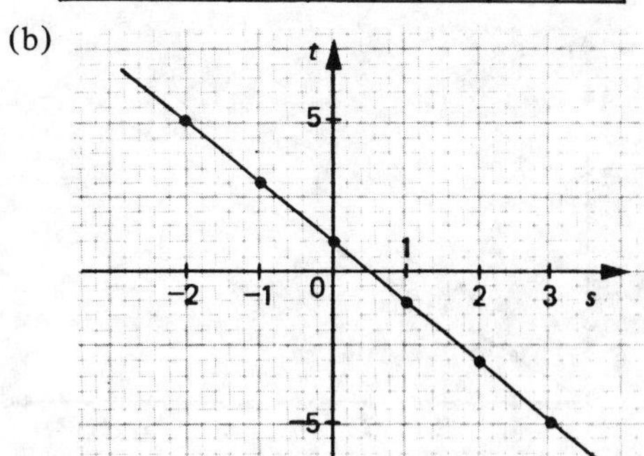

(iii) (a)

F	0	50	100	150	200	250
$F - 32$	−32	18	68	118	168	218
$C = \frac{5}{9}(F - 32)$	−17·8	10	37·8	65·5	93·3	121·1

(b)

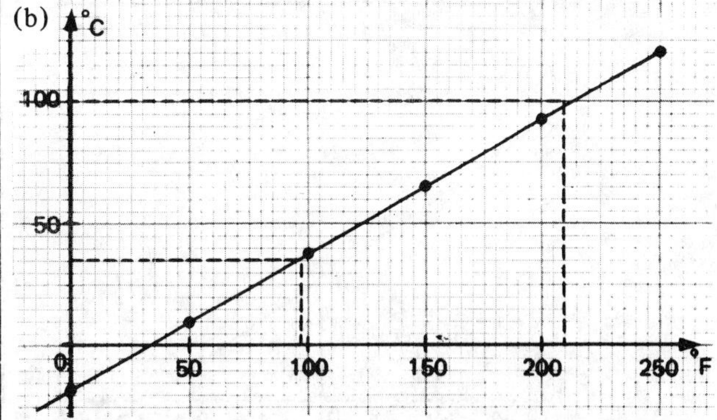

(c) 100° C ≏ 210° F. This checks, since 100° C is known to be 212° F.

(d) 98·4° F ≏ 35° C. This checks, since body temperature is 98·4° F or 36·9° C.

4 The graphs below indicate the general shapes you should get.

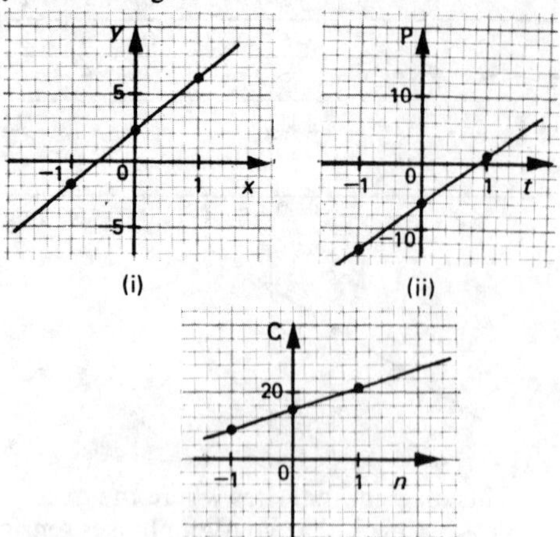

(i) (ii)

(iii)

(iii)

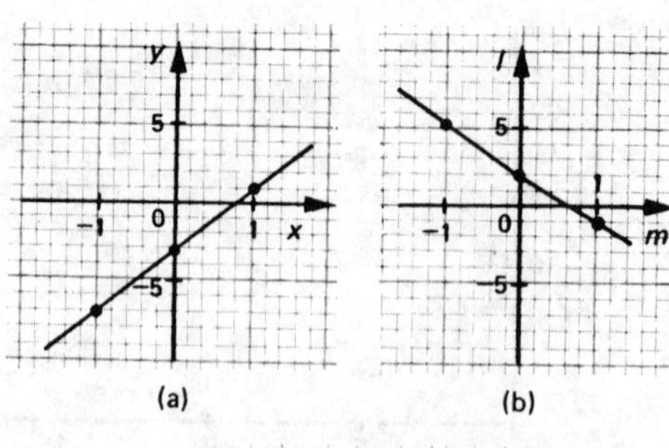

(a) (b)

(c)

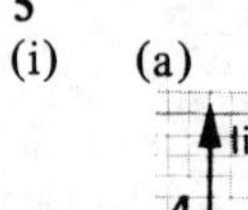

5

(i) (a)

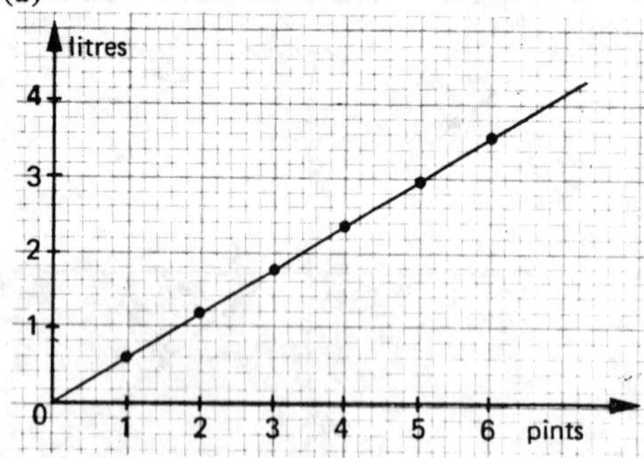

(b) 3·5 pints $\simeq$ 2·1 litres

(c) 4 litres $\simeq$ 6·7 pints

(ii) (a)

d	−3	−2	−1	0	1	2	3
$3-d$	6	5	4	3	2	1	0
$E=\frac{3-d}{3}$	2	1·67	1·33	1	0·67	0·33	0

(b)

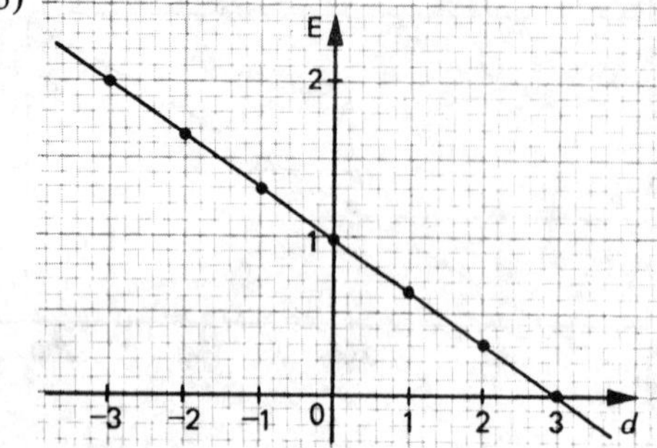

(c) When $d = 1{\cdot}5$, $E \simeq 0{\cdot}5$.

(d) When $E = 0{\cdot}1$, $d \simeq 2{\cdot}6$.

Section 3.3 Solutions

1

(i)

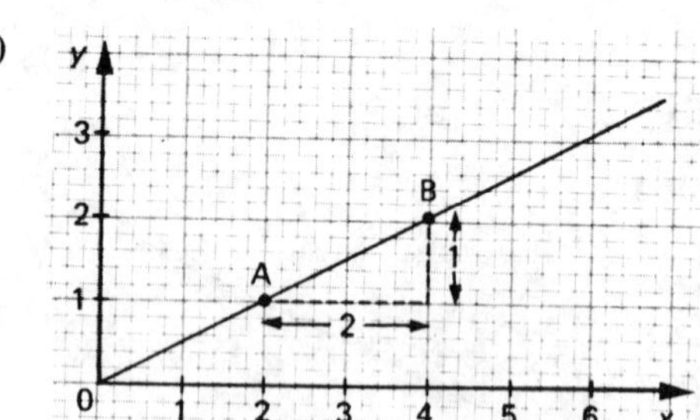

Horizontal run from A to B is 2.
Vertical rise from A to B is 1.

$$\text{Gradient} = \frac{\text{Rise}}{\text{Run}} = \frac{1}{2} = 0{\cdot}5$$

(ii)

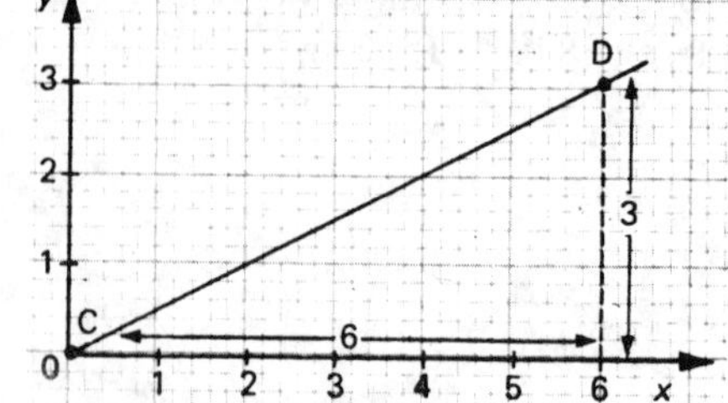

Horizontal run from C to D is 6.
Vertical rise from C to D is 3.

$$\text{Gradient} = \frac{\text{Rise}}{\text{Run}} = \frac{3}{6} = 0{\cdot}5.$$

So the gradient of the line is constant. Whatever two points we choose, the gradient will be 0·5.

2

(i) This exercise illustrates the importance of reading the scales carefully.
Horizontal run = 3; vertical rise = −2.

$$\text{Gradient} = \frac{\text{Rise}}{\text{Run}} = \frac{-2}{3} = -0{\cdot}67$$

(ii) Horizontal run = 10; vertical rise = 40.

$$\text{Gradient} = \frac{\text{Rise}}{\text{Run}} = \frac{40}{10} \text{ or } 4$$

(iii) Horizontal run = 8; vertical rise = 1.

$$\text{Gradient} = \frac{\text{Rise}}{\text{Run}} = \frac{1}{8} = 0{\cdot}13.$$

3

(i)

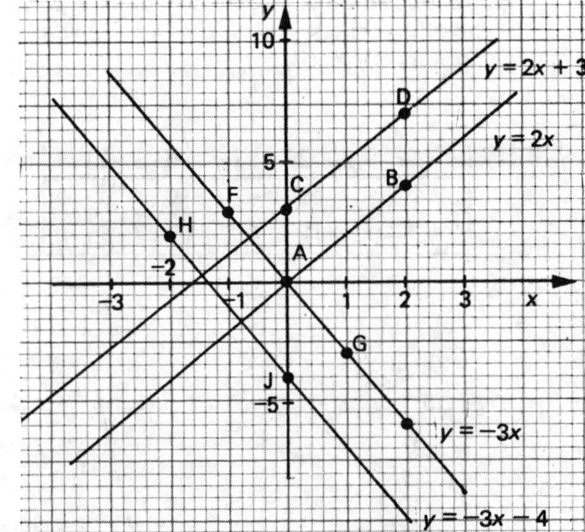

It is especially important to label all the graphs clearly to avoid confusion.

(ii) (a) Gradient $= \frac{4}{2} = 2$.
(Choosing points A and B.)
(b) Gradient $= \frac{4}{2} = 2$.
(Choosing points C and D.)
(c) Gradient $= \frac{-6}{2} = -3$.
(Choosing points F and G.)
(d) Gradient $= \frac{-6}{2} = -3$.
(Choosing points H and J.)

4

(i) Gradient is 3, intercept is −2.
(ii) Gradient is 0, intercept is 3 (since $y = 0x + 3$). In fact this graph is a horizontal line, passing through $y = 3$.
(iii) Gradient is −2, intercept is 1.
(iv) Gradient is −193, intercept is 721.

5

(i) Gradient $= \frac{2}{3} = 0{\cdot}67$, intercept is $y = 0{\cdot}6$. So the equation is $y = 0{\cdot}67x + 0{\cdot}6$ or $y = 0{\cdot}7x + 0{\cdot}6$ (rounding to one decimal place).

(ii) Gradient $= -\frac{20}{20} = -1$ (remember that you must read the distance from the scales—not the actual measurement), intercept is $N = 0$. So the equation is $N = -t$.

(iii) Gradient is $-\frac{2}{10} = -0{\cdot}2$, intercept is $C = 4$.
So the equation is $C = 4 - 0{\cdot}2n$.

6

(i) (a)

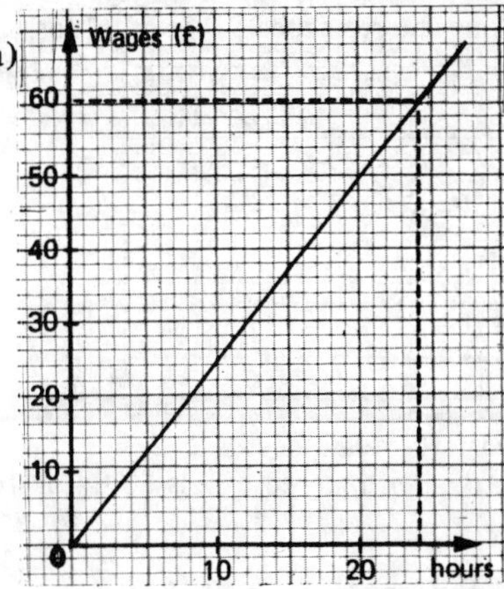

(b) From the graph, Sue earned £60 for a 24 hour week. (Notice that we must ensure that the horizontal axis runs at least as far as 24 hours.)

(ii) (a) Gradient of the graph is $\frac{1{\cdot}1}{2} = 0{\cdot}55$. The conversion rate from pints to litres is therefore 1 pint = 0·55 litres. Take care that you get the conversion the right way round.
(b) The conversion rate from litres to pints is the reciprocal of this (since if $y = kx$, $x = \frac{1}{k}y$). The rate is therefore 1 litre = 1·8 pints. Check this from the graph by finding the imperial equivalent of 1 litre.

7

(i)

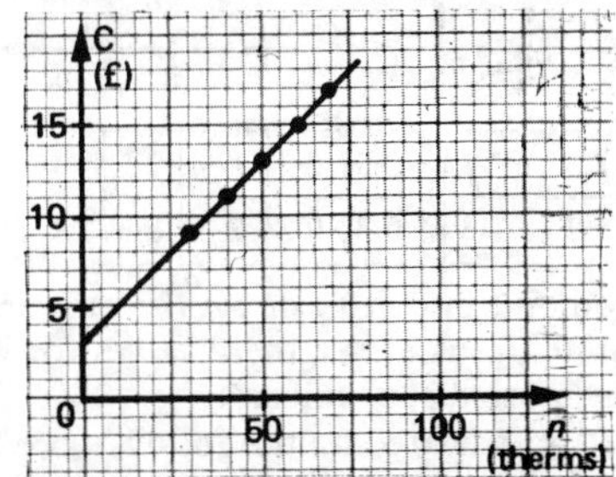

(ii) The gradient is 0·2 and the intercept is $C = £3$.
(iii) The intercept gives the quarterly charge = £3. The gradient gives the cost per therm = £0·2 or 20p per therm.

8

(i) (a) Gradient is 17, intercept is $t = 4$.
(b) Gradient is 0, intercept is $s = -3$.
(c) Gradient is 1, intercept is $t = 0$.
(d) Gradient is m, intercept is $y = c$.

(ii) (a) $y = 1 - x$ (b) $t = \frac{1}{2}s + \frac{3}{2}$
(c) $l = -\frac{1}{4}m - 2$

(iii) (a) From the graph 1 lb $\triangleq$ 0·45 kg.
(b) Taking the reciprocal, 1 kg = 2·2 lbs.

(iv)

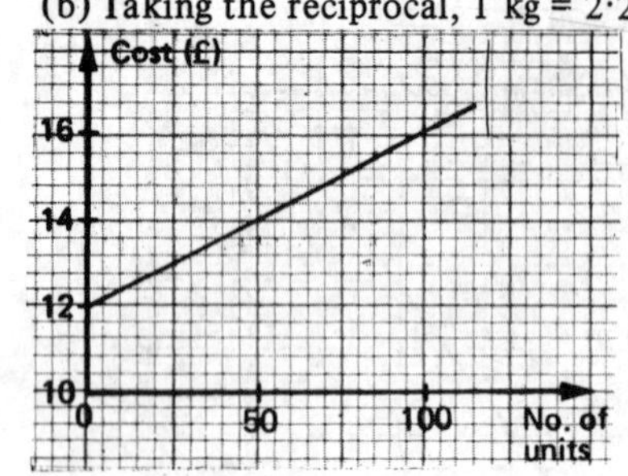

The quarterly rental cost is £12.
The charge per metered unit is £0·04 or 4p.

Section 3.4 Solutions

1

(i)

x	−3	−2	−1	0	1	2	3
$y = x^2$	9	4	1	0	1	4	9

Notice that the values for y are positive or zero.

(ii)

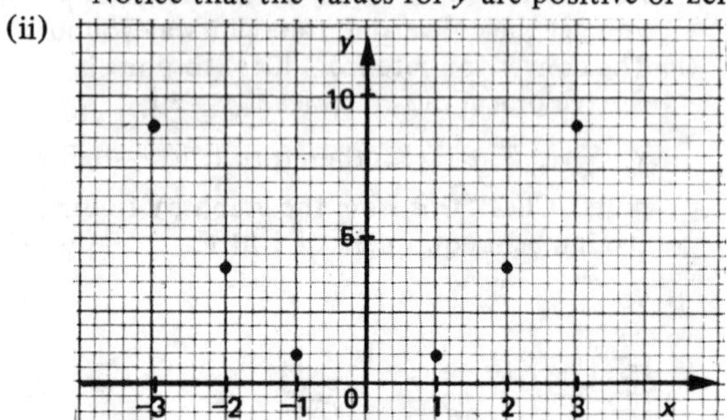

(iii)

x	−2·5	−1·5	−0·5	0·5	1·5	2·5
$y = x^2$	6·3	2·3	0·3	0·3	2·3	6·3

There is no point in writing down the numbers more accurately, since even these values for y can only be plotted approximately.
The points are indicated on the following diagram.

(iv)

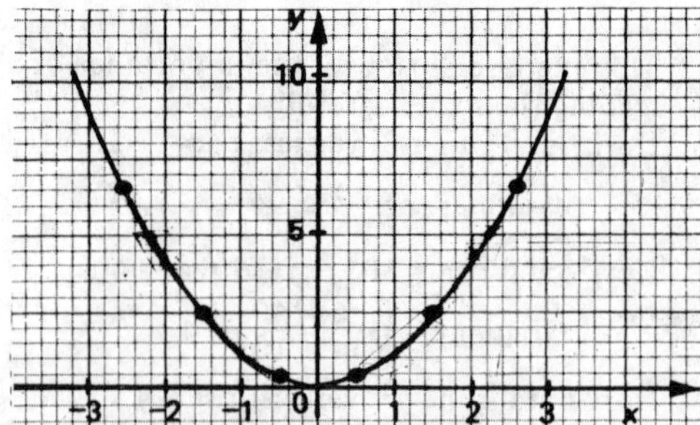

Notice that the graph is a smooth curve. There are no bumps or corners. Make sure that your graph is as smooth as possible. If you do find some odd corners—check that you have plotted the points correctly.

2

(i) (a)

x	−3	−2	−1	0	1	2	3
x^2	9	4	1	0	1	4	9
$y = 2x^2$	18	8	2	0	2	8	18

(b)

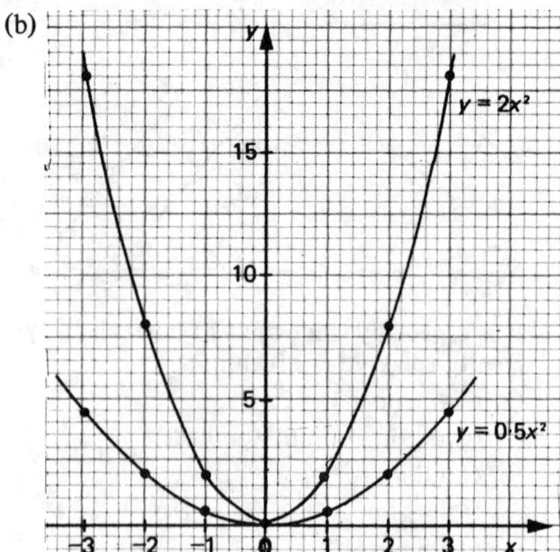

The steep parabola is labelled $y = 2x^2$. Again the curve is smooth. If you have difficulty in drawing a smooth curve try putting in some more points, using your calculator to evaluate $2x^2$ if necessary.

(ii) (a)

x	−3	−2	−1	0	1	2	3
x^2	9	4	1	0	1	4	9
$y = 0{\cdot}5x^2$	4·5	2	0·5	0	0·5	2	4·5

(b) The flatter parabola in the diagram above is labelled $y = 0{\cdot}5x^2$.

3

(i)

x	-3	-2	-1	0	1	2	3
$y = -x^2$	-9	-4	-1	0	-1	-4	-9

Notice that the values of y are negative or zero.

(ii)

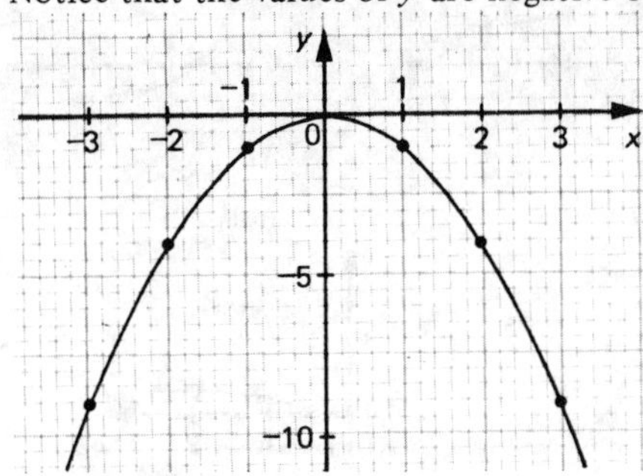

4

(i) We know that $A = kr^2$.
From the graph, when $r = 1$ cm, $A \triangleq 3$ cms.

Now if $A = kr^2$, $k = \frac{A}{r^2}$
So $k = \frac{3}{1} = 3$.
(We must indicate that this is an approximation for k since we can only read approximate values of A from the graph.)
When $r = 2$, $A \triangleq 12$.

Again $k = \frac{A}{r^2} \triangleq \frac{12}{4} \triangleq 3$.

In fact, k (or π), is approximately equal to 3·142.

(ii) (a)

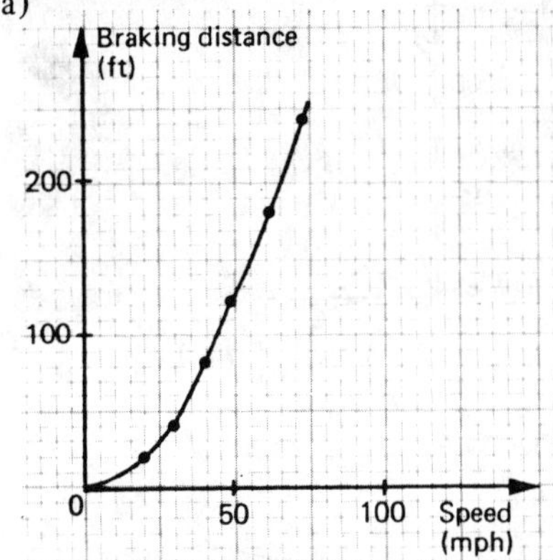

(b) From the graph the braking distance for a car travelling at 55 m.p.h. is about 140 ft.

(c) From the graph a braking distance of 60 ft corresponds to a speed of about 35 m.p.h. So a car travelling less than 35 m.p.h. will stop in less than 60 ft.

5

(i)

x	$\frac{1}{5}$	$\frac{1}{3}$	$\frac{1}{2}$	1	2	3	5
$\frac{1}{x}$	5	3	2	1	0·5	0·3	0·2

(ii) (a) $\frac{1}{10^6} = 10^{-6}$ or $\frac{1}{1,000,000}$ or 0·000001

(b) $\frac{1}{10^{-6}} = 10^6$ or 1,000,000

6

(i)

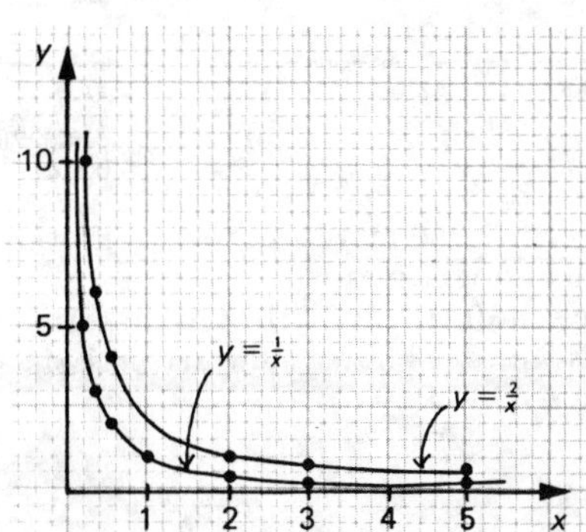

Once again the graph is a smooth curve with no bumps.

(ii) (a)

x	$\frac{1}{5}$	$\frac{1}{3}$	$\frac{1}{2}$	1	2	3	5
$y = \frac{2}{x}$	10	6	4	2	1	0·7	0·4

(b) The graph of $y = \frac{2}{x}$ is plotted on the diagram above. It also is a smooth curve. It has a similar shape to the graph of $y = 1/x$ but is steeper and lies just above $y = 1/x$.

7

(i) (a)

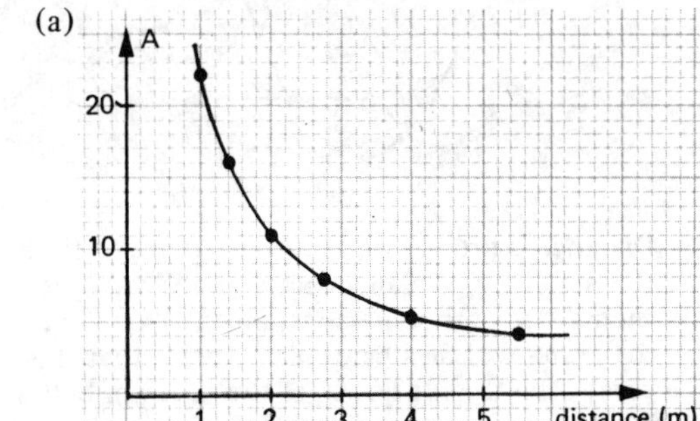

(b) From the graph, when $d = 4$ metres, the aperture setting should be about 5·5. (In fact the appropriate setting is 5·6 but we cannot read this value accurately from the graph.)

(ii) (a)

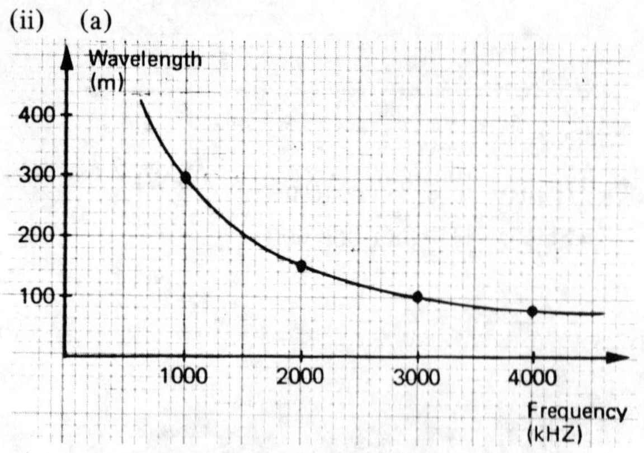

(b) 1500 kHz ≏ 200 m (This checks with the known radio settings.)

(c) 433 m ≏ 430 m ≏ 700 kHz (Again this checks with the known radio settings, since 433 m = 693 kHz.)

(iii)

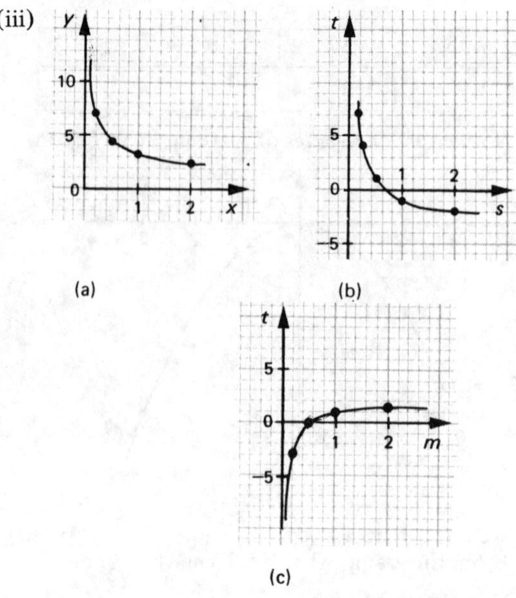

(iv) If t stands for time and n stands for the number of men employed, $t = k/n$ for some constant k. Therefore the graph will be a hyperbola.

8

(i)

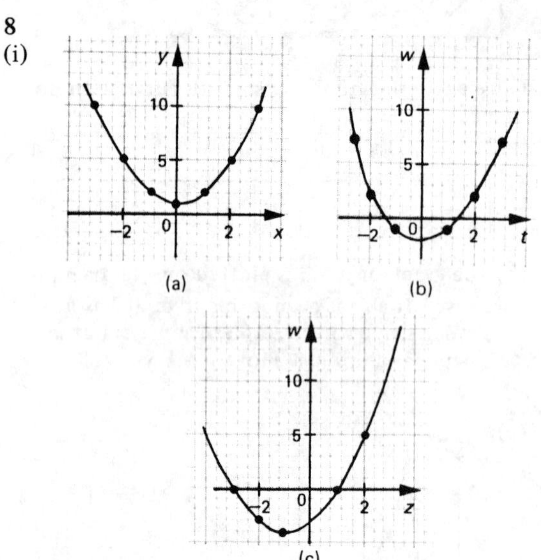

(ii) Distance is proportional to the square of time. If d stands for distance and t stands for time, then $d = kt^2$ for some constant k. Therefore, the graph will be a parabola. Notice that this parabola is upside down, as the diagram below illustrates.

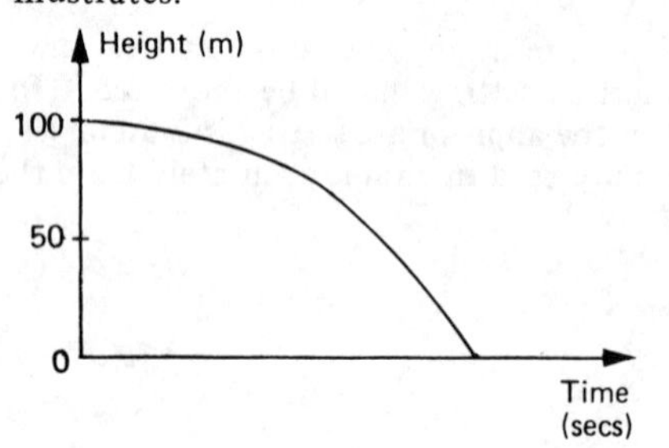

Section 3.5 Solutions

1(i)

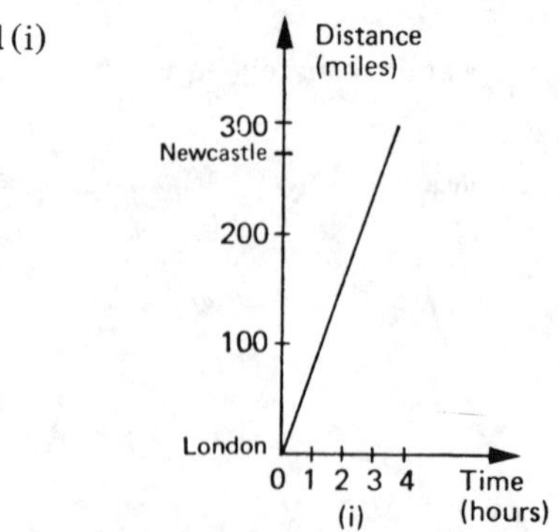

(i)

(ii)

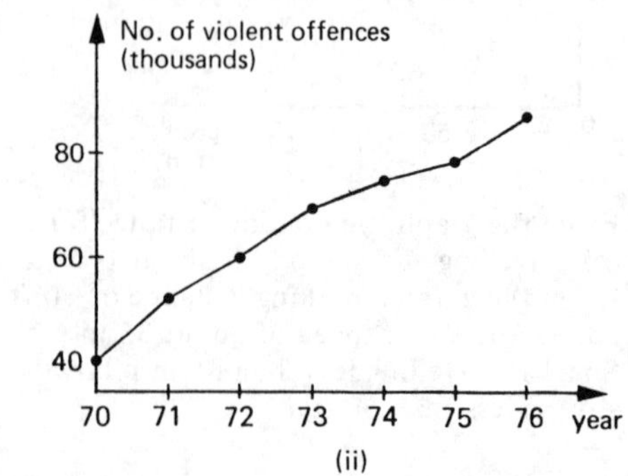

(ii)

In (ii) notice that the vertical axis is cut off at 30 000 to give the impression that the number of violent crimes in 1970 was very small.

2 In 1975 the % increase was about 50%. This suggests that about 750,000 were unemployed. In 1976 the % increase was about 130%. This suggests that about 1,150,000 were unemployed. In 1977 about 1,300,000 were unemployed (% increase of 160%).
This suggests the following graph:

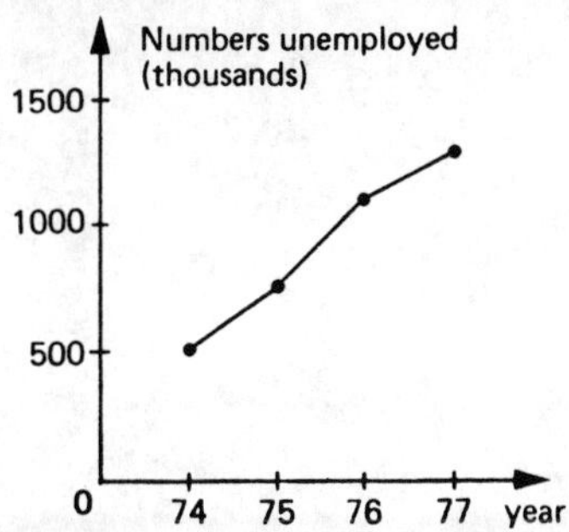

The graph has the same shape but by giving the actual numbers we can get a more realistic picture of what actually happened.

3 Now the graph gives the *rate* of increase; it gives the percentage increase in each year based upon the numbers unemployed the previous year. Thus, although the graph shows a fall, the actual number of unemployed continued to increase. If you compare this graph with the graph given above in the solution to Exercise 2 you will find that although the graphs *look* very different they give the same information.

4

(i) The graph suggests that the number of foreign visitors was falling and so we may predict that the numbers would continue to fall in subsequent years.

(ii) However, foreign travel has certainly increased and may affect predictions. It depends upon where this town is–if it is near to an area of great beauty or historical interest then the effect of increased popularity of British holidays to foreigners may be considerable. It would also depend upon the strength of the British and foreign currencies. There may be several other factors to be considered. This illustrates that the graph alone can give a restricted picture. We need to know a good deal more about the individual circumstances before we can really begin to make useful suggestions.

MODULE 4

4.1 Data in Tables

4.2 Representing Data in Pictures (1):
Graphs, Scatter Diagrams and Pie Charts

4.3 Representing Data in Pictures (2):
Bar Charts and Histograms

4.4 Averages

4.5 Tables of Figures

4.1 Data in Tables

TRY THESE QUESTIONS FIRST

1 The ages of 50 people are recorded in the margin opposite. Sort the data into a frequency table indicating the number of people aged 1-10, 11-20, 21-30 and so on.

80 15 31 22 61 40 71 49
42 23 21 69 17 33 64 19
52 39 50 58 30 86 63 47
14 62 51 29 8 4 15 70
42 74 27 72 38 57 18 39
37 41 26 73 55 5 19 28
20 35

2 The table below shows rent rebates and allowances for different categories of families.

1 The size of your family	2 Your weekly income (£)	3 Your rent (per week, after taking off rates, etc) £4.00 or less	£5.00	£6.00	£7.00	£8.00	£9.00	£10.00
		up to		Your rebate or allowance (£)				
Couple (or single parent) and one dependent child	35	4.00	5.00	6.00	6.95	7.55	8.15	8.75
	40	3.90	4.50	5.10	5.70	6.30	6.90	7.50
	45	2.65	3.25	3.85	4.45	5.05	5.65	6.25
	50	1.72	2.32	2.92	3.52	4.12	4.72	5.32
	60	–	0.62	1.22	1.82	2.42	3.02	3.62
	70	–	–	–	–	0.72	1.32	1.92
Couple (or single parent) and two dependent children	40	4.00	5.00	6.00	7.00	7.94	8.54	9.14
	45	4.00	4.89	5.49	6.09	6.69	7.29	7.89
	50	3.04	3.64	4.24	4.84	5.44	6.04	6.64
	55	1.98	2.58	3.18	3.78	4.38	4.98	5.58
	65	0.28	0.88	1.48	2.08	2.68	3.28	3.88
	75	–	–	–	0.38	0.98	1.58	2.18

How much allowance could a family with two children and a weekly income of £55 a week claim, if the rent was £9 a week?

3 The table below gives the car rental charges for one particular firm.

	Mileage rate			Unlimited mileage	
Vehicle	Daily	Weekly	Mileage rate	Daily	Weekly
Fiesta	£ 7	£42	6p	£10	£60
Golf	£ 8	£48	6p	£11	£66
Cortina 1·6 litre	£10	£60	6p	£13	£78
Cortina 2·0 litre	£12	£72	6p	£15	£90

Use the table to work out the cheapest cost of renting a 1·6 litre Cortina for two weeks if you intend to travel 500 miles.

4 The table below indicates weekly earnings in 1977. It is based on a sample of 7190 men.

'Gross weekly earnings' is the amount earned before tax is deducted.

Average Gross Weekly Earnings (£)								Great Britain 1977
	less than 30	30-39	40-49	50-59	60-69	70-99	more than 100	total = 100%
Has been unemployed %	9	15	21	21	16	13	6	597
Not unemployed %	2	5	13	19	18	20	13	6593
Total %	3	6	13	19	18	28	12	7190

(a) What is the actual number of people earning between £60 and £69 a week who were not unemployed?
(b) Explain why the total for the 'has been unemployed' row does not equal 100%.

4.1(i) COLLECTING DATA

After an experiment or survey you will usually end up with a lot of information from which to draw conclusions. Such information is called *data*. Data can also be extracted from newspapers or books.

***Data** is the plural of **datum** which comes from the Latin **dare** meaning 'given'.*

The most convenient way to present data is in a table. Sometimes information can be recorded in a table as it is collected. For example, many motorists keep a log book to record petrol consumption, in which the information is recorded as soon as it is collected. Experiments often require regular measurements. Again, information can be recorded as it is collected. For example, the table below resulted from an experiment to determine how quickly a cup of tea cooled down.

Date	Gallons	Mileage	Comments
4.1.81	9·4	22695	(full)
10.1.81	9·1	22916	
20.1.81	8·6	23212	(oil)
24.1.81	9·2	23320	
27.1.81	8·5	23595	(full)

Time (mins)	0	5	10	15	20	25	30	35	40	45	50
Temp (°C)	90	65	60	50	36	35	30	26	25	22	20

Tables can be read horizontally or vertically. The column or row headings should indicate what is being measured and the units of measurement.

Date	Gallons	Mileage	Comments
4.1.81	9·4	22695	(full)
10.1.81	9·1	22916	
20.1.81	8·6	23212	(oil)
24.1.81	9·2	23320	
27.1.81	8·5	23595	(full)

Time (mins)
Temp (°C)

Petrol is measured in gallons. Time is measured in minutes. Temperature is measured in degrees Centigrade.

Some conclusions can be drawn directly from the table.

CHECK YOUR ANSWERS

1 *Section 4.1(i)*

Age group	Frequency
1-10	3
11-20	8
21-30	8
31-40	8
41-50	6
51-60	5
61-70	6
71-80	5
over 80	1
Total	50

2 £4·98 a week.

3 At the unlimited mileage rate the cost would be £156. Alternatively the weekly cost is £60 per week plus 6p a mile. The total cost would then be £150. The answer is therefore £150. *Section 4.1(ii)*

4 (a) The table indicates that 18% of 6593 earned between £60 and £69 per week. The answer is therefore 1187 (rounded to the nearest whole number). *Section 4.1(iii)*

(b) The 'has been unemployed' row totals 101%. This is probably because the percentages have been rounded off to whole numbers.

EXAMPLE

Jane travelled to work by car. The table opposite indicates her journey times over a two week period.

(i) What was the shortest journey time?

(ii) What was the longest journey time?

	Mon	Tue	Wed	Thu	Fri
Time (mins)	14	16	20	13	15
	15	17	14	10	17

SOLUTION

(i) Reading across the table, the lowest number is 10. So the shortest journey time was 10 minutes.

This occurred on the second Thursday.

(ii) Similarly the longest journey time was 20 minutes.

This was on the first Wednesday.

You may want to suggest reasons why Jane's journey times vary as they do. She may have got caught in a traffic jam on the first Wednesday, or travelled by a different route. Similarly the shortest journey time could have resulted from travelling by a different route or may have been at a different time of day. Notice that we only *suggest* reasons. Without any other information we cannot say anything definite.

*You can state factual conclusions, but only **suggest** reasons. Interpretation of the information often depends on your own experience or some other information not included in the table.*

TRY SOME YOURSELF

1(i) The table below indicates the cooling rate of tea.

Time (mins)	0	5	10	15	20	25	30	35	40	45	50
Temp (°C)	90	65	60	50	36	35	30	26	25	22	20

(a) How long does it take for the tea to cool to 50°C?
(b) By how much does the temperature drop in the first twenty minutes? By how much in the second twenty minutes?
(c) What can you suggest about the pattern of the cooling?

(ii) This table indicates the amount of money Tony and Sue spent each week on groceries.
(a) What is the maximum amount spent?
(b) What is the minimum amount spent?
(c) Suggest reasons why such large or small amounts could arise.

Week ending	Cost (£)
11.2.81	14·90
18.2.81	10·27
25.2.81	3·22
2.3.81	12·81
9.3.81	9·12
16.3.81	19·14
23.3.81	11·17

It is sometimes convenient to sort data into different categories as they are collected. For example, a survey was set up to investigate the types of traffic using a small road on a housing estate. The survey recorded the number of cars, buses, lorries etc. passing a specific point in a half-hour interval.

This is particularly the case if you want to record the number of times something happened.

The resulting information is displayed in a *frequency table.*

Vehicle	Frequency
Car	27
Bus	10
Lorry	8
Bicycle	12
Others	3
Total	60

The frequency indicates the number of cars, buses etc. passing in half an hour.

The total number of vehicles passing in half an hour was 60.

The total represents the ***sample size.*** *Here, 60 vehicles passed the point in half an hour; the sample size was 60.*

When drawing up a frequency table you might find that a *tally count* helps you to record the frequencies quickly. The next example indicates how a tally count is used to sort data into groups of ten.

Vehicle	Tally	Frequency
Car	𝍸 𝍸 𝍸 𝍸 𝍸 II	27
Bus	𝍸 𝍸	10

EXAMPLE

The geography marks (out of 100) of a class of 40 children are given in the margin. Sort the data into a table indicating the number of children who scored 1-10, 11-20, 21-30 etc.

63 29 64 54 59 98 74 34
71 90 73 69 19 21 54 87
61 62 14 75 8 23 21 29
84 38 47 81 64 30 53 40
47 51 59 39 71 36 53 49

SOLUTION

The table below sets out the different categories. It has three columns, one to record the different categories, one for the tally count and one for the frequencies.

Score	Tally	Frequency
1-10	I	1
11-20	II	2
21-30	~~IIII~~ I	6
31-40	~~IIII~~	5
41-50	III	3
51-60	~~IIII~~ II	7
61-70	~~IIII~~ I	6
71-80	~~IIII~~	5
81-90	IIII	4
91-100	I	1
	Total	40

The tally count allows us to work methodically through the data, recording each mark in its category as it is read. The frequencies are then written down straight from the tally column. Notice that 30 belongs in the 21-30 category.

The total is a useful check. We were told, originally, that there were 40 children. We can quickly see whether a mark has been missed.

TRY SOME YOURSELF

2(i) The marks scored in a mathematics examination by 40 children are given in the margin. Sort the data into a table giving the number who scored 1-10, 11-20, 21-30 and so on.

43 25 68 72 81 54 61 27
62 63 14 79 87 61 72 3
49 21 11 85 56 47 41 19
71 35 3 91 59 43 81 28
84 24 81 99 54 31 30 80

(ii) The information below indicates the various sizes of ladies shoes sold on 17th January 1981 in Bloggs's shoe shop.

4 4½ 5 6 6½ 7 4½ 5 5 6 5½
7½ 4½ 5 5½ 6 6½ 7 4½ 5 5 7
4½ 5½ 6 6½ 7 4 5 5 7 6 5
6½ 6 5½ 5 5 5 5½

Use a tally count. Check the total. Does it match?

(a) Sort the data into a frequency table.
(b) Which size(s) of shoes accounted for the largest number of sales?

4.1(ii) READING AND INTERPRETING MORE COMPLICATED TABLES

So far we have looked at the *construction* of tables to summarise data. You will often find that information is already presented in tables and that you are required to read values from the table and interpret that information in some given context. In many cases this is straightforward but some tables are more complicated and need more thorough examination in order to interpret the data. Simple tables involve only two rows or two columns; more complex tables can involve many rows and columns, which may also include subheadings. The most difficult part lies in identifying the row (or rows) and column (or columns) which give the information you want. To give you practice at reading and interpreting data from tables like this we are going to look at some common examples.

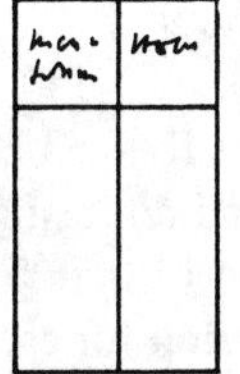
2 columns

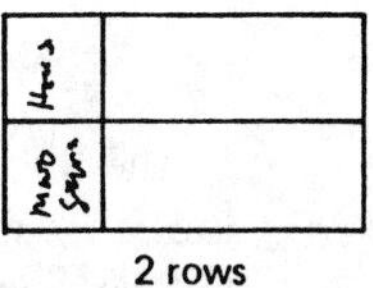
2 rows

The table below is taken from a holiday brochure.

Inclusive prices per person in £'s (excluding airport taxes—see page 193)											
Hotel		Pueblo Full board									
Departure airport		GATWICK		LUTON		BIRMINGHAM		MANCHESTER		GLASGOW	
No. of nights in hotel		7	14	7	14	7	14	7	14	7	14
Aircraft		B707	B707	B720	B720	B737	B737	B727	B727	B727	B727
Departure day/approx. time		1530	1530	0810	0810	0835	0835	0840	0840	1635	1635
Arrive back day/approx. time		2120	2120	1355	1355	1455	1455	1435	1435	2320	2320
First departure		26 Apr	26 Apr	26 Apr	26 Apr	26 Apr	26 Apr	26 Apr	26 Apr	17 May	17 May
Last departure		18 Oct	11 Oct	18 Oct	11 Oct	18 Oct	11 Oct	18 Oct	11 Oct	27 Sep	20 Sep
Holiday number		T2255									
Departure between	26 Apr	98	133	101	136	103	138	108	143	–	–
	28 Apr-5 May	124	158	127	161	129	163	134	168	–	–
	6-18 May	120	154	123	157	125	159	130	164	145	179
	19-26 May	130	174	133	177	135	179	140	184	155	199
	27 May-15 June	137	182	140	185	142	187	147	192	162	207
	16 June-13 July	147	199	150	202	152	204	157	209	172	224
	14 July-5 Aug	153	207	156	210	158	212	163	217	178	232
	6-31 Aug	157	211	160	214	162	216	167	221	182	236
	1-14 Sep	151	204	154	207	156	209	161	214	176	229
	15-28 Sep	145	187	148	190	150	192	155	197	170	212
	29 Sep-18 Oct	115	149	118	152	120	154	125	159	–	–
Children's Reduction 2-11 yrs inc (see page 6)	23 Apr-15 June & from 1 Sept	FREE 50% when allocation of free holidays full									
	16 June-31 Aug	20%									
Transfer airport/hotel approx.		1hr5mins.									
Flying time—Gatwick/Alicante 2hrs10mins; Luton/Alicante 2hrs15mins; Birmingham/Alicante 2hrs20mins; Manchester/Alicante 2hrs. Important—please read the general information pages of this brochure.											

This table contains several rows and columns. You might find it useful to use a ruler to read off values from the same column or row.

Here, the columns and the rows are divided into subheadings. For example,

Inclusive prices per person in £'s (excluding airport taxes—see page 193)										
Hotel	Pueblo Full board									
Departure airport	GATWICK		LUTON		BIRMINGHAM		MANCHESTER		GLASGOW	
No. of nights in hotel	7	14	7	14	7	14	7	14	7	14

EXAMPLE

The Robertsons have two children, Tim aged 8 and Helen aged 11. They are leaving from Luton for the Hotel Pueblo on 6 August for two weeks. What information does the table above give about the holiday?

SOLUTION

First identify the relevant column and row.

The Robertsons are leaving from Luton and staying for two weeks or fourteen nights.

They are leaving on the 6th of August. This indicates which row to look at.

Hotel		
Departure airport		LUTON
No. of nights in hotel		14
Aircraft		B720
Departure day/approx. time		0810
Arrive back day/approx. time		1355
First departure		26 Apr
Last departure		11 Oct
Holiday number		T2255
Departure between	26 Apr	136
	28 Apr-5 May	161
	6-18 May	157
	19-26 May	177
	27 May-15 June	185
	16 June-13 July	202
	14 July-5 Aug	210
	6-31 Aug	214
	1-14 Sep	207
	15-28 Sep	190
	29 Sep-18 Oct	152

So the Robertsons fly on a B720 leaving at approximately 08.10 on the 6th August and returning at about 13.55 two weeks later. The holiday number is T2255. The holiday costs £214 for each adult with a 20% reduction for each child, that's a total of £770·40. The flight time from Luton to Alicante is 2 hours 15 minutes and it takes about 1 hour 5 minutes to transfer from the airport to the hotel. (That's quite a lot of information from one table!)

Since both children are in the '2-11 inclusive' category, they qualify for 20% reductions from mid June till the end of August. The reader is also directed to pages containing general information.

TRY SOME YOURSELF

3(i) (a) Paul and Linda Potter wish to take their 12 year old daughter to the Hotel Pueblo for a week's holiday, leaving from Birmingham on the 6th October. What information is given in the table in the example above about such a holiday?

(b) Which was the most expensive holiday? Which was the cheapest?

First identify the ***column*** *then the row.*

(ii) The table below is part of an insurance table for private vehicles.

		COMPREHENSIVE								THIRD PARTY FIRE & THEFT		THIRD PARTY ONLY	
Vehicle age		**4 Years**		**3 Years**		**2 Years**		**0/1 Year**					
Drivers		Any	Insd/ Spouse	Any	Insd/ Spouse	Any	Insd/ Spouse	Any	Insd/ Spouse	Any	Insd/ Spouse	Any	Insd/ Spouse
Driver age	NCB years	(£)	(£)	(£)	(£)	(£)	(£)	(£)	(£)	(£)	(£)	(£)	(£)
31-35	0	150	129	157	134	163	139	169	145	46	40	42	37
	1	114	98	119	102	124	106	129	110	37	33	33	29
	2	100	86	104	90	108	93	112	96	33	30	29	26
	3	79	68	82	70	85	73	88	76	28	25	24	22
	4	64	56	67	58	69	60	72	62	24	22	21	19
27-30	0	171	146	178	152	185	158	192	165	61	53	57	49
	1	130	111	135	116	141	121	146	125	48	42	44	39
	2	113	97	118	101	123	105	127	109	43	38	39	34
	3	89	77	92	80	96	83	100	86	35	32	32	28
	4	72	63	75	65	78	67	81	70	30	27	27	24
23-26	0		198		207		215		223		73		69
	1		150		157		163		169		57		54
	2		131		137		142		147		51		47
	3		103		107		111		115		42		38
	4		83		87		90		93		35		32

(a) How much would it cost a 28 year old driver for fully comprehensive insurance for himself and his wife to drive a 2 year old car if he has 2 years no claim bonus (NCB)?

Identify the row and column.

(b) For a driver with a car 3 years old, with 1 year's no claim bonus, what is the most he could pay for fully comprehensive insurance for himself only, and what is the least?

It is not always possible to get the information you want *directly* from the table. Sometimes you have to make some calculations first. This is often the case when the table indicates various *rates of payment.* The table below is taken from an insurance table for householders.

STANDARD RATES per £100 (minimum policy premium £10·00)	London Postal	London Met.	Glasgow	Liverpool/ Manchester	Elsewhere
• BUILDINGS	20p	20p	15p	15p	15p
• CONTENTS (minimum sum insured £2000)	50p	40p	50p	40p	30p
• ALL RISKS (minimum premium £2·00)					
up to £1000	£2·00	£1·50	£2·00	£1·50	£1·00
£1001 to £3000	£2·50	£2·00	£2·50	£2·00	£1·00
£3001 to £5000	£3·00	£3·00	£3·00	£3·00	£1·50

The ***rate*** *indicates the amount per £100.*

EXAMPLE

How much would it cost for Sarah, who lives in Newcastle, to insure her house (worth £15,000) and contents (worth £5000)? She also wants to insure a camera (worth £150) for all risks. What is the total cost?

SOLUTION

	Elsewhere
• BUILDINGS	15p
• CONTENTS	30p
(minimum sum insured £2000)	
• ALL RISKS (minimum premium £2·00)	
up to £1000	£1·00
£1001 to £3000	£1·00
£3001 to £5000	£1·50

First find the appropriate row and column. Sarah lives in Newcastle–so the column is 'Elsewhere'.

Now start by examining the row marked 'Buildings'. The *rate* is 15p per £100, or £0·15 per £100. So Sarah has to pay £0·15 × $\frac{15\,000}{100}$ = £22·50.

The 'Contents' row indicates that the *rate* is 30p per £100 or £0·30 per £100. So Sarah has to pay £0·30 × $\frac{5\,000}{100}$ = £15.

The total so far is £22·50 + £15 = £37·50.

It's a good idea to change everything to pounds.

Finally, the camera is worth £150. The 'All risks' rate is £1 per £100. So the table indicates that the cost should be £1 × $\frac{150}{100}$ = £1·50. But the minimum premium is £2. So the actual cost would be £2.

Thus the total amount due is £37·50 + £2 = £39·50.

TRY SOME YOURSELF

4(i) (a) Use the same table to find how much it costs to insure a house worth £41,000 and contents worth £7000 if the owner lives in Glasgow?
(b) Which is the most expensive area(s) to insure a house in?
(c) Which is the cheapest area(s) to insure house contents?

(ii) The table below gives the car rental charges for one particular firm.

Vehicle	Mileage rate			Unlimited mileage	
	Daily	Weekly	Mileage rate	Daily	Weekly
Fiesta	£ 7	£42	6p	£10	£60
Golf VW	£ 8	£48	6p	£11	£66
Marina Estate (1·3 litre)	£ 9	£54	6p	£12	£72
Cortina (1·6 litre)	£10	£60	6p	£13	£78
Cortina (2·0 litre)	£12	£72	6p	£15	£90

How much would it cost to rent a Marina Estate for 4 days travelling 625 miles at
(a) The mileage rate
(b) The unlimited mileage rate?

4.1(iii) PERCENTAGE TABLES

Tables often give information in percentages. The table below indicates household sizes in the 1970s.

Household size: 1971 to 1977					
Households					Great Britain
No. of persons in household (all ages)	1971	1973	1975	1976	1977
	%	%	%	%	%
1	17	19	20	21	21
2	31	32	32	32	33
3	19	18	18	17	17
4	18	18	17	17	18
5	8	8	8	8	7
6 or more	6	5	5	5	4
Base (h'hlds) 100%	**11988**	**11651**	**12097**	**12120**	**11979**

For example, in 1973 19% of households comprised only 1 person, 32% consisted of 2 people, 18% consisted of 3 people and so on. The figures for 1973 were calculated from a ***base*** *of 11,651 households.*

In 1973 the actual number of households consisting of 2 people is given by calculating 32% of 11,651, which is 3728·32 or 3728.

We can only really count whole numbers of households so 3728·32 must be rounded to the nearest whole number.

Since each column should include all possible households in the sample the total of all percentages in the column should be 100% and indeed, for 1973,

$$19 + 32 + 18 + 18 + 8 + 5 = 100$$

However, the column total is not always exactly equal to 100%. All the percentages have been rounded to whole numbers and this can sometimes introduce rounding errors. For example, the total of 1971 column is

$$17 + 31 + 19 + 18 + 8 + 6 = 99$$

Rounding errors are usually very small and the total is always very close to 100%.

TRY SOME YOURSELF

5 From the table given in the example above,
(i) What was the 100% base for the 1977 survey?
(ii) What was the actual number of households consisting of four people in the 1977 survey?
(iii) What is the percentage total of the 1977 column?

Sometimes the total percentages for both rows and columns are indicated. The table below indicates the percentages of families with different numbers of dependent children.

		Families with Dependent Children Number of Dependent Children				
		1	2	3	4 or more	TOTAL
Type of Family		%	%	%	%	%
Married Couple	%	32.8	34.9	15.4	8.2	91.3
Lone Mother	%	3.9	2.1	1.1	0.6	7.7
Lone Father	%	0.8	0.2	0	0.1	1.1
TOTAL	%	**37.5**	**37.2**	**16.5**	**8.9**	**100**

Base = 100% = 4855

The percentages are totalled across each row.

The percentages are totalled down each column.

In examples like this the row totals *and* the column totals should add up to 100%, although rounding errors might mean that the total is not exactly equal to 100%. In this case the total for both row and column is in fact 100·1%. The total is always the same for both rows and columns.

					Total %
					91·3
					7·7
					1·1
Total %	37·5	37·2	16·5	8·9	100

Base (=100%) = 4855

The survey was based upon a sample of 4855 households. Thus 100% = 4855.

TRY SOME YOURSELF

6(i) Use the table above to calculate the actual number of families consisting of a married couple with 2 dependent children.

(ii) The table below illustrates male cigarette smoking habits in various age groups.

Age		Males			
		No. of cigarettes smoked per day			
		none	under 20	over 20	total %
16-24	%	10	4	2	16
25-34	%	9	4	5	18
35-49	%	13	5	7	25
50-59	%	10	3	5	18
over 60	%	14	5	3	22
total	%	56	21	22	100

Base = 10,480

(a) What percentage of males in the survey smoked over 20 cigarettes a day?

(b) What was the actual number of male non-smokers aged between 16 and 24?

(c) How many men aged over 60 were interviewed in the survey?

(d) Explain why the percentage totals do not actually add up to 100%.

After you have worked through this section you should be able to

a Record data systematically in a suitable table

b Construct a frequency table, using a tally count if necessary

c Read and interpret information from tables, recognising that any such interpretation can only be a suggestion, particularly if no other information is available

d Read and interpret more complicated tables which involve a number of rows or columns and which may involve some calculation

e Read and interpret tables which give information in percentages

Finally, here are some exercises if you want more practice.

TRY SOME MORE YOURSELF

7(i) The daily midday temperatures (in degrees Centigrade) for June one year are given in the margin. Sort the data into a frequency table.

17 18 17 16 20 21
23 20 17 19 26 27
28 24 21 16 15 16
19 18 21 17 18 20
24 20 19 18 20 21

(ii) Look back at the table given in Exercise 3(ii). Jane is 35 and has 4 years no claim bonus. How much would it cost her to insure her car for third party fire and theft for herself only?

(iii) Look back at the table given in Exercise 4(ii). How far would Jim need to travel in a Fiesta hired for a week to make it worth his while to hire the car at the unlimited mileage rate?

(iv) The table below also illustrates male cigarette smoking habits in various age groups.

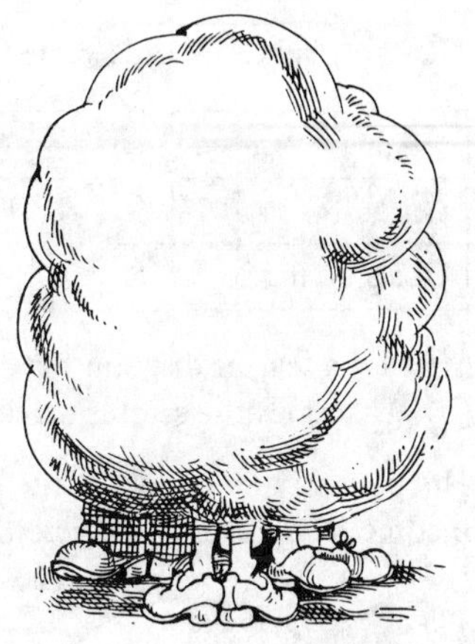

CIGARETTE SMOKING STATUS BY AGE

Age		Males			
		No. of cigarettes smoked per day			
		none	under 20	over 20	total (= 100%)
16-24	%	60	24	15	1730
25-34	%	51	21	28	1950
35-49	%	52	20	28	2560
50-59	%	52	20	27	1830
over 60	%	62	23	15	2410
all aged over 16	%	54	22	23	10,480

(a) Write down the percentage of men aged between 35 and 49 who smoked less than 20 cigarettes a day.
(b) What was the actual number of men aged over 60 who were non-smokers?
(c) Identify the rows, if any, in which the percentage totals are exactly 100%.
(d) Which age groups contain the heaviest smokers?

4.2 Representing Data in Pictures (1): Graphs, Scatter Diagrams and Pie Charts

TRY THESE QUESTIONS FIRST

1 What relationship is suggested by the following data?

Length of bus journey (miles)	0·5	1	1·5	2	2·5	3	3·5	4
Cost (£)	0·10	0·15	0·18	0·25	0·30	0·40	0·45	0·60

2 The graphs opposite illustrate domestic fuel consumption. Compare the graphs showing solid fuel consumption and gas consumption.

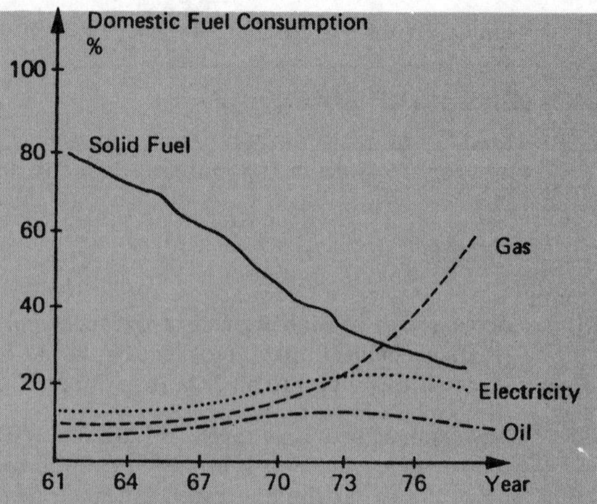

3 The table below records some football results.

Team	A	B	C	D	E	F	G	H	I	J
Goals for	9	9	6	11	19	4	15	10	4	14
Goals against	9	8	7	4	2	10	2	8	5	5

(a) Plot a scatter diagram for this data.
(b) What does the scatter diagram indicate?

4 The pie charts below show the proportions of world population and income for various countries. Summarise the information.

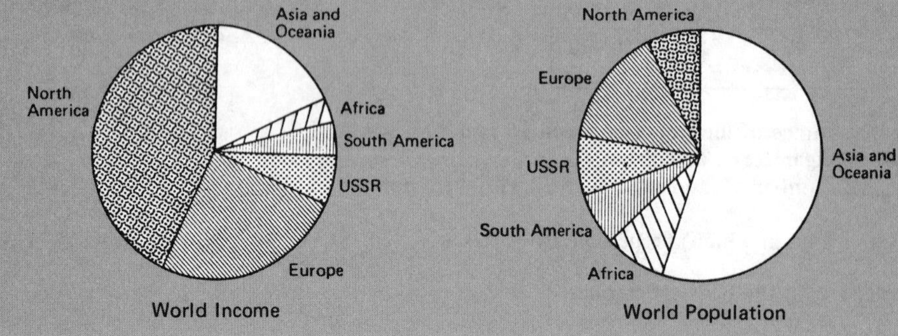

4.2(i) RELATIONSHIPS

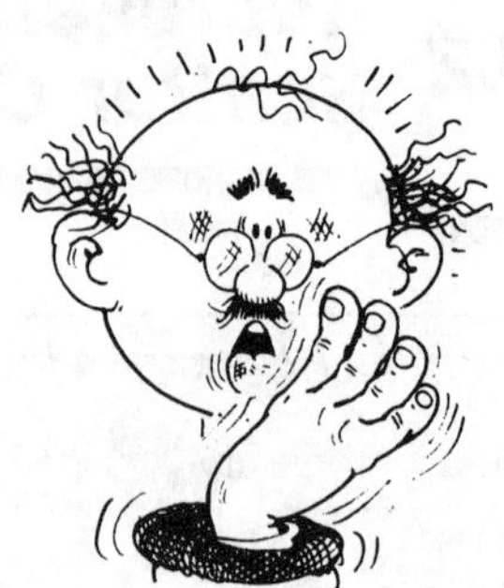

Graphs help us to visualise relationships and can often be drawn straight from a table of data. The table below resulted from an experiment to investigate how quickly a cup of tea cooled down.

Time (mins)	0	5	10	15	20	25	30	35	40	45	50
Temp (°C)	90	65	60	50	36	35	30	26	25	22	20

The resulting graph illustrates the relationship between temperature and time. It looks something like this:

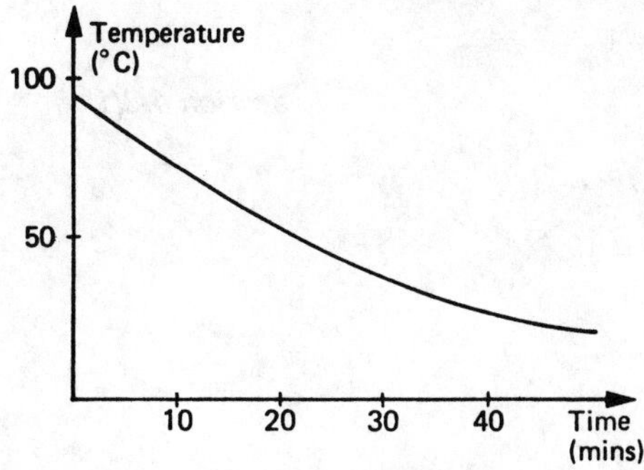

The temperature drops quickly at first then levels off.
The graph provides more information than the table, since it allows us to read off values in between the points given in the table.

Temperature and time are the *variables* in this graph. We have indicated that time is measured along the horizontal axis and temperature is measured up the vertical axis. It's often quite difficult to decide which variable should be measured along which axis. It doesn't really matter as long as the axes are clearly labelled, although it is usual, when time is involved, to measure time along the horizontal axis.

Similarly, when distance is involved (but not time) it is usual to measure distance along the horizontal axis.

The example above involves both time and distance, so time should be measured along the horizontal axis.

In Module 3, we discussed the characteristics of straight line graphs and looked briefly at parabolas and hyperbolas. These graphs were all smooth shapes and could all be categorised easily. More generally, graphs do not fit such standard descriptions. Nevertheless, the graph of a relationship between two variables always indicates:

(i) How quickly or slowly one variable changes with respect to the other (the rate of change)
(ii) The general pattern, thus allowing interpolation and extrapolation.

These features can all be interpreted in the context of the particular example.

Although you will need to proceed very cautiously here!

EXAMPLE

The diagram opposite is a geological cross-section of part of Wales. It plots height above sea level against distance from the coast. What does the graph indicate?

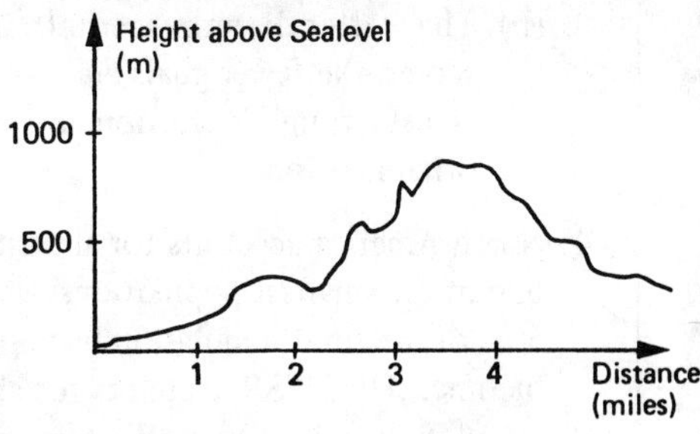

SOLUTION

Starting at sea level the height changes slowly at first–so the terrain is quite flat. Then it rises after about 1 mile and increases irregularly to about 800 metres, before falling again. This suggests that the geological cross-section is taken through a mountain range close to the coast.

CHECK YOUR ANSWERS

1 *Section 4.2(i)*

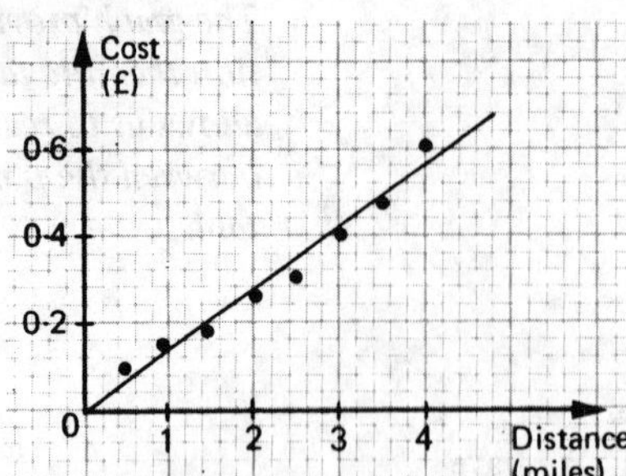

The graph is approximately a straight line. It indicates that the cost is roughly proportional to distance.

2 Solid fuel consumption decreased steadily between 1961 and 1977. Gas consumption increased slowly until 1970 then more rapidly in the 1970s. This was probably due to the introduction of natural gas. Whereas solid fuel accounted for about 80% of fuel consumption in 1961, in 1976 it accounted for only about 30% of the market. On the other hand, gas jumped from under 10% to about 40% of the market in this period. *Section 4.2(ii)*

3 (a) *Section 4.2(iii)*

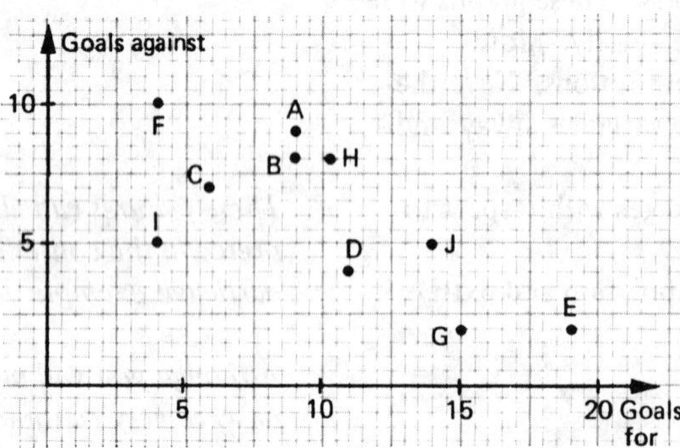

(b) The scatter diagram seems to indicate that the more goals a team scores, the fewer goals are scored against them; of course, this is just a rough indication and should be checked using some other method.

4 North America accounts for almost half the world's income but has one of the smallest populations. Asia and Oceania account for over 50% of the total population but only about 20% of the world's income. The USSR accounts for about the same proportion of population and income. Western civilisations (Europe and North America) contribute only about 25% of the world's population but account for nearly 75% of the world's income. *Section 4.2(iv)*

TRY SOME YOURSELF

1(i) The graph opposite indicates the price of a car on the second-hand market. Summarise the information given by the graph.

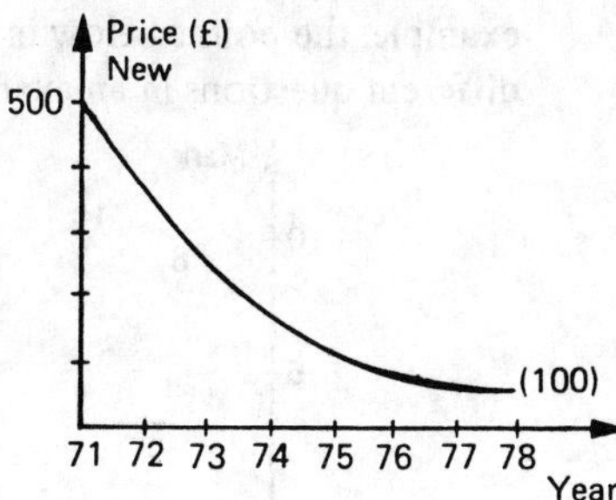

(ii) What does the graph below indicate?

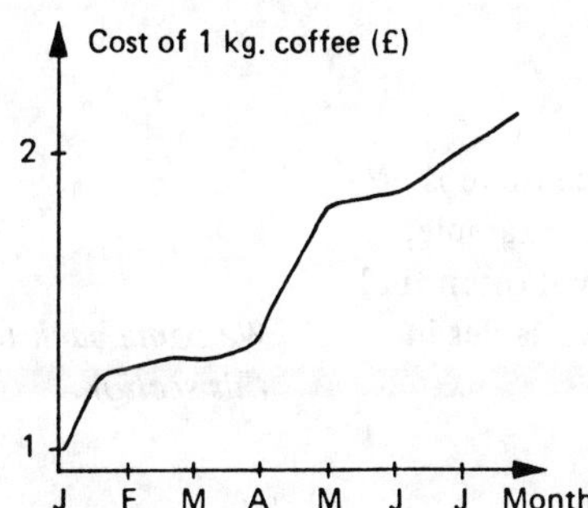

Usually, when a graph is plotted from a table of data, the points do not lie *exactly* on a curve. For example, the experiment on the cooling rate of tea actually produced the following points:

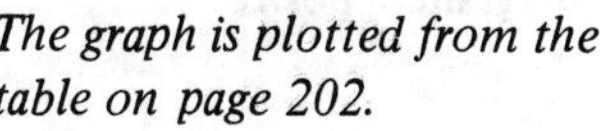
The graph is plotted from the table on page 202.

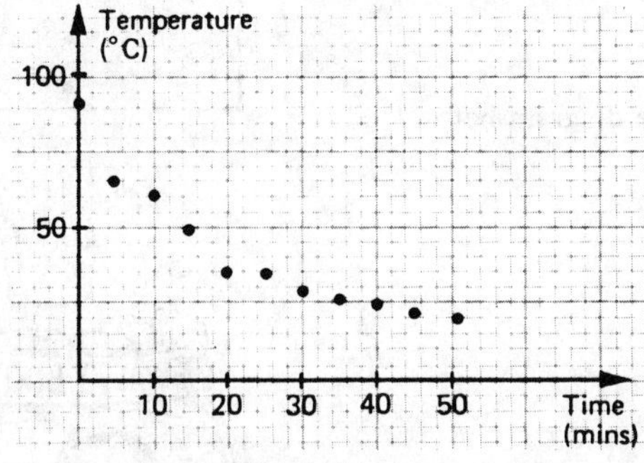

The points almost lie on a smooth curve–but not exactly.

In this case the graph is completed by drawing the smoothest curve possible. This is often just done by trial and error although you may later find that if the points seem to lie on a straight line, there is an algebraic method which derives the best straight line through the given points. The graph illustrating the cooling rate of tea can therefore be completed as follows:

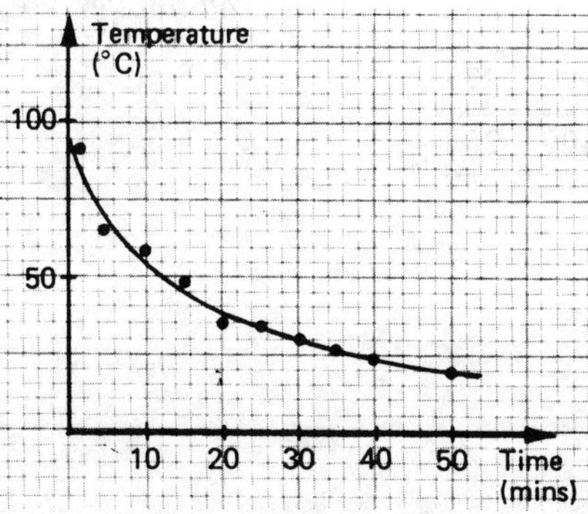

Occasionally it does not make sense to join the points up at all. For example, the points below indicate the marks Jonathan scored on different questions in an examination.

Up until now we have considered only ***continuous*** *variables. Time and distance both increase gradually. The points in between the whole numbers have physical meaning–such as 1·5 seconds or 1·5 metres.*

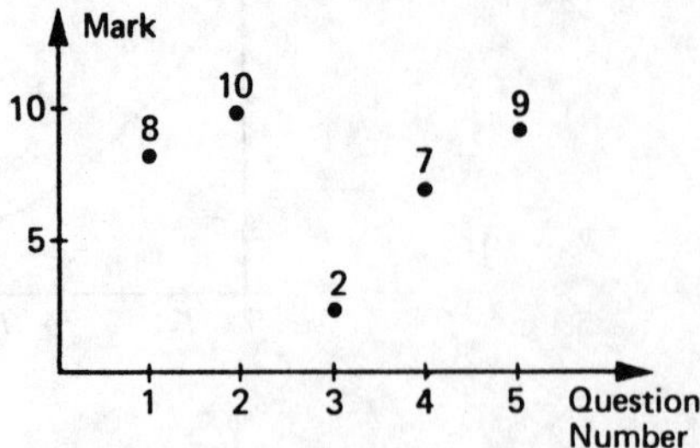

In this example the variables are said to be *discrete* since there is no meaning for points in between the whole numbers. For example, there is no such thing as Question 1·5! However, you will often find that graphs are drawn through points relating discrete variables in order to indicate a trend or pattern.

We come back to this later in this section.

TRY SOME YOURSELF

2(i) Plot the best smooth curve you can through the points on the diagram opposite.

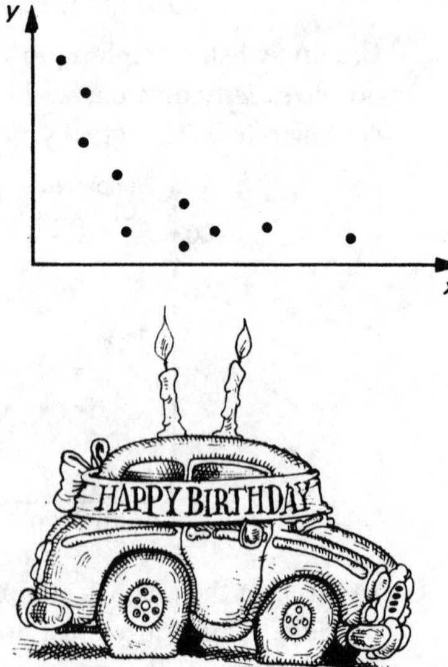

(ii) Would it make sense to join up the points in the diagram below? Give reasons for your answer.

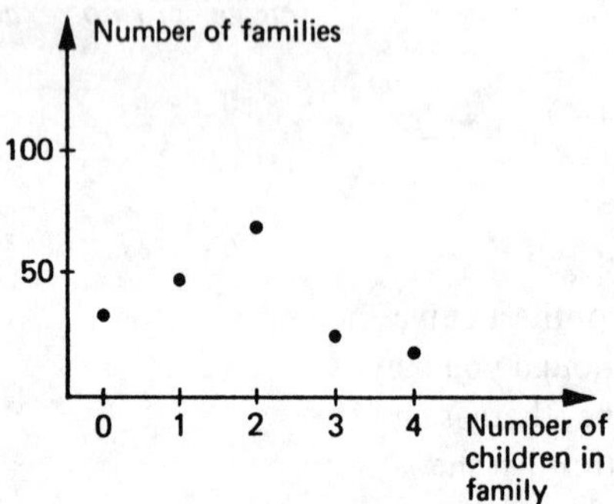

(iii) The table below gives the price of a car related to its age. Plot a graph to illustrate the relationship.

Draw the best straight line you can through the points.

Year	1972	1973	1974	1975	1976	1977	1978	1979
Price (£)	1500	1200	1000	975	800	625	400	200

4.2(ii) MULTIPLE GRAPHS

Sometimes several relationships are illustrated on the same graph so that the reader can *compare* the information. The diagram below indicates the percentages of men with various qualifications relative to age.

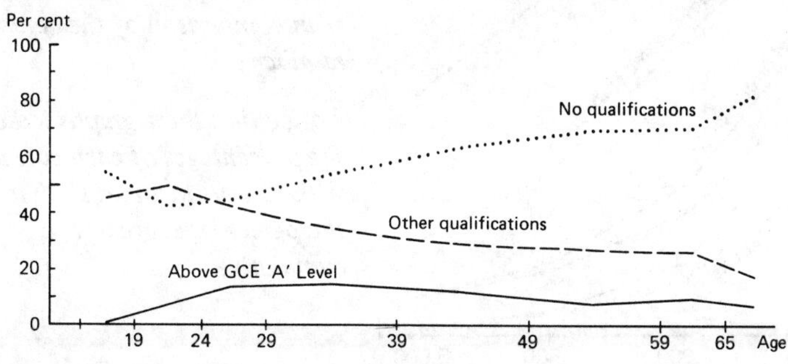

Notice that the vertical axis measures percentage, but we're not told the sample size. You should always be a bit suspicious if the sample size is not indicated. It's OK if the percentages are based upon large samples (e.g. 100% = 1000), but dubious if the samples are very small.

These graphs can be analysed separately. But it is also possible to compare the information from several graphs. For example, the following features can be identified:

- A higher percentage of young people have some form of qualification than old age pensioners. Notice how the graph plotting 'no qualifications' increases with age whereas the graph plotting 'other qualifications' decreases with age. This perhaps reflects better educational opportunities now than at the beginning of the century.

Here two graphs are compared and the information put together.

- The percentage of men with qualifications above A-level remains fairly constant for men aged over 26.

This information is given by just one graph.

- At the age of 39,
  About 12% have qualifications over A level
  About 30% have some other qualification
  About 60% have no qualifications.

Here, the information from all three graphs is put together and compared.

Notice that the total percentage is about

12 + 30 + 60 = 102%

The discrepancy arises because the percentages can only be read approximately from the graph.

This graph shows the cumulative percentages. At each age point the percentage total of all three graphs should be approximately equal to 100%. Check this for yourself.

TRY SOME YOURSELF

3 The diagram below illustrates the relationship between annual income and qualifications in 1977.

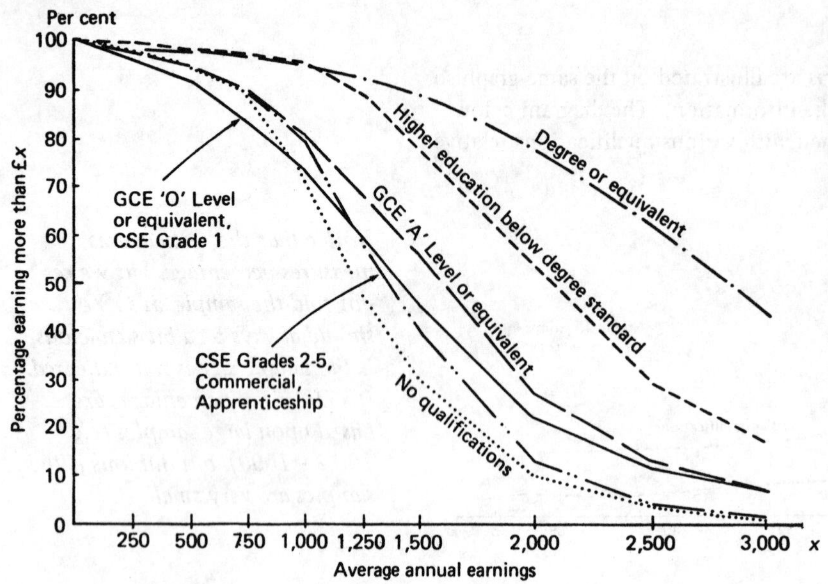

Study the graphs carefully before answering the questions. Once again the graphs indicate percentage, or proportion. As long as the surveys are based on large samples, comparison of different percentages can be more useful than comparison of the actual number.

Notice that these graphs indicate the percentages of **each** *category with various incomes so that the percentages are* **not** *cumulative.*

(i) What percentage of people with a degree (or equivalent) earned more than £3000 per annum?
(ii) Summarise the information given by the graph showing 'no qualifications'.
(iii) Compare the percentages in each category earning over £2000 per annum.
(iv) Compare the graphs showing 'degree or equivalent' and 'no qualifications'. What conclusions can you draw?

4.2(iii) SCATTER DIAGRAMS

We indicated earlier that although discrete variables cannot really give a continuous curve, we are often interested in the trend or pattern of the points. The table below gives the marks on three questions in a mathematics examination for 23 Open University students.

Student	A	B	C	D	E	F	G	H	I	J	K	L
Question 1	10	6	10	4	7	10	10	10	9	6	10	8
Question 2	10	8	8	3	9	9	9	5	1	3	9	10
Question 3	10	10	10	3	10	7	7	10	2	1	10	9
Student	**M**	**N**	**O**	**P**	**Q**	**R**	**S**	**T**	**U**	**V**	**W**	
Question 1	10	10	10	9	8	4	6	4	8	9	5	
Question 2	9	9	10	9	10	7	10	2	6	0	5	
Question 3	8	10	10	10	10	6	10	1	7	2	4	

In this examination, the questions were all marked out of 10.

We may wish to investigate whether there is a tendency for students who do well on one question to do well on another. The questions may not be related to each other at all—one question may be on algebra, another on geometry, and another on statistics. Nevertheless, by plotting points on a diagram we can get some idea of the trend or pattern.

For example, we may want to compare the marks for Question 1 with those for Question 2. We can *plot* the marks for Question 1 against those for Question 2 on a *scatter diagram.*

For example, student A scored 10 on Question 1 and 10 on Question 2.

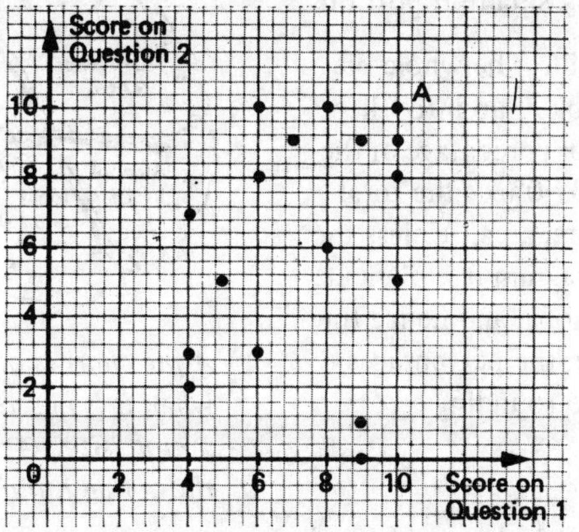

Student A scored (10, 10). So did Student O. These points coincide. You may wish to label the points to identify each student, especially when points coincide.

Even though scatter diagrams cannot give precise information, they can give a 'feel' for the relationship and are often a first step. There are many statistical techniques which allow more specific analysis than this, and you may find yourself using some of these at some time. Even then a scatter diagram is still a good start to the investigation.

The points seem to lie in a band, suggesting that students who score high marks on Question 1 are likely to score high marks on Question 2. However, this is a very rough interpretation and should be checked using some other method.

TRY SOME YOURSELF

4(i) From the table above, plot a scatter diagram to compare the marks for Question 2 with those for Question 3. What does your diagram seem to indicate?

(ii) The table below shows the list prices in 1980 for a number of different cars. The first row gives the price of a new car. The second row gives the price of a five year old car.

It doesn't matter which question score is indicated along which axis, as long as you label the diagram clearly.

Make of car	A	B	C	D	E	F	G	H	I	J
Price new	2500	4000	3500	2500	3000	3000	3800	2000	3000	3200
Price for 5 year old car	500	1100	1100	600	1200	800	1500	500	800	1100

(a) Plot a scatter diagram from this table.
(b) What does your diagram indicate?

4.2(iv) PIE CHARTS

We mentioned earlier that the reader often has to compare data. Multiple graphs are one way of comparing data, particularly when they indicate percentages or proportion.

Provided that the actual numbers are large.

Pie charts also compare proportion. In particular they allow quick identification of very large proportions and very small proportions.

This pie chart summarises a weekly shopping bill. The whole pie represents 100% of the bill. The pie is then broken up into 'slices' and the area of each 'slice' represents a fraction or percentage of the total bill. For example, groceries account for 37·0% of the total bill. The area of this 'slice' is 37/100 of the total area. Unless the percentages or fractions are indicated on the diagram it is difficult to give any precise information, although fractions, and hence the percentages, *can* be estimated. In this example, the percentages actually add up to 100·2%. Again this is due to rounding errors.

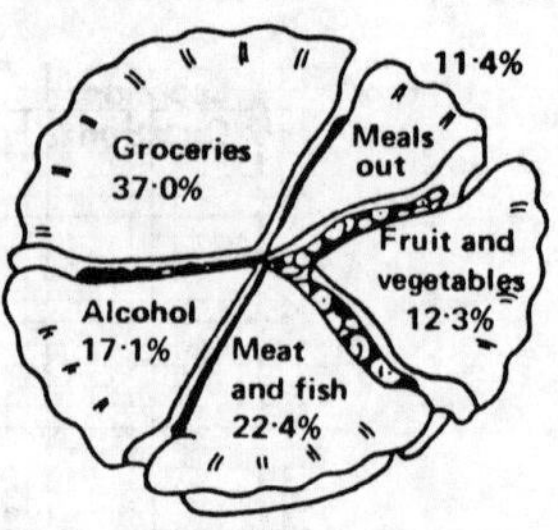

A pie chart is only really useful when it is based upon a large sample size. Again, it is dubious if the actual numbers are very small.

Pie charts can be constructed by dividing a circle into equal 'slices' and then shading in fractions.

This circle is divided into 10 parts.

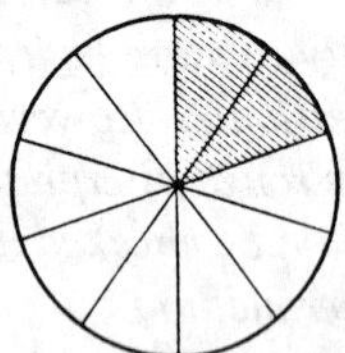

The shaded area is $\frac{2}{10}$ or 20% of the total circle.

For more accurate information the pie chart can be broken up into a greater number of equal parts.

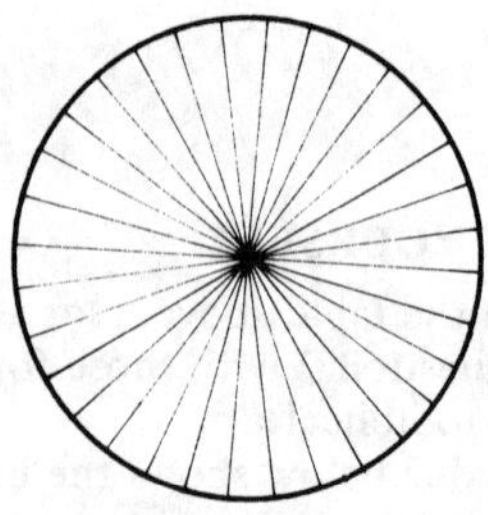

EXAMPLE

This pie chart indicates how people first heard about their present job.

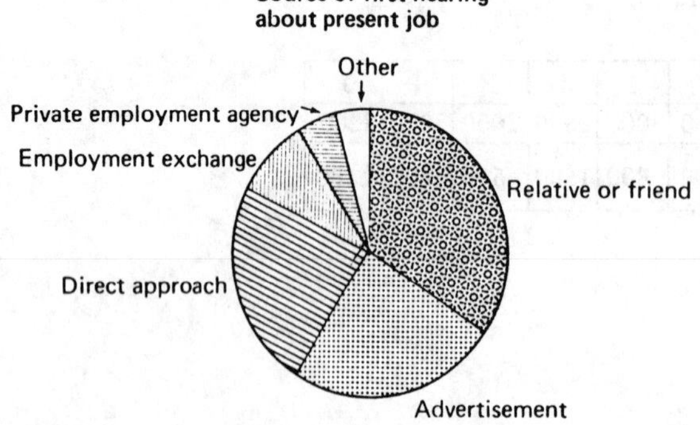

No percentages are indicated on the pie chart so we will need to make some estimates. Remember that 50% corresponds to half the circle, 25% to a quarter of it and so on.

What does the pie chart indicate?

SOLUTION

Most people (about 35%) heard about their job from a relative or a friend. About 25% took the initiative from an advertisement or direct approach. Only about 15% heard of their job through an employment exchange or private agency.

The 'others' category covers all other sources of information.

TRY SOME YOURSELF

5(i) The table opposite breaks up Tom's activities for the day. Shade in the pie chart below to illustrate how the day is broken up. Label each category on the chart.

Activity	Time (hrs)
Sleeping	10
Working	8
Eating	2
Watching TV	1
Pub	3
Total	24

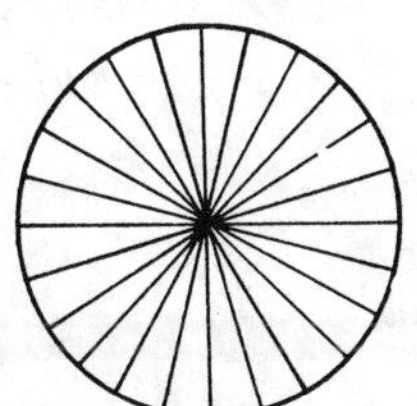

(ii) The pie chart below shows the proportions of people living in different kinds of accommodation.

You will need to estimate the percentages.

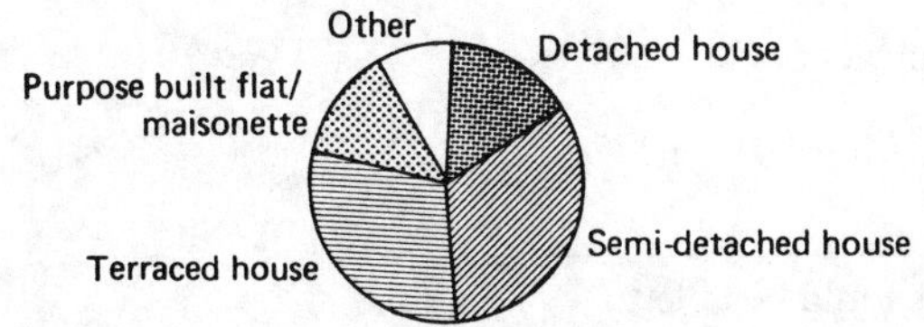

What does the pie chart indicate?

Pie charts give a good visual comparison of proportions arising from different sets of similar data. The pie charts below summarise the means by which people consult their GP.

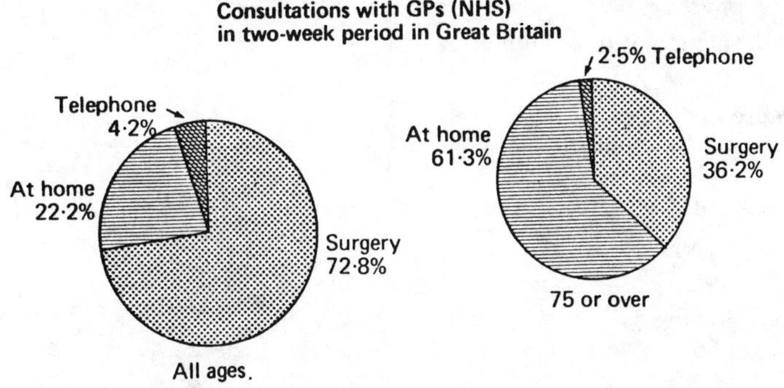

The pie chart on the left summarises the information for people of all ages; the pie chart on the right summarises the information for elderly people.

Taking the population as a whole, most people (72·8%) consult their doctor at his surgery. However, elderly people are more likely to be visited at home.

TRY SOME YOURSELF

6(i) The diagram below shows the proportions of people living in various kinds of accommodation. Separate pie charts are drawn for people in different types of occupation.

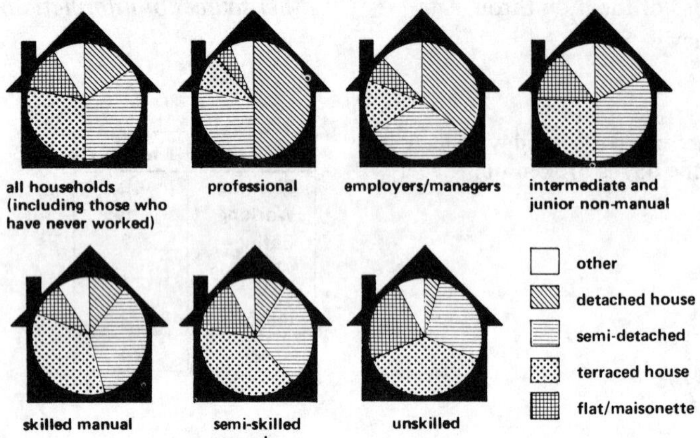

The key shows the different shading used to represent each type of accommodation.

(i) Look at the pie chart labelled 'professional'. Which type of accommodation accounts for the largest proportion?
(ii) Look at the pie chart labelled 'unskilled'. Which type of accommodation accounts for the largest proportion?
(iii) Compare the pie chart labelled 'employers/managers' with that labelled 'skilled manual'.
(iv) Overall, which is the most common type of accommodation?

After you have worked through this section you should be able to

a Fit a reasonably smooth curve to a given set of points
b Identify some of the distinguishing features of a relationship from a graph
c Compare the information given by several graphs on the same diagram, including graphs which indicate percentages
d Plot a scatter diagram
e Read and compare information from a pie chart
f Construct a simple pie chart by dividing a circle into a number of equal parts
g Compare information from two or more pie charts

Finally, here are some exercises if you want more practice.

TRY SOME MORE YOURSELF

7(i) What relationship is indicated by the following data, which gives the prices of different sized packets of soap powder?

Weight of box (kg)	0·32	0·56	0·78	0·87	2·8
Cost (£)	0·30	0·50	0·60	0·70	2·30

(ii) The graphs below relate household size to the age of the head of the household. Summarise the information.

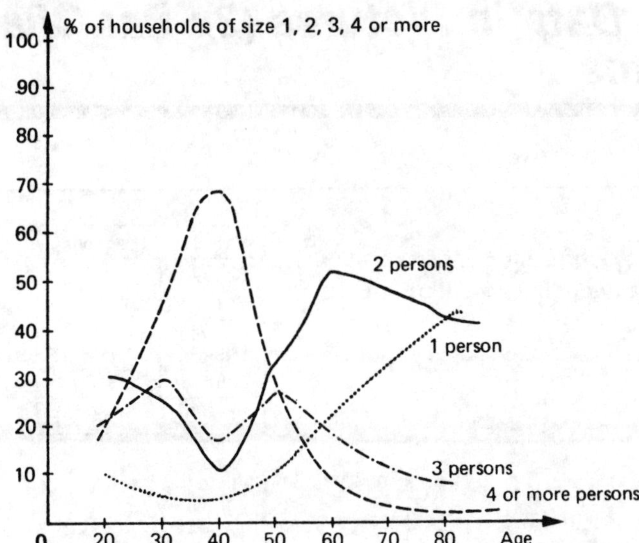

(iii) The table below records the number of accidents per 1000 employees in a number of factories. Plot a scatter diagram to illustrate these data and indicate the trend.

Factory	A	B	C	D	E	F	G
Number of accidents/ 1000 employees	2	3	8	1	5	4	2
Age of factory (yrs)	12	14	20	5	15	8	10

(iv) The pie charts below illustrate domestic fuel consumption in 1951 and 1977. What are the major differences between the two pie charts?

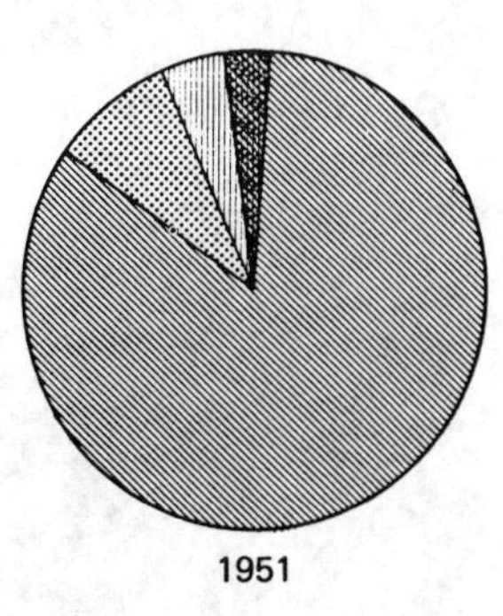

1951

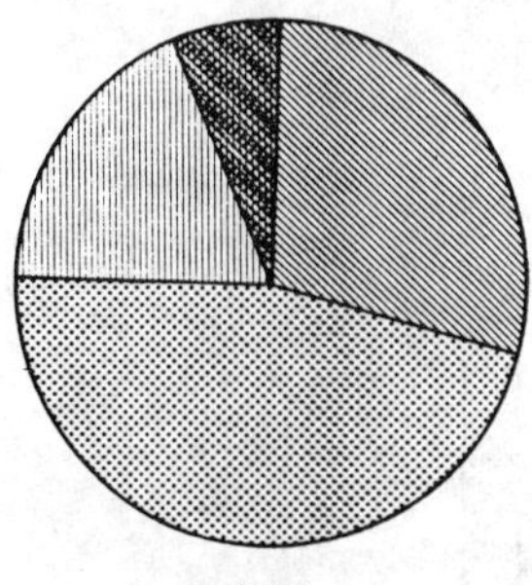

1977

Solid Fuel
Gas
Electricity
Oil

4.3 Representing Data in Pictures (2): Bar Charts and Histograms

TRY THESE QUESTIONS FIRST

1 The frequency table opposite records the different numbers of vehicles passing a specific point in a half hour interval. Illustrate the information on a bar chart.

Vehicle	Frequency
Car	27
Bus	10
Lorry	8
Bicycle	12
Other	3
Total	60

2 The histogram opposite shows the age distribution of the male population of the United Kingdom.
(a) What is the width of each interval?
(b) How many men are aged between 50 and 54?

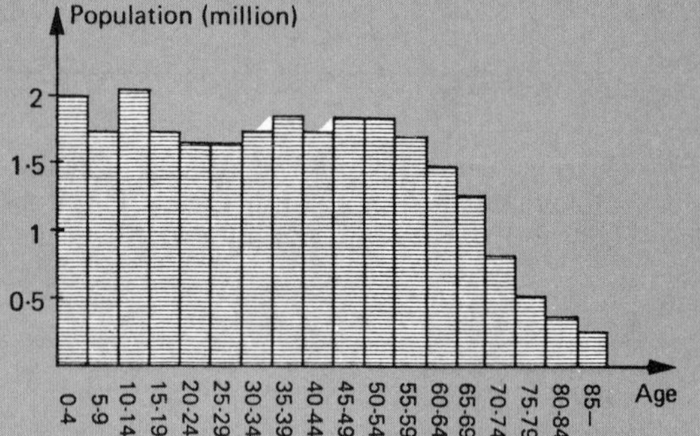

3 The bar chart below indicates the various numbers of offences of different crimes in 1969, 1970 and 1971.

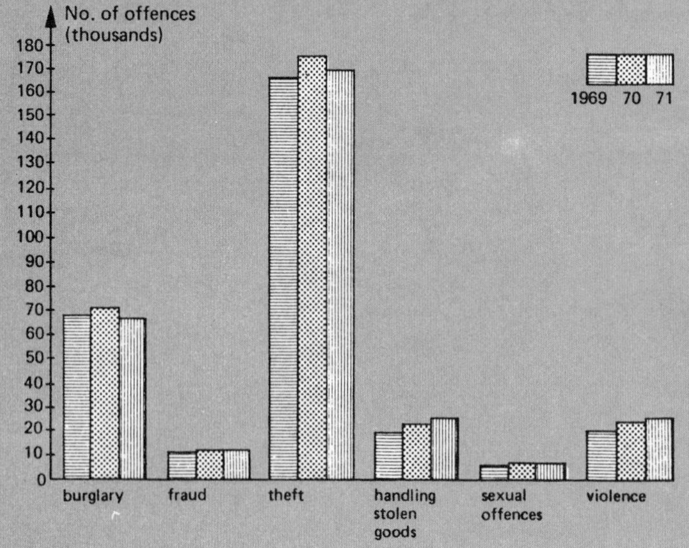

Compare the numbers of violent crimes and the numbers of cases of theft.

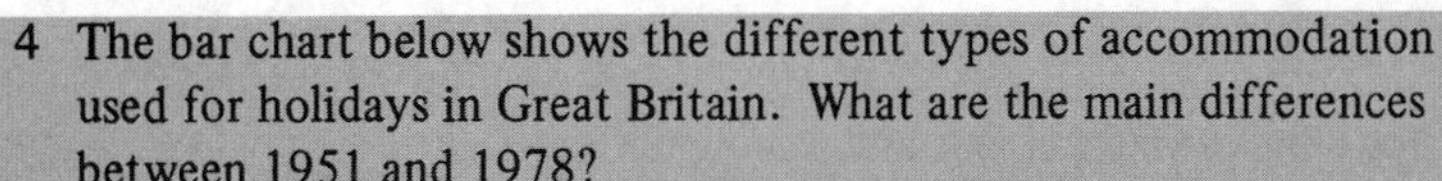

4 The bar chart below shows the different types of accommodation used for holidays in Great Britain. What are the main differences between 1951 and 1978?

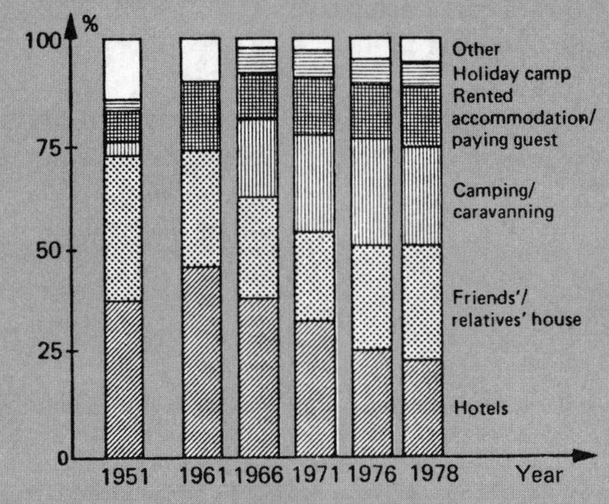

4.3(i) BAR CHARTS

Bar charts are commonly used to represent data since they allow quick assimilation of the information and immediate comparison.

This bar chart compares investment in 'things' with more conventional forms of investment. The height of each bar or column represents the percentage increase in value between 1950 and 1978. For example, shares increased in value by over 1000%. Roman coins proved to be the best investment since the height of that column is the greatest. They increased in value by over 3500%. Building society shares, with the shortest bar, proved to be the worst investment and did not even keep pace with inflation.

The next example shows how a bar chart can be constructed from a frequency table.

Increase in value %

Investment made in 1950 cashed in 1978

4000
3000
2000
1000
0

Increase needed to keep pace with inflation

Shares | Building society ordinary shares | Rare GB stamps | Roman coins | Georgian silver-ware

EXAMPLE

This frequency table indicates the various numbers of children in a sample of 30 households. Draw a bar chart to illustrate the information.

Number of children	Frequency
1	4
2	7
3	11
4	4
5	3
6	0
7	1
Total	30

SOLUTION

The number of children is measured along the horizontal axis and the frequency up the vertical axis. The height of each bar or column represents the frequency. Notice that the bars do not touch each other; this is because the numbers of children are discrete.

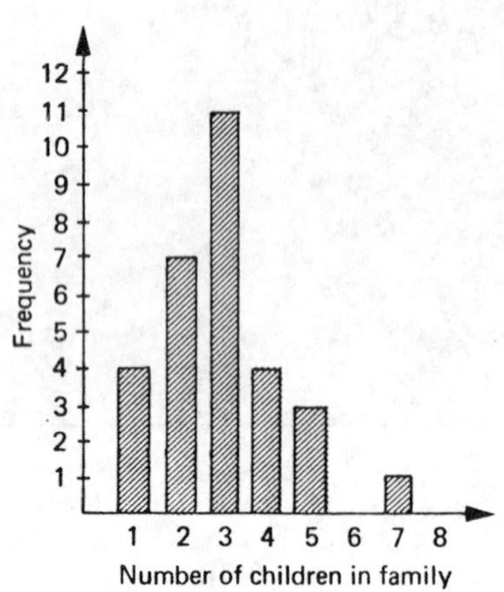

This bar chart could have been drawn the other way round. It doesn't matter as long as the axes are labelled clearly.

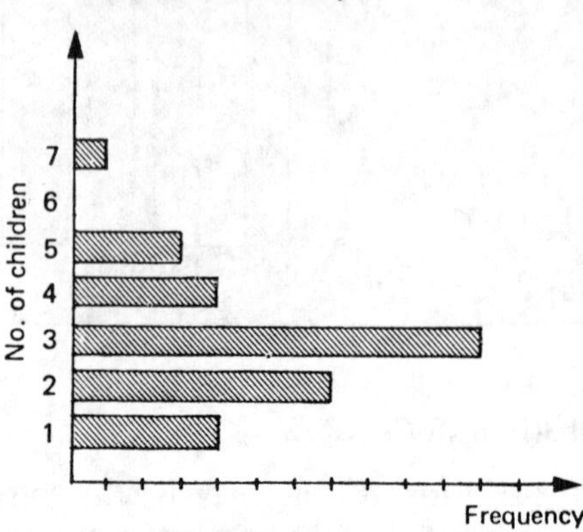

CHECK YOUR ANSWERS

1 *Section 4.3(i)*

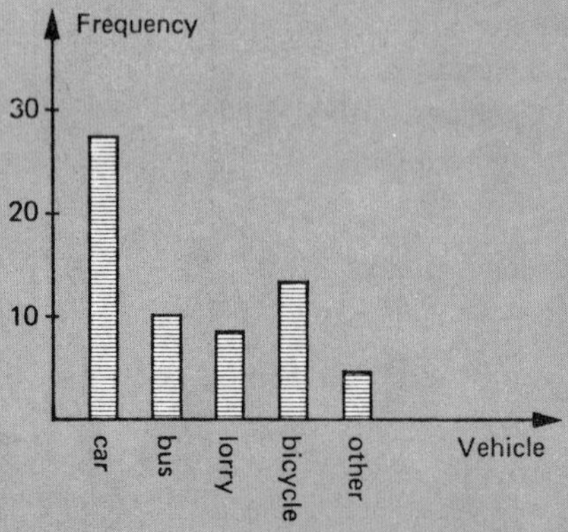

2 (a) 5 years. *Section 4.3(ii)*
(b) About 1·7 million.

3 Theft is by far the most common crime. The number of cases of theft in each year was about eight times the number of violent crimes. The number of cases of theft was at a maximum in 1970, decreasing slightly in 1971 to just over 160,000. The number of violent crimes increased from about 20,000 in 1969 to about 25,000 in 1971. *Section 4.3(iii)*

4 In 1951, about 35% of all holidays were spent in hotels, whereas in 1978 this proportion dropped to under 25%. Camping/caravanning has become more popular, its share rising from under 5% to about 25%. Fewer people stayed with friends or relatives in 1978. There seems to be more 'guest house' type accommodation in 1978 and holiday camps seem to be more widespread. The percentage in the category labelled 'other' has halved.

Section 4.3(iv)

TRY SOME YOURSELF

1(i) The frequency table opposite indicates the various numbers of people who carry out their main weekly shopping on different days.
(a) Illustrate this information on a bar chart.
(b) Which is the most popular shopping day?
(c) Which is the least popular shopping day?

Day	Frequency
Monday	2
Tuesday	5
Wednesday	10
Thursday	24
Friday	29
Saturday	30
Total	100

You may like to draw the bar chart on graph paper. This will make it easier to mark off the correct heights.

(ii) The bar chart below indicates the numbers of cars sold by one firm in successive years.

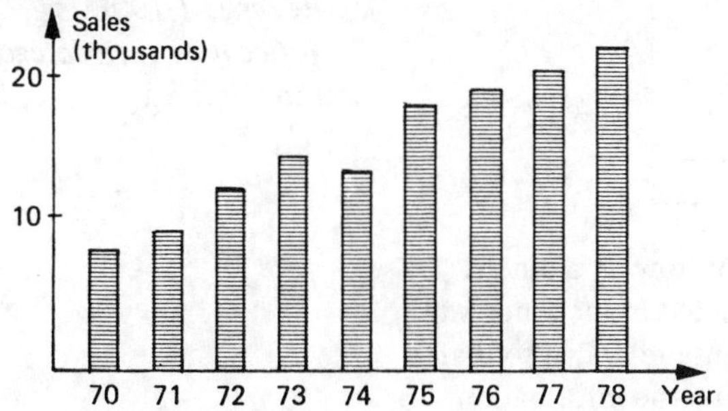

(a) What were the sales figures for 1972?
(b) What were the sales figures for 1977?
(c) What does the bar chart suggest about the pattern of sales?

Like graphs, bar charts can be misleading. The two bar charts below illustrate the same information but look very different because they have been drawn to different scales.

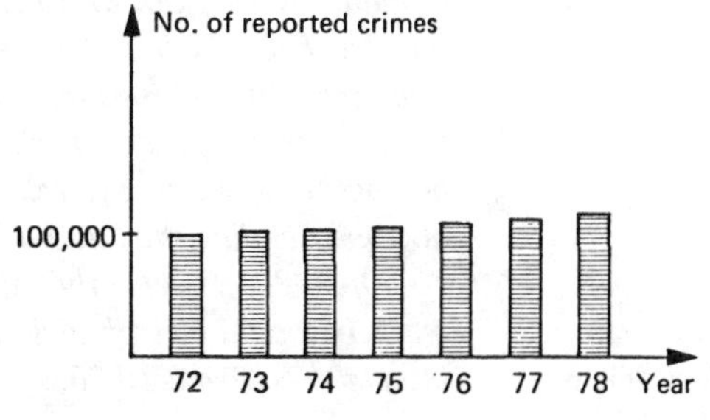

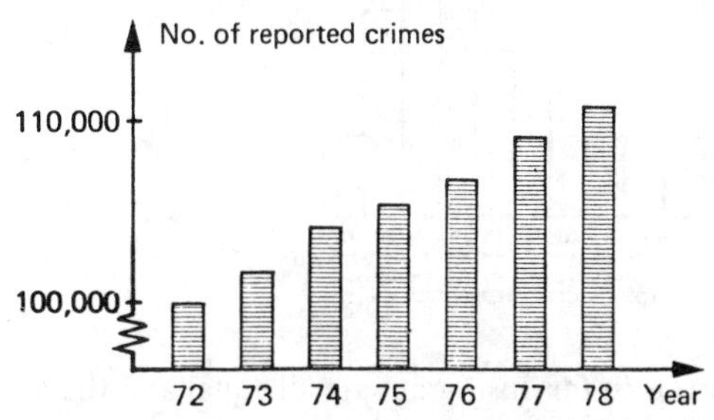

The bar chart on the left suggests that the number of reported crimes has remained steady, whereas the bar chart on the right suggests a dramatic increase.

The scales on a bar chart are often selected deliberately to mislead the unsuspecting reader, so you should always inspect such diagrams carefully.

4.3(ii) HISTOGRAMS

Bar charts are drawn to illustrate discrete forms of data. Similar diagrams can be constructed to represent data distributed across a range of values. The table below records examination scores for 80 students.

For example, days of the week, numbers of children, and different types of vehicle are all discrete.

Range of marks	Frequency
1-10	2
11-20	2
21-30	4
31-40	6
41-50	7
51-60	8
61-70	15
71-80	22
81-90	10
91-100	4
Total	80

The frequency table indicates how the marks are distributed across the range 1-100. Notice how each category runs into the next.

A *histogram* provides a visual representation of how data are distributed across a range of values. It is constructed in the same way as a bar chart, but this time the bars or columns are drawn so that they touch. The following histogram represents the information given in the frequency table above.

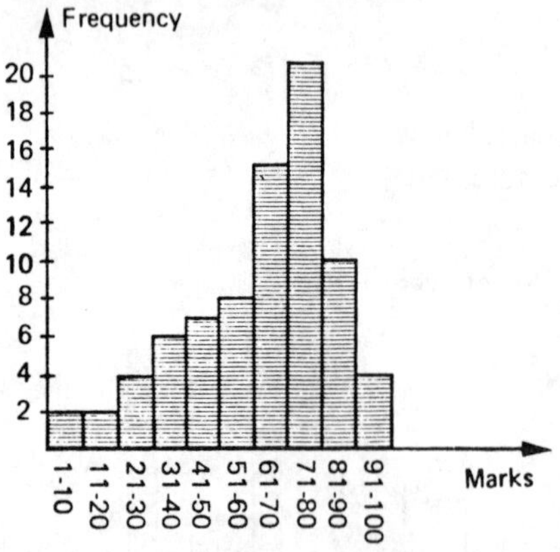

The horizontal axis is divided into intervals to represent each group of data: 1-10, 11-20, 21-30 and so on. The bars touch each other since each interval runs into the next. Notice that each mark belongs to only one interval. It's no good dividing the line 1-10, 10-20, 20-30 . . . since then the mark 10 would be included in two intervals: 1-10 and 10-20.

The intervals in this histogram are all equal and the *height* of each bar or column indicates the frequency of data in any interval.

However, not all histograms are based on equal intervals. In the histogram below some intervals are wider than others.

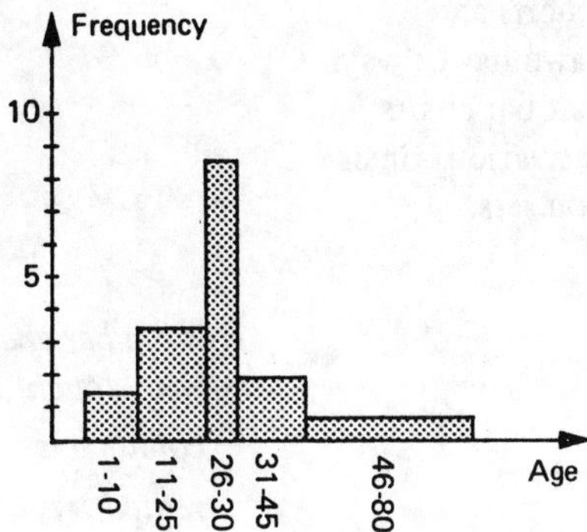

We include this example as a warning.
It is very difficult to compare data when the intervals are unequal. You must compare the ***area*** *of each rectangle to make any useful comparisons.*

However, most histograms are based upon equal intervals. Because the width of each column or rectangle is the same, the area of each bar is proportional to its height. Therefore the different categories can be compared by height alone.

Area = Height x Width

TRY SOME YOURSELF

2(i) The frequency table opposite resulted from a survey of 30 people who were asked 'How long did it take to get from Central London to Heathrow Airport?'.
(a) Illustrate the information in a histogram.
(b) What is the width of each interval?
(c) Which interval has the highest column?

Length of journey (mins)	Frequency
20-24	3
25-29	6
30-34	11
35-39	5
40-44	3
45-49	1
50-54	1
Total	30

Again you may find it easier to draw your histogram on graph paper.

(ii) The histogram below shows the various ages of a sample of the UK population in 1976.

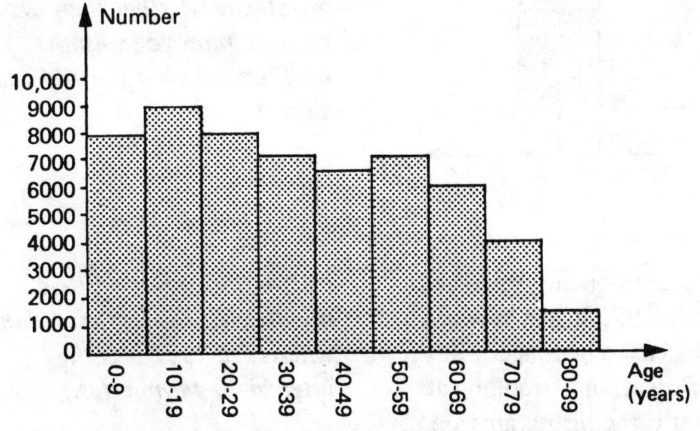

The histogram does not show the number of people aged over 90, probably because there were so few that the information would not be detected on this scale.

(a) What is the width of each interval?
(b) Which age group contains the largest number of people?
(c) What is the ratio of the number of people in their seventies to the number of people in their thirties?
(d) Use the histogram to estimate the sample size.

You will need to add up ***all*** *the frequencies to find the total number of people.*

Sometimes it is difficult to distinguish between bar charts and histograms, since although the bar charts we have drawn have always had 'non-touching' bars you will occasionally find that bar charts look like histograms. The bar chart below records the various times taken to repair different categories of colour television sets.

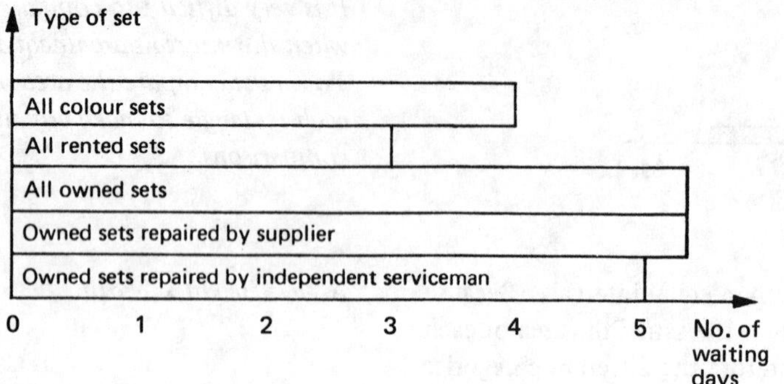

This is a bar chart since, although the bars touch, the data:

colour sets

rented sets

are discrete; the intervals do not run into each other.

Again, this example serves as a warning to read the axis labelling carefully.

4.3(iii) COMPARING BAR CHARTS AND HISTOGRAMS

Sometimes you may need to compare the data given in two or more bar charts or histograms. The two histograms below resulted from two surveys to investigate the time taken to travel from Central London to Heathrow Airport. The first was in 1968; the second in 1972.

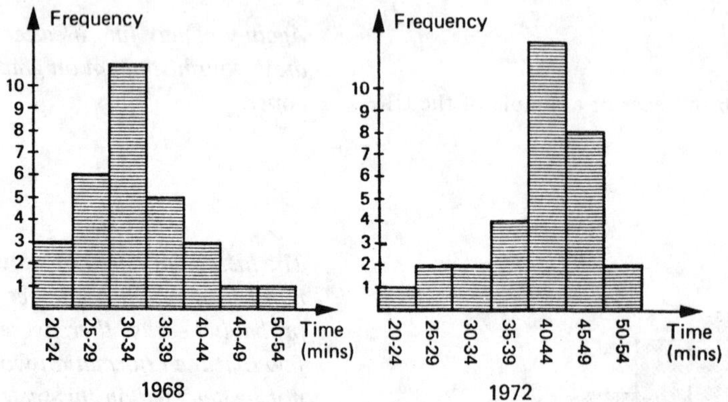

These surveys were organised to investigate whether there was a need to improve transport facilities between the city and the airport.

These histograms show that in 1968 the most frequently reported time was between 30 and 34 minutes, but in 1972 it was between 40 and 44 minutes. So the journey times seem to be longer. But it is not particularly easy to compare the information by looking at the histograms separately; it is much clearer if the histograms are combined onto the same diagram.

In 1968 the interval 30-34 minutes has the highest column, whereas in 1972 it is the interval 40-44 minutes.

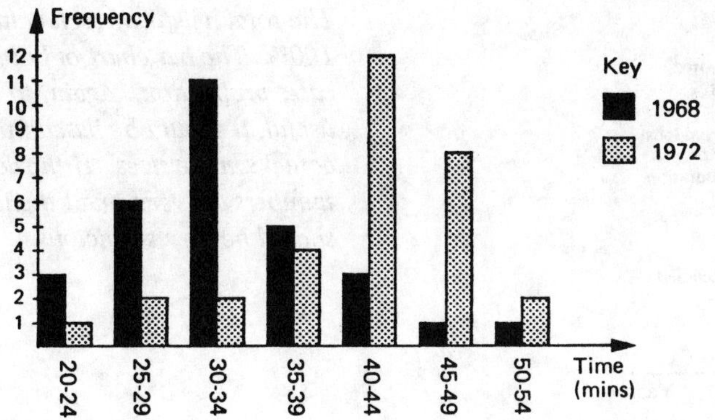

The key indicates the shading for each year. Although two histograms are combined, we've left a space between each interval just to make the comparison easier.

This *comparative bar chart* illustrates the differences between 1968 and 1972. The black bars appear to be clustered to the left, whereas the shaded bars are grouped to the right.

This indicates that the journey times tended, overall, to be shorter in 1968 than they were in 1972.

So there seemed to be some evidence to suggest that travel facilities should be improved. In the 1970s the underground network was subsequently extended to Heathrow Airport.

A comparative bar chart therefore allows quick comparison of several sets of similar data.

TRY SOME YOURSELF

3 The bar chart opposite illustrates cigarette smoking habits of men in the United Kingdom.
- (i) How many men aged between 35 and 49 smoked more than 20 cigarettes a day?
- (ii) How many men over 60 were non-smokers?
- (iii) Compare the bar charts of men aged between 16 and 24 and men aged between 35 and 49.

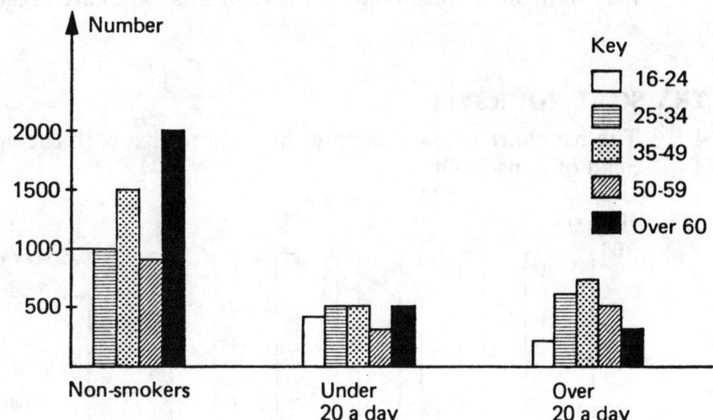

4.3(iv) STACKED BAR CHARTS TO SHOW PROPORTION

Bar charts can also be stacked on top of each other to compare proportions. The bar chart overleaf indicates the proportions of different types of living accommodation in the UK between 1950 and 1977.

The total height of each column is 100%. The bar chart only indicates proportion. Again, to be useful, it should be based upon actual sample sizes. If the actual numbers are very small then you should be very suspicious.

This bar chart shows that the proportion of owner occupiers increased from about 30% in 1959 to nearly 50% in 1977. The proportion of accommodation rented from the local authority also increased, from about 15% to about 35%. The proportion of privately rented accommodation or 'other' decreased significantly from nearly 60% to about 15%. This does not necessarily mean that fewer people lived in privately rented accommodation in 1977, just that a smaller proportion of the total population were thus housed. The population has increased substantially since 1950, so although the percentage dropped dramatically, the actual numbers may have stayed about the same, or if they did drop the decrease may have been considerably less than this bar chart suggests.

For example,

3000 = 60% of 5000

3000 = 30% of 10,000

TRY SOME YOURSELF

4 The bar chart below compares household size with the age of the head of household.

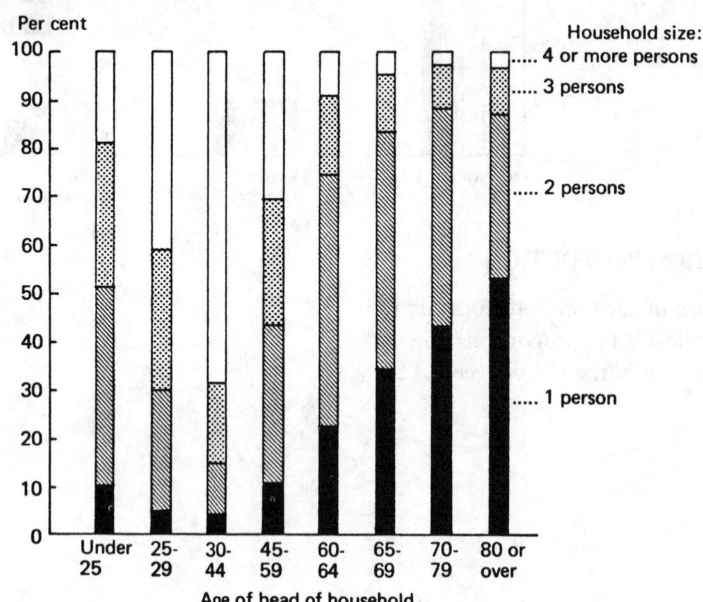

(i) What percentage of under 25s live alone?
(ii) What percentage of over 80s live alone?
(iii) What does the bar chart suggest about people living alone?

After you have worked through this section you should be able to

a Construct a bar chart from a frequency table
b Read and compare information from a bar chart
c Construct a histogram from a frequency table
d Compare information from a histogram based on equal intervals
e Combine two bar charts or histograms together and compare the resulting patterns
f Compare information from a bar chart indicating proportions

Finally, here are some exercises if you want more practice.

TRY SOME MORE YOURSELF

5(i) The frequency table below indicates the daily calorie consumption per person in different countries. Illustrate the information on a bar chart.

Country	Daily calorie consumption per person
India	1810
Ceylon	2180
France	3150
UK	3220
Denmark	3300

(ii) The frequency table opposite indicates the number of floods over 14 ft at Shrewsbury over several years.
(a) Illustrate the information on a histogram.
(b) What is the length of each time interval?

Years	Frequency
1910-1919	1
1920-1929	4
1930-1939	1
1940-1949	4
1950-1959	2
1960-1969	4
Total	16

(iii) The bar chart overleaf indicates different British occupations in the years between 1841 and 1951.
(a) What does it indicate about white collar/public administration workers?
(b) What does it indicate about the numbers involved in the mining industry?
(c) Summarise the differences between the bar charts for mining and white collar/public administration.

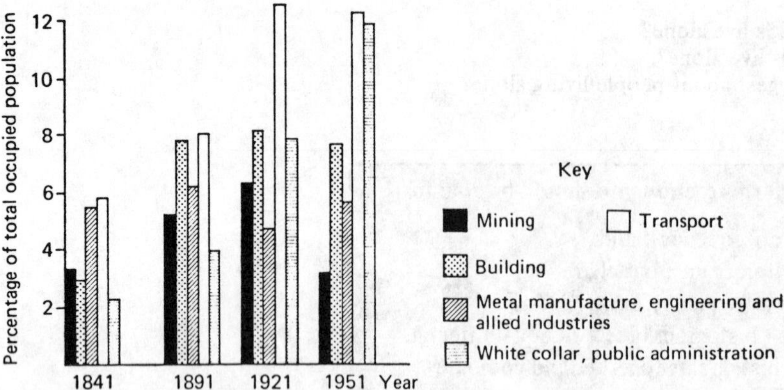

(iv) The bar chart below indicates different types of living accommodation according to the age of the head of household. Summarise the information.

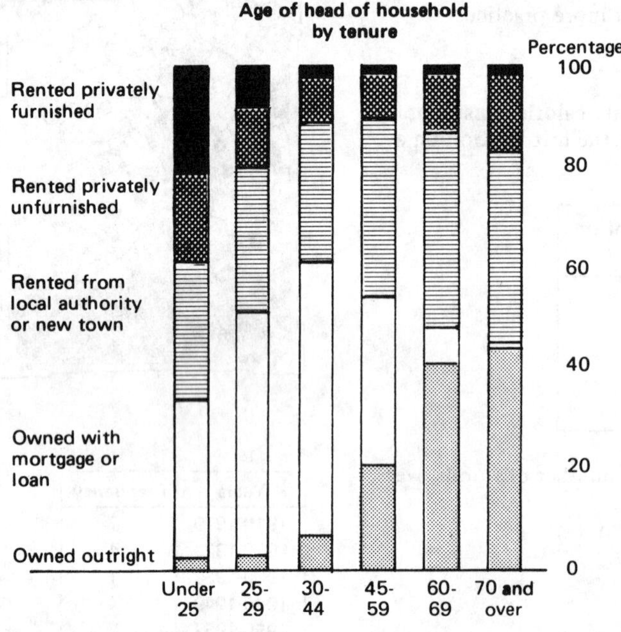

4.4 Averages

TRY THESE QUESTIONS FIRST

The frequency table opposite indicates the numbers of children in 20 families.

1 Find the mean number of children in a family.
2 Find the modal number of children in a family.
3 Find the median number of children in a family.

No of children	Frequency
1	6
2	7
3	3
4	2
5	1
6	1
Total	20

4 The bar chart below indicates the heights of 40 children.

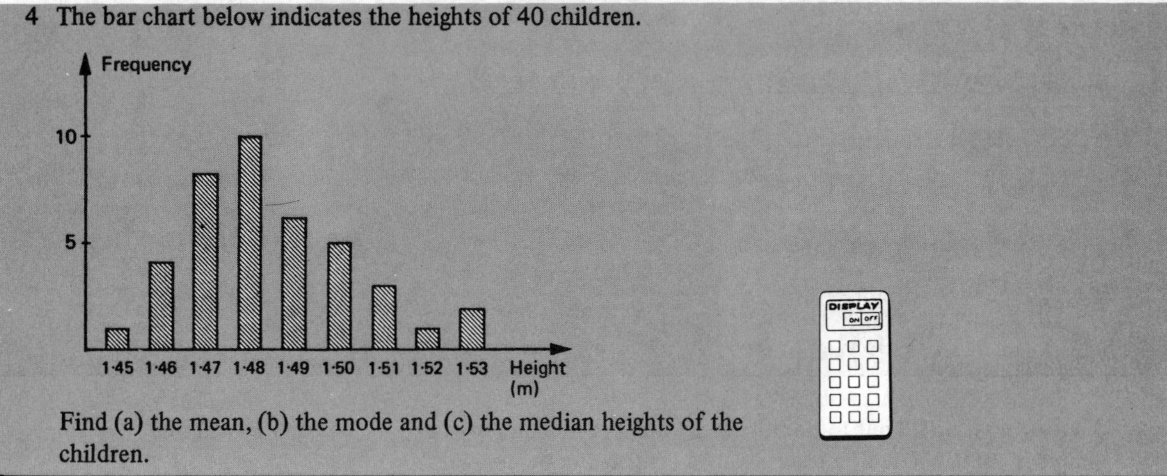

Find (a) the mean, (b) the mode and (c) the median heights of the children.

You may have heard expressions like

'On an average day . . . '

or 'The average number of children in a family is 2·4.'

or 'The average dress size is size 14.'

But what does the word *'average'* mean? In fact there are several types of average, and we discuss all of these in this section.

2·4 children

4.4(i) ARITHMETIC MEAN

Average often means 'the mean of a set of numbers'. The *arithmetic mean* of the five numbers

1, 3, 7, 10, 15

is found by adding all the numbers together

$$1 + 3 + 7 + 10 + 15 = 36$$

and dividing by the total number of numbers (five). So the mean is

$$\frac{1 + 3 + 7 + 10 + 15}{5} = 7{\cdot}2$$

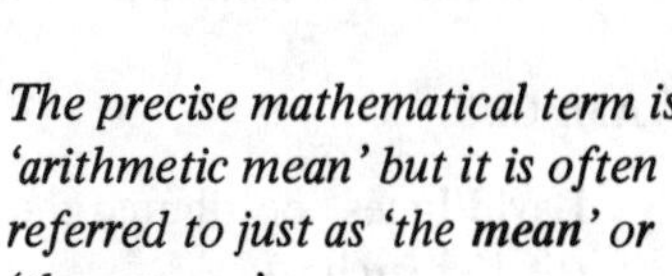

The precise mathematical term is 'arithmetic mean' but it is often referred to just as **'the mean'** *or* **'the average'**.

A calculator sequence for this is

The same process is used to find the arithmetic mean of any set of numbers, no matter how many. This is when a calculator is particularly useful.

CHECK YOUR ANSWERS

1 The mean = $\frac{(6 \times 1) + (7 \times 2) + (3 \times 3) + (2 \times 4) + (1 \times 5) + (1 \times 6)}{20}$ *Section 4.4(i)*

= 2·4 children

2 The modal number of children is 2. *Section 4.4(ii)*

3 The median is 2 children. *Section 4.4(iii)*

4 (a) The mode is 1·48 metres. *Section 4.4(iv)*
(b) The mean height is 1·49 metres.
(c) The median height is 1·48 metres.

A survey was carried out to find the mean number of children in a sample of 20 families. The resulting data are listed in the margin.

Number of children

1 1 1 2 2 4
5 1 2 2 1 1
3 6 2 2 4 3
2 3

The mean is calculated by finding the total number of children and dividing by 20.

$$\text{The mean} = \frac{\text{Total number of children}}{\text{Total number of families}}$$

$$= \frac{1+1+1+2+2+4+5+1+\ldots+3+2+3}{20}$$

$$= 2{\cdot}4$$

Although 2·4 children has no real meaning it does provide useful information for clothes manufacturers, education authorities, etc.

EXAMPLE

David Jones contributed the following scores in a cricket season (of 13 matches):

17, 53, 15, 81, 19, 3, 23, 31, 43, 1, 8, 22, 4

What was his average score?

SOLUTION

We want to find the mean score over 13 matches. The total score over the season was

17 + 53 + 15 + 81 + 19 + 3 + 23 + 31 + 43 + 1 + 8 + 22 + 4
= 320

$$\text{The mean} = \frac{\text{Total score}}{\text{Number of matches}} = \frac{320}{13} = 24{\cdot}6 \simeq 25$$

Here the mean is rounded to the nearest whole number.

So Jones's average score over the season was 25.

TRY SOME YOURSELF

1(i) Jane had these examination results one summer. What was her average mark?

English	*52*	*French*	*56*
Maths	*49*	*Geography*	*63*
History	*64*	*Biology*	*66*

(ii) **The following table gives the daily midday temperatures in August for one year. Calculate the mean midday temperature for the month.**

Date	Temp (°C)	Date	Temp (°C)	Date	Temp (°C)
1	19	11	19	21	19
2	18	12	20	22	18
3	18	13	21	23	21
4	17	14	18	24	22
5	19	15	22	25	21
6	20	16	23	26	19
7	21	17	22	27	19
8	20	18	21	28	18
9	22	19	19	29	20
10	19	20	20	30	21
				31	19

It's easy to lose your place on a large table like this. You may find it useful to use a ruler as a guide. Alternatively you may want to cross off each number after you have entered it into the calculator.

This last exercise indicates that you may often have some idea of the answer before you start the calculation. The temperatures all lie between 17°C and 23°C. So the average temperature must also lie between 17°C and 23°C.

If you got an answer less than 17°C or more than 23°C you should have known that you made a mistake, maybe by missing out one number, or counting one twice.

Calculating the mean from a frequency table

EXAMPLE

The table opposite indicates the different numbers of children in a sample of 30 families. Find the mean number of children in a family.

No. of children	Frequency
1	4
2	7
3	11
4	4
5	3
6	0
7	1
Total	30

SOLUTION

$$\text{The mean} = \frac{\text{Total number of children}}{\text{Total number of families}}$$

From the table, the total number of families is 30.

Finding the total number of *children* requires a little thought. Notice that 4 families each have 1 child, making 4 children all together. The total number of children is therefore

$(4 \times 1) + (7 \times 2) + (11 \times 3) + (4 \times 4) + (3 \times 5) + (0 \times 6) + (1 \times 7)$

$= 89$

The total number of children in each category is found by multiplying the number of children by the frequency.

The mean = $\frac{89}{30}$ = 2·97 (Rounded to 2 decimal places.)

Check this using your calculator.

You may have noticed that we obtained a mean of 2·97 children in this example whereas previously (on page 226) we calculated a mean of 2·4 children. The difference arises because the examples are quite distinct. The example above concerns a sample of 30 families; the earlier example concerns a different sample of 20 families. Since the samples were quite distinct, the results are also unrelated.

This illustrates that the type of sample considerably influences the mean. For example, if the survey includes only large families, then the mean will also be large.

TRY SOME YOURSELF

2(i) The table opposite gives the frequency of sales of different sizes of ladies shoes for one day. Calculate the mean shoe size sold that day.

Size	Frequency	Size	Frequency
3	1	5½	10
3½	3	6	9
4	4	6½	5
4½	10	7	3
5	11	7½	1
		Total	57

(ii) The frequency table below indicates the length of time taken by 30 children to travel to school each morning. What was the mean time taken?

Time (mins)	Frequency
5	3
10	5
15	10
20	6
25	5
30	1
Total	30

4.4(ii) MODE

When we talk of the 'average family' we do not usually mean a family with 2·4 children. An 'average' family' is usually taken to be father, mother and two children—the *most common* type of family.

Similarly, an 'average day' is taken to be an 'ordinary' day, the sort of day that occurs *most often*, and this gives the definition for the second type of average—the mode.

The *mode* of a set of numbers is the number which appears *most often.*

An 'average' family

EXAMPLE

Find the mode of the numbers

2, 1, 3, 5, 2, 5, 9, 1, 7, 1, 9, 2, 3, 5, 9, 2

SOLUTION

The mode is found by counting the number of times each number occurs. By inspection, 1 occurs 3 times, 2 occurs 4 times, 3 occurs twice, etc. So the mode is 2.

The number 2 occurs most often.

In fact the easiest way to find the mode is to draw up a frequency table. The mode is the number which occurs most often, that is, the number with the *highest frequency*. From the table, the mode is therefore 2. (Notice that we want the actual number, *not* the frequency.)

Number	Frequency
1	3
2	4
3	2
5	3
7	1
9	3
Total	16

The mode is often not the same number as the mean. We return to the example on page 227, where a sample of 30 families produced this frequency table.

From the table, the highest frequency is 11. So the mode is 3 (this is also referred to as the modal number of children). The mode (3) and the mean (2·97) are very close but they are not the same.

Notice that the mode is the *number of children* (not the associated frequency). Referring to the mode as the 'modal number of children' avoids any possible confusion.

No. of children	Frequency
1	4
2	7
3	11
4	4
5	3
6	0
7	1
Total	30

TRY SOME YOURSELF

3(i) Find the mode of the following set of numbers:

3 1 2 3 4 5 2 1 5 2 4 5 1 1 5
2 5 3 1 4 5 6 4 1 2 4 6 1 5 6
6 4 4 1 5 2 2 4 3 3 6 1

Construct a frequency table. Remember to include the total as a check.

(ii) This table indicates the number of cars, buses etc. passing in a half hour interval. Which vehicle passed most often?

Vehicle	Frequency
Car	21
Bus	5
Lorry	17
Motorbike	11
Other	20
Total	74

(iii) This table shows the different lengths of time that 70 people have spent in their current job. Find the modal time spent in the current job.

No. of years	Frequency	No. of years	Frequency
1	3	9	7
2	2	10	4
3	4	11	5
4	0	12	4
5	5	13	2
6	10	14	0
7	15	15	1
8	8	Total	70

4.4(iii) MEDIAN

The third type of average refers to the *number in the middle*–the *median.*

For example, consider the numbers below:

5 4 7 16 17 2 11 1 9

Listing the numbers in ascending order from left to right gives

1 2 4 5 7 9 11 16 17

4 numbers | 4 numbers

7 is the number in the middle

There are 9 numbers altogether. The middle number is the fifth number, 7.

So 7 is the median of this set of numbers.

In this example the total number of numbers was 9, an odd number. Finding the median is a little more difficult when dealing with an even number of numbers, as the next example illustrates.

EXAMPLE

Find the median of the following numbers.

10 7 2 5 1 7 9 4 3 2

SOLUTION

Rearranging the numbers in ascending order gives

1 2 2 3 4 | 5 7 7 9 10

Since there are 10 numbers, the 'middle' lies between the fifth and sixth numbers (4 and 5). So the median lies half-way between 4 and 5. In fact the median is the *mean* of 4 and 5 which is $\frac{4+5}{2} = 4{\cdot}5$.

When finding the median the numbers must always be rearranged first into ascending order. In general, for an even number of numbers the median ***is given by the mean of the middle two numbers.***

TRY SOME YOURSELF

4(i) Find the median of

1 3 7 9 11 13 15 17 19

(ii) Find the median of

5 3 4 3 3 4 5 3 4 3 4 4

First arrange the numbers in ascending order.

(iii) Find the median of

7 11 10 14 3 4 8 12

The median can also be calculated from a frequency table.

EXAMPLE

The frequency table opposite indicates the ages of 20 children. What is the median age?

Age (yrs)	Frequency
2	1
3	2
4	4
5	7
6	4
7	2
Total	20

SOLUTION

If we rewrite the data in ascending order we get

$$\underbrace{2}_{1}\quad \underbrace{33}_{2}\quad \underbrace{4444}_{4}\quad \underbrace{5555555}_{7}\quad \underbrace{6666}_{4}\quad \underbrace{77}_{2}$$

There are 20 numbers altogether so we need to find the middle two numbers–the 10th and 11th numbers. Counting from the left, the 10th number is 5 and the 11th number is also 5.

The median = mean of middle two numbers = $\frac{5+5}{2}$ = 5

The median *can* be written down straight from the table. We need to find the 10th and 11th 'ages'. Since the ages are already listed in ascending order we can count down to the 10th number using the frequency column. Both the 10th and 11th numbers are 5. So the median is 5.

Adding the frequencies up to age 4 gives

$1 + 2 + 4 = 7$

So the 7th number is 4.

Adding the frequencies up to age 5 gives

$1 + 2 + 4 + 7 = 14$

So the 8th-14th numbers are all 5.

It's quite easy to make the mistake of finding the median *frequency* rather than the median age. If you feel unsure try writing out the data as we did in the example above just to get a 'feel' for what to do.

TRY SOME YOURSELF

5 The frequency table opposite indicates the numbers of children in 30 families. What is the median number of children in a family?

No. of children	Frequency
1	4
2	7
3	11
4	4
5	3
6	0
7	1
Total	30

4.4(iv) FINDING AVERAGES FROM BAR CHARTS AND HISTOGRAMS

It's quite straightforward to find the mode from a bar chart or a histogram.

The bar chart opposite indicates the numbers of different vehicles passing in half an hour. The mode or modal vehicle is the vehicle with the highest frequency or highest column. In this example the highest column corresponds to 'car' so the modal vehicle is the car.

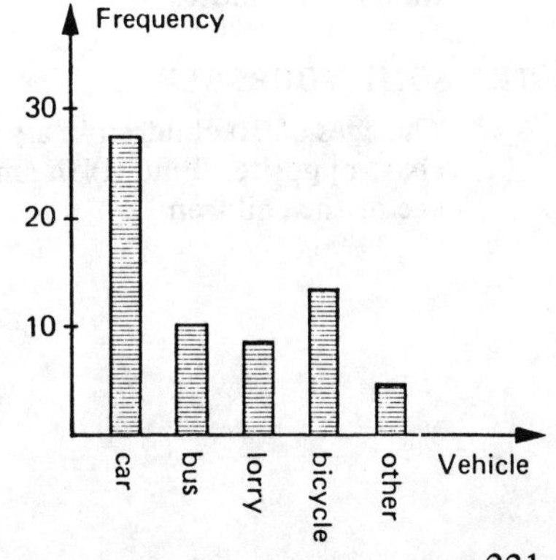

This histogram indicates the distribution of marks in an exam. The modal interval is 51-60 because it occurs with the highest frequency —it has the highest column.

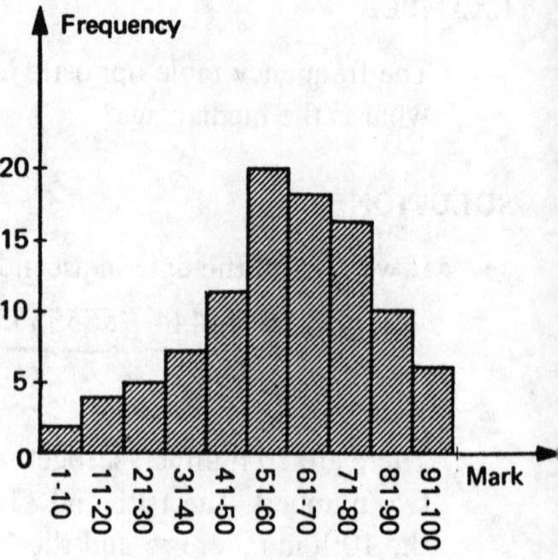

However, it is not so easy to find the mean or median straight from the diagram and indeed it is better to rewrite the information in a table before carrying out the calculation.

EXAMPLE

The bar chart below indicates the lengths of time taken by 30 children to travel to school each day. Find the mode, mean and median time taken.

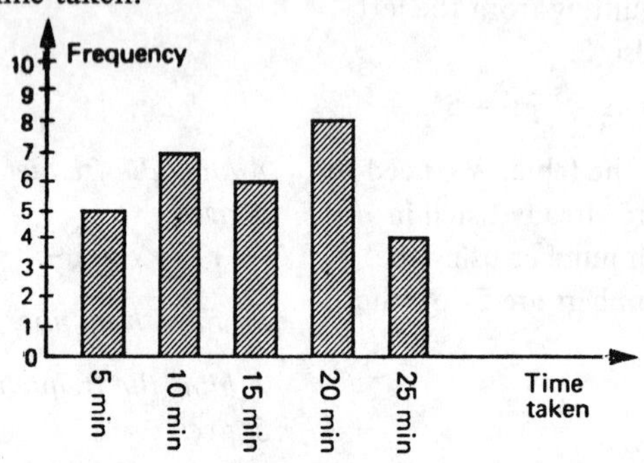

SOLUTION

The modal time is the time with the highest column. From the bar chart this is 20 minutes.

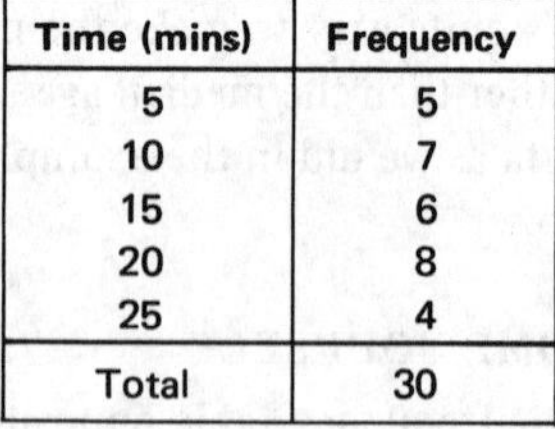

Time (mins)	Frequency
5	5
10	7
15	6
20	8
25	4
Total	30

To find the mean and median, the information must be rewritten in a table. The frequencies are read from the bar chart. Now

$$\text{The mean} = \frac{(5 \times 5) + (7 \times 10) + (6 \times 15) + (8 \times 20) + (4 \times 25)}{30}$$

$$= 14{\cdot}8 \text{ minutes} \simeq 15 \text{ minutes}$$

There are 30 numbers altogether. The median is given by the mean of the middle two (the 15th and 16th).

From the table the 12th number corresponds to 10 minutes and the 13th to 18th numbers all correspond to 15 minutes. So the median time is 15 minutes.

TRY SOME YOURSELF

6 The ages of 30 children in a youth club are indicated on the bar chart opposite. Find (i) the mode, (ii) the mean and (iii) the median age of the children.

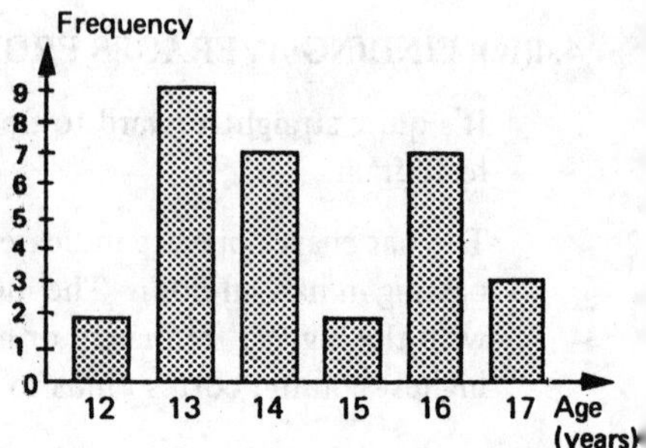

Since the mean, mode and median can often be completely different there can be considerable confusion if the word 'average' is used loosely without indicating which type of average in intended. For example, the following statements appeared in three different newspapers:

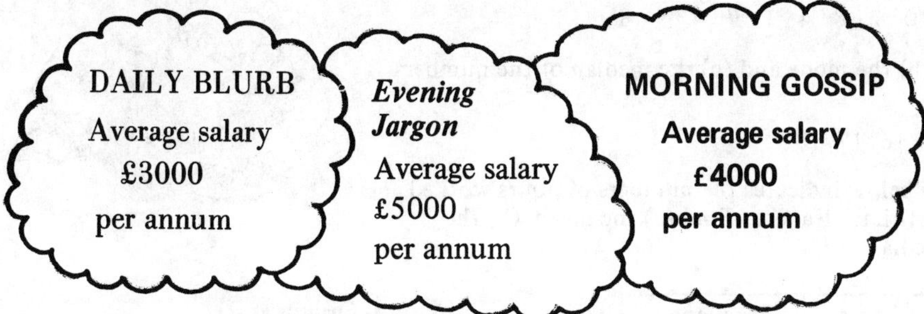

In fact the statements were based on the same information, illustrated by the histogram below.

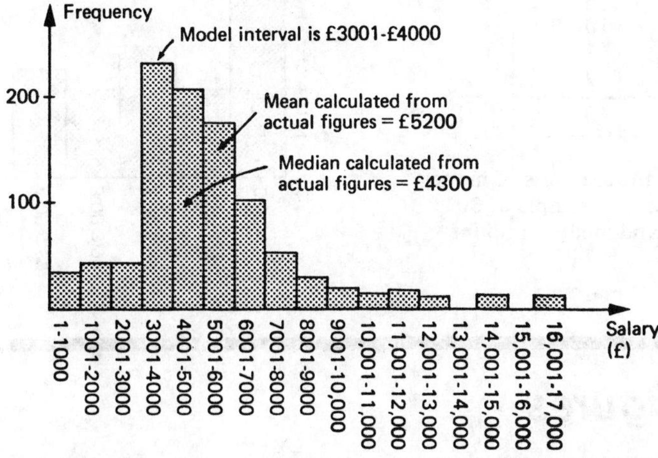

The modal interval is £3001-£4000 per annum. The mean, calculated from the actual figures, is between £5000 and £6000 a year. The median, also calculated from the actual figures, is between £4000 and £5000 a year.

The word 'average' is often used to deliberately mislead the reader. You should, therefore, always be a bit suspicious if it is used without a specific meaning.

After working through this section you should be able to

a Find the arithmetic mean of a set of numbers or from a frequency table
b Find the mode of a set of numbers or from a frequency table
c Find the median of a set of numbers or from a frequency table
d Find the mean, mode and median if the information is given in a bar chart or histogram

Finally, here are some exercises if you want more practice.

TRY SOME MORE YOURSELF

7(i) Find (a) the mean, (b) the mode and (c) the median of the numbers below:

2 2 4 7 10 2 8 4 1 7 4 5 8 4

(ii) Find (a) the mean, (b) the mode and (c) the median of the numbers below:

14 17 13 21 16 17 18

(iii) The frequency table below indicates the numbers of hours worked each week by 40 part-time cleaners. Find (a) the mean, (b) the mode and (c) the median.

No. of hours	Frequency
2	3
5	14
8	8
12	10
15	3
20	2
Total	40

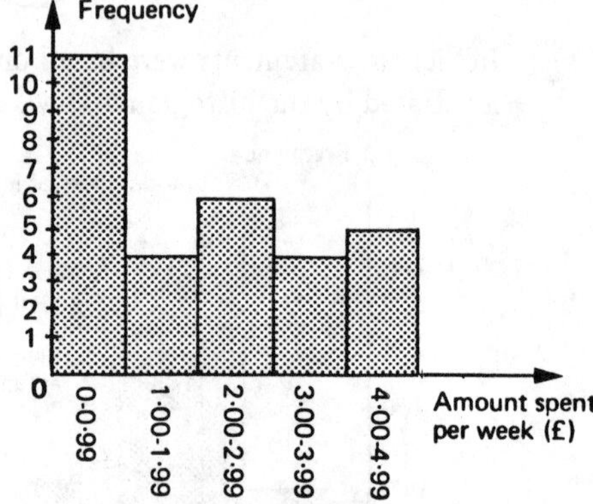

(iv) The histogram opposite indicates the amounts of money spent on entertainment each week by a sample of 30 teenagers. Find the mean, mode and median amount spent each week.

4.5 Tables of Figures

TRY THESE QUESTIONS FIRST

The table below evaluates various expressions for given values of n.

n	n^2	n^3	$\sqrt{n}$	$\sqrt{10n}$	$\sqrt[3]{n}$	$\sqrt[3]{10n}$	$\sqrt[3]{100n}$	$1/n$
17	289	4 913	4·123 106	13·038 40	2·571 282	5·539 658	11·934 83	0·058 824
18	324	5 832	4·242 641	13·416 41	2·620 741	5·646 216	12·164 40	0·055 556
19	361	6 859	4·358 899	13·784 05	2·668 402	5·748 897	12·385 62	0·052 632
20	400	8 000	4·472 136	14·142 14	2·714 418	5·848 035	12·599 21	0·050 000
21	441	9 261	4·582 576	14·491 38	2·758 924	5·943 922	12·805 79	0·047 619
22	484	10 648	4·690 416	14·832 40	2·802 039	6·036 811	13·005 91	0·045 455
23	529	12 167	4·795 832	15·165 75	2·843 867	6·126 926	13·200 06	0·043 478
24	576	13 824	4·898 979	15·491 93	2·884 499	6·214 465	13·388 66	0·041 667
25	625	15 625	5·000 000	15·811 39	2·924 018	6·299 605	13·572 09	0·040 000
26	676	17 576	5·099 020	16·124 52	2·962 496	6·382 504	13·750 69	0·038 462
27	729	19 683	5·196 152	16·431 68	3·000 000	6·463 304	13·924 77	0·037 037
28	784	21 952	5·291 503	16·733 20	3·036 589	6·542 133	14·094 60	0·035 714

1 From the table find (i) $(26)^3$ (ii) $\sqrt[3]{260}$.

2 Use the table to find an approximation for $\sqrt{225}$.

3 The table opposite indicates the numbers of children in parallel classes in Redmont School (R) and Greengage School (G). From the table find (i) R_4 (ii) G_6.

4 Use the same table to find

(i) $\sum_{i=1}^{i=7} G_i$

(ii) $\sum_{i=3}^{i=6} R_i$.

Class i	R_i	G_i
1	27	31
2	25	30
3	28	29
4	27	28
5	25	29
6	27	27
7	29	27

4.5(i) TABLES OF FIGURES

Statistical tables can be large and complicated and reading them requires careful interpretation. This is not the place to discuss statistical tables, but we include some common tables to give you practice in reading large tables of figures.

Before the advent of calculators, tables of squares, square roots, powers and reciprocals were often used to speed up calculations.

The table below is taken from a larger table giving squares, cubes and square roots of numbers between 1 and 100.

No.	Square	Cube	Square root	Reciprocals of square roots	
x	x^2	x^3	$\sqrt{x}$	$\frac{1}{\sqrt{x}}$	$\frac{1}{\sqrt{10x}}$
50	25 00	125 000	7·0711	0·1414	0·0447
51	26 01	132 651	7·1414	0·1400	0·0443
52	27 04	140 608	7·2111	0·1387	0·0439
53	28 09	148 877	7·2801	0·1374	0·0434
54	29 16	157 464	7·3485	0·1361	0·0430
55	30 25	166 375	7·4162	0·1348	0·0426
56	31 36	175 616	7·4833	0·1336	0·0423
57	32 49	185 193	7·5498	0·1325	0·0419
58	33 64	195 112	7·6158	0·1313	0·0415
59	34 81	205 379	7·6811	0·1302	0·0412
60	36 00	216 000	7·7460	0·1291	0·0408
61	37 21	226 081	7·8102	0·1280	0·0405
62	38 44	238 328	7·8740	0·1270	0·0402
63	39 69	250 047	7·9373	0·1260	0·0398
64	40 96	262 144	8·0000	0·1250	0·0395
65	42 25	274 625	8·0623	0·1240	0·0392
66	43 56	287 496	8·1240	0·1231	0·0389
67	44 89	300 763	8·1854	0·1222	0·0386
68	46 24	314 432	8·2462	0·1213	0·0383
69	47 61	328 509	8·3066	0·1204	0·0381
70	49 00	343 000	8·3666	0·1195	0·0378
71	50 41	357 911	8·4261	0·1187	0·0375
72	51 84	373 248	8·4853	0·1179	0·0373
73	53 29	389 017	8·5440	0·1170	0·0370
74	54 76	405 224	8·6023	0·1162	0·0368

The column on the far left indicates the number x. In this section of the table x ranges from 50 to 74.

CHECK YOUR ANSWERS

1 (i) $(26)^3 = 17{,}576$ (ii) $\sqrt[3]{260} = \sqrt[3]{10 \times 26} \doteq 6{\cdot}382504$ *Section 4.5(i)*

2 225 is $10 \times 22{\cdot}5$. $22{\cdot}5$ is half-way between 22 and 23. So an approximate value for $\sqrt{225}$ is half way between $14{\cdot}83240$ ($= \sqrt{220}$) and $15{\cdot}16575$ ($= \sqrt{230}$): *Section 4.5(ii)*

$$= \frac{14{\cdot}83249 + 15{\cdot}16575}{2} = 14{\cdot}99912 \simeq 15$$

(In fact $\sqrt{225} = 15$.)

3 (i) $R_4 = 27$ (ii) $G_6 = 27$ *Section 4.5(iii)*

4 (i) $\sum_{i=1}^{i=7} G_i = 31 + 30 + 29 + 28 + 29 + 27 + 27 = 201$ *Section 4.5(iv)*

(ii) $\sum_{i=3}^{i=6} R_i = R_3 + R_4 + R_5 + R_6$

$= 28 + 27 + 25 + 27 = 107$

We have shaded in the row and column which give $\sqrt{62} = 7{\cdot}8740$. The other columns give the following approximations when $x = 62$.

$(62)^2 = 3844$

$(62)^3 = 238{,}328$

$\dfrac{1}{\sqrt{62}} = 0{\cdot}1270$

$\dfrac{1}{\sqrt{620}} = 0{\cdot}0402$

Again, you may find it helpful to use a ruler as a guide.
Here $x = 62$.
The x^2 column gives $(62)^2$.
The x^3 column gives $(62)^3$.
The $\dfrac{1}{\sqrt{x}}$ column gives $\dfrac{1}{\sqrt{62}}$.
The $\dfrac{1}{\sqrt{10x}}$ column gives $\dfrac{1}{\sqrt{620}}$.

TRY SOME YOURSELF

1(i) Use the same table to find
(a) $(57)^2$ (b) $(73)^3$ (c) $\dfrac{1}{\sqrt{67}}$ (d) $\dfrac{1}{\sqrt{590}}$.

Check your answers with a calculator.

(ii) This table is rather more complicated. It evaluates some algebraic expressions for different numerical values.

p	p^2	$\sqrt{p}$	$p(1-p)$	$\sqrt{p(1-p)}$	$1-p^2$	$\sqrt{1-p^2}$	p
0·71	·5041	·8426	·2059	·4538	·4959	·7042	0·71
0·72	·5184	·8485	·2016	·4490	·4816	·6940	0·72
0·73	·5329	·8544	·1971	·4440	·4571	·6834	0·73
0·74	·5476	·8602	·1924	·4386	·4524	·6726	0·74
0·75	·5625	·8660	·1875	·4330	·4375	·6614	0·75
0·76	·5776	·8717	·1824	·4271	·4224	·6499	0·76
0·77	·5929	·8775	·1771	·4208	·4071	·6380	0·77
0·78	·6084	·8832	·1716	·4142	·3916	·6258	0·78
0·79	·6241	·8888	·1559	·4073	·3759	·6181	0·79
0·80	·6400	·8944	·1600	·4000	·3600	·6000	0·80
0·81	·6561	·9000	·1539	·3923	·3439	·5864	0·81
0·82	·6724	·9055	·1476	·3842	·3276	·5724	0·82
0·83	·6889	·9110	·1411	·3756	·3111	·5578	0·83
0·84	·7056	·9165	·1344	·3666	·2944	·5426	0·84
0·85	·7225	·9220	·1275	·3671	·2775	·5268	0·85
0·86	·7396	·9274	·1204	·3470	·2604	·5103	0·86
0·87	·7569	·9327	·1131	·3363	·2431	·4931	0·87
0·88	·7744	·9381	·1055	·3250	·2256	·4750	0·88
0·89	·7921	·9434	·0979	·3129	·2079	·4560	0·89
0·90	·8100	·9487	·0900	·3000	·1900	·4359	0·90

Use this table to find

(a) The value of $p(1 - p)$ when $p = 0{\cdot}84$.

(b) The value of $\sqrt{p(1 - p)}$ when $p = 0{\cdot}85$.

(c) When $p = 0{\cdot}74$, which is bigger, $\sqrt{p(1 - p)}$ or $\sqrt{1 - p^2}$?

The rows and columns sometimes combine to give more detailed information. The table below is an extract from a table of square roots for numbers between 10 and 100.

SQUARE ROOTS FROM 10 to 100

	0	1	2	3	4	5	6	7	8	9
55	7·416	7·423	7·430	7·436	7·443	7·450	7·457	7·463	7·470	7·477
56	7·483	7·490	7·497	7·503	7·510	7·517	7·523	7·530	7·537	7·543
57	7·550	7·556	7·563	7·570	7·576	7·583	7·589	7·596	7·603	7·609
58	7·616	7·622	7·629	7·635	7·642	7·649	7·655	7·662	7·668	7·675
59	7·681	7·688	7·694	7·701	7·707	7·714	7·720	7·727	7·733	7·740
60	7·746	7·752	7·759	7·765	7·772	7·778	7·785	7·791	7·797	7·804
61	7·810	7·817	7·823	7·829	7·836	7·842	7·849	7·855	7·861	7·868
62	7·874	7·880	7·887	7·893	7·899	7·906	7·912	7·918	7·925	7·931
63	7·937	7·944	7·950	7·956	7·962	7·969	7·975	7·981	7·987	7·994
64	8·000	8·006	8·012	8·019	8·025	8·031	8·037	8·044	8·050	8·056
65	8·062	8·068	8·075	8·081	8·087	8·093	8·099	8·106	8·112	8·118
66	8·124	8·130	8·136	8·142	8·149	8·155	8·161	8·167	8·173	8·179
67	8·185	8·191	8·198	8·204	8·210	8·216	8·222	8·228	8·234	8·240
68	8·246	8·252	8·258	8·264	8·270	8·276	8·283	8·289	8·295	8·301
69	8·307	8·313	8·319	8·325	8·331	8·337	8·343	8·349	8·355	8·361
70	8·367	8·373	8·379	8·385	8·390	8·396	8·402	8·408	8·414	8·420
71	8·426	8·432	8·438	8·444	8·450	8·456	8·462	8·468	8·473	8·479
72	8·485	8·491	8·497	8·503	8·509	8·515	8·521	8·526	8·532	8·538
73	8·544	8·550	8·556	8·562	8·567	8·573	8·579	8·585	8·591	8·597
74	8·602	8·608	8·614	8·620	8·626	8·631	8·637	8·643	8·649	8·654
75	8·660	8·666	8·672	8·678	8·683	8·689	8·695	8·701	8·706	8·712
76	8·718	8·724	8·729	8·735	8·741	8·746	8·752	8·758	8·764	8·769
77	8·775	8·781	8·786	8·792	8·798	8·803	8·809	8·815	8·820	8·826
78	8·832	8·837	8·843	8·849	8·854	8·860	8·866	8·871	8·877	8·883
79	8·888	8·894	8·899	8·905	8·911	8·916	8·922	8·927	8·933	8·939

In this case, the column on the left hand side gives the whole numbers (ranging from 55 to 79).

The other columns give decimal parts, so that the table gives the square roots of
55·0, 55·1, 55·2, 55·3 . . .
56·0, 56·1, 56·2, 56·3 . . .
etc.

For example, this table indicates that

$$\sqrt{64} = \sqrt{64{\cdot}0} = 8{\cdot}000$$

and

$$\sqrt{64{\cdot}3} = 8{\cdot}019$$

In each case, the square root is given to three decimal places.

TRY SOME YOURSELF

2 Use the same table to find
(i) $\sqrt{59}$ (ii) $\sqrt{59{\cdot}9}$ (iii) $\sqrt{75{\cdot}6}$

Check your answers with a calculator.

4.5(ii) READING BETWEEN THE LINES

Quite often the tables do not give exactly the information you want. It is sometimes possible to *interpolate* for values in between the ones which are given.

The following table is a handy conversion chart from pounds to dollars. This is based on 1980 figures.

£	1	2	3	4	5	6	7	8	9	10
$	2·3	4·6	6·9	9·2	11·5	13·8	16·1	18·4	20·7	23

Of course, it's easy to interpolate using the graph.

The table converts only whole numbers of pounds to dollars. However, it is possible to use the table to find, for example, the dollar equivalent of £2·50.

Since £2·50 is half-way between £2 and £3, its dollar equivalent must be half-way between \$4·6 and \$6·9:

£2·50 = £2 + £0·50

So

$$£2{\cdot}50 = \$4{\cdot}6 + \frac{\$6{\cdot}9 - 4{\cdot}6}{2}$$

= £2

Half the difference between £2 and £3.

= \$5·175

Similarly the dollar equivalent of £2·25 can also be evaluated from the table. £2·25 is a quarter of the way between £2 and £3.

£2·25 = £2 + £0·25

So

$$£2{\cdot}25 = \$4{\cdot}6 + \frac{\$6{\cdot}9 - 4{\cdot}6}{4}$$

= £2

Quarter of the difference between £2 and £3.

= \$5·75

In the example above the interpolated values were quite accurate. That's because the graph was a straight line. More generally, the interpolated values only give an approximation.

EXAMPLE

The table below indicates the square roots of numbers between 1 and 10.

Number	1	2	3	4	5	6	7	8	9	10
Square root	1	1·41	1·73	2	2·24	2·45	2·65	2·83	3	3·16

The square roots are given to two decimal places.

Use the table to find an approximate value for $\sqrt{7 \cdot 5}$.

SOLUTION

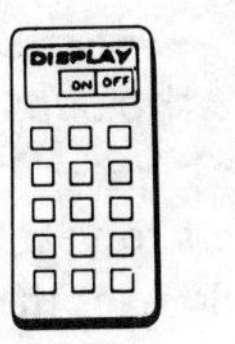

Check:

$\sqrt{7 \cdot 4} = 2 \cdot 7386$

$= 2 \cdot 74$

(Rounded to 2 decimal places.)

7·5 is half-way between 7 and 8. So the number half-way between 2·65 and 2·83 should give a reasonable approximation:

$$\sqrt{7 \cdot 5} \simeq 2 \cdot 65 + \frac{2 \cdot 83 - 2 \cdot 65}{2} = 2 \cdot 74$$

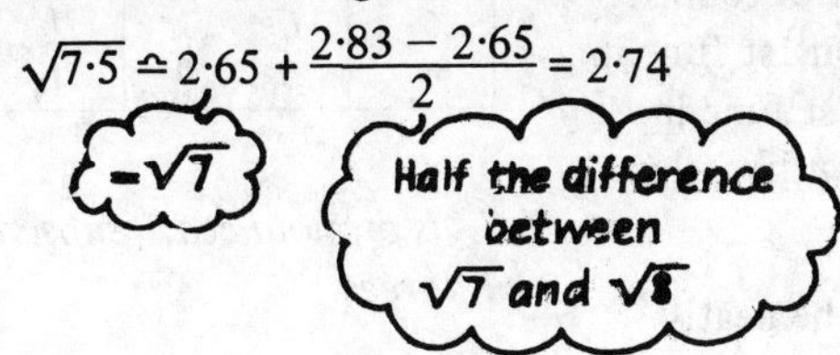

TRY SOME YOURSELF

3(i) Use the conversion table from pounds to dollars to find the dollar equivalent of
(a) £3·50 (b) £8·50.

(ii) Use the square root table given in the example above to find
(a) $\sqrt{4 \cdot 5}$ (b) $\sqrt{6 \cdot 5}$

4.5(iii) SUBSCRIPT NOTATION IN TABLES

Tables are often used to record information in successive stages. The table below resulted from a study of 144 robins who had left the nest one year on June 1st. It indicates the number of robins who remained alive on that date in subsequent years.

Year	Alive on June 1st	Died in following 12 months
0	144	111
1	33	15
2	18	14
3	4	2
4	2	1
5	1	1
6	0	0

The exact year is not given in this table. The table starts in year 0 when 144 robins first left the nest.

For example, at the end of one year 33 robins remained alive. In the following 12 months 15 robins died.

On June 1st in year 1 thirty three robins were alive.

This type of sequential information can be expressed neatly using *subscripts.*

Suppose

L represents the numbers alive on June 1st,

then L_0 represents the number alive on June 1st in the year 0

L_1 represents the number alive on June 1st in year 1

etc.

Similarly, D stands for the number who died in the following 12 months after June 1st.

Using this subscript notation, the table can be rewritten as shown in the margin. From this table, $L_3 = 4$.

The *subscript* 3 refers to year 3. L refers to the number of robins alive on 1st June. So L_3 is the number of robins alive on 1st June in year 3. In general, L_i is the number of robins alive on 1st June in year i. D_i is the number of robins who die in the 12 months following 1st June in year i.

Year	L	D
0	L_0 144	D_0 111
1	L_1 33	D_1 15
2	L_2 18	D_2 14
3	L_3 4	D_3 2
4	L_4 2	D_4 1
5	L_5 1	D_5 1
6	L_6 0	D_6 0

L_3 is pronounced L subscript 3 or L three.

Subscripts therefore allow quite lengthy expressions to be neatly condensed.

EXAMPLE

The table opposite records the numbers of television sets (T) and radios (R) sold by one firm in successive months in 1980. Find
(a) T_4
(b) The number of radios sold in August.

i	T_i	R_i
1	148	137
2	167	125
3	130	108
4	133	158
5	118	172
6	123	134
7	103	131
8	111	102
9	104	94
10	106	95
11	101	92
12	94	84

SOLUTION

(a) T_4 is given by the 4th row in column T_i. From the table, $T_4 = 133$.
(b) August is the 8th month of the year. So we need to find R_8. From the table, $R_8 = 102$.

TRY SOME YOURSELF

4 Use the table given in the example above to find
(i) R_{12}
(ii) The number of television sets sold in March
(iii) $T_2 + R_2$.
(iv) What does T_i stand for?

4.5(iv) ADDING UP DATA USING SUBSCRIPTS

The table opposite indicates the numbers of boys and girls born in successive weeks in a large hospital.

Week i	*Boys* B_i	*Girls* G_i
1	4	2
2	2	4
3	1	7
4	9	4
5	3	1
6	0	2

The total number of boys born in the 6 week period was

$$4 + 2 + 1 + 9 + 3 + 0 = 19$$

In subscript notation this is equivalent to

$$B_1 + B_2 + B_3 + B_4 + B_5 + B_6$$

This sum can be abbreviated to

$$\sum_{i=1}^{i=6} B_i$$

This notation is commonly used in statistics. Σ is the capital Greek letter 'sigma'. It means ***'add up'***.

Thus

$$\sum_{i=1}^{i=6} B_i = B_1 + B_2 + B_3 + B_4 + B_5 + B_6 = 19$$

Similarly

$$\sum_{i=1}^{i=6} G_i = G_1 + G_2 + G_3 + G_4 + G_5 + G_6$$

$$= 2 + 4 + 7 + 4 + 1 + 2$$

$$= 20$$

So

$\sum_{i=1}^{i=6} B_i$ *means* ***add up*** *all the* B_i*s between* $i = 1$ *and* $i = 6$.

$\sum_{i=1}^{i=6} B_i$ *is pronounced 'sigma* B_i *from 1 to 6'.*

The 'sigma' notation can also be used for subtotals:

$\sum_{i=1}^{i=3} B_i$ means *add up* the B_is between $i = 1$ and $i = 3$.

From the table above,

$$\sum_{i=1}^{i=3} B_i = B_1 + B_2 + B_3 = 4 + 2 + 1 = 7$$

and

$$\sum_{i=3}^{i=6} B_i = B_3 + B_4 + B_5 + B_6 = 1 + 9 + 3 + 0 = 13$$

Thus the 'sigma' notation is a convenient, shorthand way of writing down long sums using subscript notation.

EXAMPLE

The table opposite indicates the numbers of new and second hand cars sold by one garage over a number of years. Find the mean number of new cars sold each year.

Year i	New cars N_i	Second hand cars S_i
1	100	150
2	170	110
3	200	210
4	150	160
5	170	150
6	120	100
7	190	200
8	160	180

SOLUTION

$$\text{The mean} = \frac{\text{Total number of new cars sold}}{\text{Number of years}}$$

$$= \frac{\sum_{i=1}^{i=8} N_i}{8} = 157{\cdot}5 \simeq 158 \text{ cars.}$$

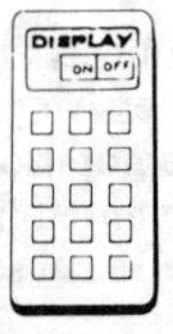

Use your calculator to check this yourself.

TRY SOME YOURSELF

5(i) Use the table in the example above to find

(a) $\sum_{i=1}^{i=4} N_i$ (b) $\sum_{i=5}^{i=8} S_i$

(c) The mean number of second hand cars sold each year.

(ii) Go back to the table indicating sales of television and radio sets on page 240. Use the table to find

(a) $\sum_{i=4}^{i=10} T_i$ (b) $\sum_{i=2}^{i=5} R_i$ (c) $\sum_{i=7}^{i=12} R_i$

(d) The mean number of TV sets sold each month.

The examples on' sigma' notation which we have included have all used the letter i as the subscript. The letters n and r are also commonly used with this notation. For example,

$$\sum_{n=1}^{n=3} D_n = D_1 + D_2 + D_3$$

Add up all the D_ns between $n = 1$ and $n = 3$.

and

$$\sum_{r=1}^{r=4} F_r = F_1 + F_2 + F_3 + F_4$$

Add up all the F_rs between $r = 1$ and $r = 4$.

In fact, as long as the notation is clear, any letter can be used.

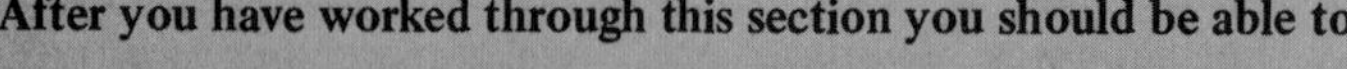

After you have worked through this section you should be able to

a Read common tables of squares, square roots, reciprocals etc.
b Interpolate from tables to find values half-way between given points
c Interpret information given in subscript notation
d Use sigma notation to add up values from a table

Finally, here are some exercises if you want more practice.

TRY SOME MORE YOURSELF

6(i) Go back to the tables on page 235 and find

(a) $(68)^2$ (b) $\sqrt{70}$

(c) An approximate value for $\sqrt{59{\cdot}5}$

(d) An approximate value for $\frac{1}{\sqrt{655}}$

(ii) Go back to the table on page 237 and find

(a) $\sqrt{61}$ (b) $\sqrt{65{\cdot}5}$

(c) An approximate value for $\sqrt{65{\cdot}55}$.

(iii) The table opposite indicates the numbers of trees planted on a housing estate in successive years.

(a) Choose a suitable notation to represent the number of deciduous trees planted in the year i.

(b) Rewrite the table in subscript form using this notation.

Year	Deciduous trees	Conifers
1970	10	3
1971	12	5
1972	19	11
1973	31	13
1974	43	8
1975	41	8
1976	36	6
1977	30	4

(iv) Go back to the table on page 240 and find

(a) T_5

(b) The number of radio sets sold in October

(c) The number of radio sets sold in January

(d) $\sum_{i=3}^{i=6} T_i$

(e) $\sum_{i=4}^{i=9} R_i$

(f) The mean number of radio sets sold each month.

Section 4.1 Solutions

1

(i) (a) 15 minutes.

(b) The temperature drops to 36° C–a total of 54° centigrade. In the second 20 minutes, the temperature dropped from 36° C to 25° C–a total of 11° C.

(c) The tea cools off quickly at the beginning then cools more gradually. (This is certainly my experience when drinking tea!)

(ii) (a) The maximum amount was £19·14.

(b) The minimum was £3·22.

(c) The reasons might be because Tony and Sue entertained during the week of 16.3.81. Similarly, they may have gone away during the week of 25.2.81 or eaten out more often. There are many reasons why the amounts vary. We can only suggest possibilities. You may have thought of other possibilities, though it's important to remember that these are only suggestions.

2

(i)

Score	Tally	Frequency
1-10	||	2
11-20	|||	3
21-30	~~||||~~ |	6
31-40	||	2
41-50	~~||||~~	5
51-60	||||	4
61-70	~~||||~~	5
71-80	~~||||~~	5
81-90	~~||||~~ |	6
91-100	||	2
	Total	40

Check that the figures add up to the given total. (Here there are 40 children.)

(ii) (a) First we must select the headings for the table. We could sort the data into completely separate categories–4, 4½, 5, 5½, etc.–but we have sorted them as shown in the table below. If you have sorted them into separate sizes do not worry–it is probably still correct–as long as the total is 40.

Size	Tally	Frequency
4–4½	~~||||~~ ||	7
5–5½	~~||||~~ ~~||||~~ ~~||||~~ ||	17
6–6½	~~||||~~ ~~||||~~	10
7–7½	~~||||~~ |	6
	Total	40

The hardest part is deciding how to group the data. This is difficult at first but becomes easier as you gain more experience.

(b) From the table, sizes 5–5½ accounted for the largest number of sales.
If you sorted the data into separate sizes you should have found that the answer was size 5.

3

(i) (a) The appropriate column is Birmingham (7 nights),

Hotel	
Departure airport	BIRMINGHAM
No. of nights in hotel	7
Aircraft	B737
Departure day/approx. time	0835
Arrive back day/approx. time	1455
First departure	26 Apr
Last departure	18 Oct
Holiday number	T2255
Departure between 26 Apr	103
28 Apr-5 May	129
6-18 May	125
19-26 May	135
27 May-15 June	142
16 June-13 July	152
14 July-5 Aug	158
6-31 Aug	162
1-14 Sep	156
15-28 Sep	150
29 Sep-18 Oct	120

The Potters will fly by a B737 leaving Birmingham at about 08.35 arriving back at about 14.45 one week later. The cost is £120 per adult. Since their daughter is 12 she is not entitled to any reduction. The flight time is 2 hours 20 minutes and it takes about 1 hour 5 minutes to transfer from the airport to the hotel. The total cost of the holiday is therefore £360.

(b) The most expensive holiday costs £236 for 2 weeks starting between the 6th and 31st of August, leaving from Glasgow.
The cheapest holiday is for 1 week leaving from Gatwick on 26th April at a cost of £98.
To glean information like this from a table it is necessary to go through the data systematically. Even then it is easy to make mistakes. Unfortunately there is no quick method; you just have to be careful.

(ii) (a) The appropriate row and columns are shown below:

age of car → 2 Years

2 years no claim bonus

Driver age	NCB years					2 Years Any	Insd/Spouse
						(£)	(£)
						163	139
						124	106
						108	93
						85	73
						69	60
27-30	0	171	146	178	152	185	158
	1	130	111	135	116	141	121
	2	113	97	118	101	123	105
	3	89	77	92	80	96	83
	4	72	63	75	65	78	67

Insured and spouse only

The cost therefore is £105.

(b) From the table, the younger the driver, the more he must pay. So the maximum occurs when the driver is between 23 and 26, when it costs £157. The cheapest premium occurs if the driver is aged between 31 and 35, when he must pay £102.

4

(i) (a) The rate for the house itself is 15p or £0·15 per £100 since the house is in Glasgow.
So the cost is

$$\frac{£0{\cdot}15 \times 41{,}000}{100} = £61{\cdot}50$$

The contents are rated at 50p or £0·5 per £100.
So the cost to insure the contents is

$$£0{\cdot}50 \times \frac{7000}{100} = £35$$

The total cost is £61·50 + £35 = £96·50.

(b) The most expensive area to insure a house is in London, where the rate is 20p per £100.

(c) The cheapest areas to insure house contents are in the 'elsewhere' category where the rate is 30p per £100.

(ii) (a) The daily rate is £9, so 4 days costs 4 x £9 = £36.
The mileage rate is 6p (or £0·06) per mile, so 625 miles costs £625 x 0·06 = £37·50.
The total cost is therefore £36 + £37·50 = £73·50.

(b) The unlimited mileage daily rate is £12, so 4 days costs 4 x £12 = £48.
It is therefore much cheaper to hire the car at the unlimited mileage rate.

5 (a) From the table the 100% base for 1977 is 11,979.

(b) 18% of the households surveyed contained 4 people. The actual number was 18% of 11,979, which is
2156·22 or 2156 (Rounded to the nearest whole number.)
Therefore 2156 households comprised 4 people in 1977.

(c) The percentages for the 1977 column total
21 + 33 + 17 + 18 + 7 + 4 = 100%

6

(i) From the table, 34·9% of families consisted of a married couple with 2 dependent children. The total number of families interviewed was 4855.
34·9% of 4855 = 1694·39 = 1694
(Rounded to the nearest whole number.)
So the actual number of families consisting of a married couple and two children was 1694.

(ii) (a) 22%. Notice that this is the column *total.*

(b) 10% of the survey consisted of male non-smokers aged between 16 and 24. The total number interviewed was 10,480.
10% of 10,480 = 1048
Therefore, there were 1048 non-smokers aged between 16 and 24.

(c) 22% of the survey was aged over 60 (row total).
Therefore the actual number was 22% of 10,480 = 2305·6 or 2306. So the number of men interviewed who were over 60 was 2306.

(d) The percentage totals actually add up to 99%. This is probably due to rounding errors.

7

(i) Grouping the temperatures into groups of two, the frequency table is

Temperature	Frequency
15-16	4
17-18	8
19-20	8
21-22	4
23-24	3
25-26	1
27-28	2
Total	30

(ii) £22

(iii) The weekly unlimited mileage rate is £60. The alternative is to pay £42 and use up £18 of petrol.
£18 means (18 ÷ 0·06) miles = 300 miles.
So if he intends to travel over 300 miles he should hire the car at the unlimited mileage rate.

(iv) (a) 20%
(b) 1494
(c) '25-34', '35-49' and 'over 60'.
(d) The age group 25-34 contains both the highest percentage of smokers and the highest percentage of heavy smokers.

Section 4.2 Solutions

1

(i) The cost reduces with age. It falls off quickly at first, losing half its value in the first two years (its value is about £200 in 1973), then drops more gradually.

(ii) The cost of coffee increased over the six month period from £1 per kg to £2 per kg. It remained steady between February and April, rose significantly during the month of April then increased steadily until July.

2 Your graph should look something like this:

(i)

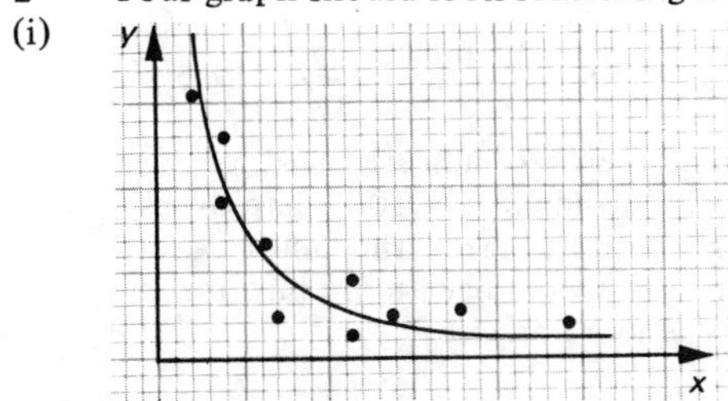

(ii) No. The numbers of children are *discrete*. There is no meaning to 1·5 children.

(iii) Your graph should look something like this:

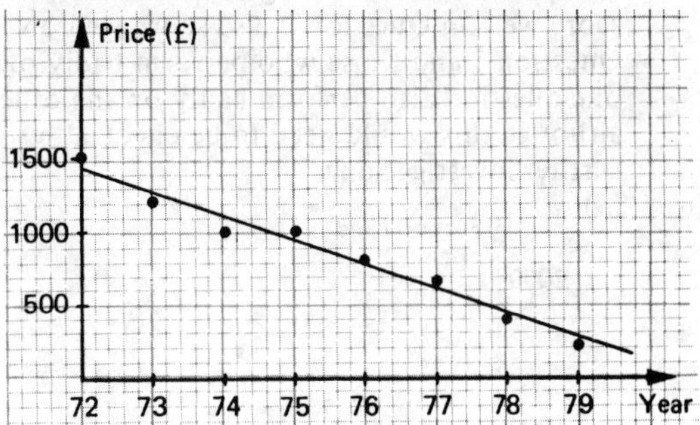

3

(i) Just under 50%.

(ii) Less than 5% earned over £3000 a year and only about 10% earned more than £2000 a year. This means that about 90% earned *less than* 2000 a year. About 70% earned over £1000 a year.

(iii) About 10% of men with no qualifications earned over £2000 per year. About 12% of men with CSE or commercial apprenticeships earned over £2000 per year. The remaining percentages are about 20% of men with O-level or equivalent, 27% of men with A-level or equivalent, 50% of men with education below degree standard and about 80% of men with a degree or equivalent.

(iv) Most men with a degree earned over £2000 (about 80%).
Most men with no qualifications earned under £2000 (about 90%).
About 50% of men with a degree earned over £3000 whereas only about 3% of men with no qualifications fell into this category.
The two graphs differ significantly over £1000. The graph showing 'no qualifications' drops quickly whereas the graph showing 'degree or equivalent' remains steady and falls off very gradually.

4

(i)

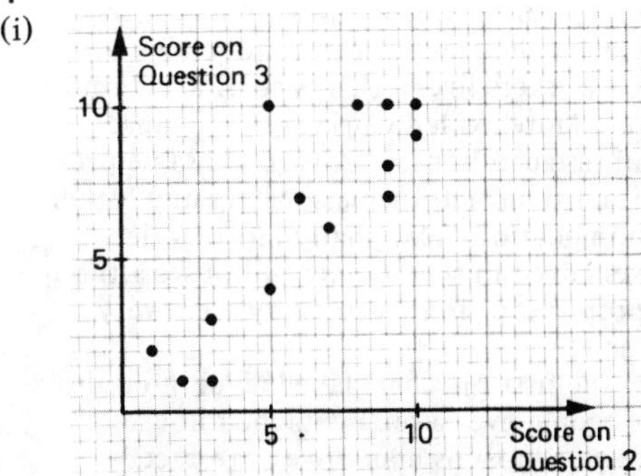

The pattern seems to suggest that those who score well on Question 2 also score well on Question 3 and those who perform badly on one question also perform badly on the other, although this would need to be checked by further investigation.

(ii) (a)

(b) The scatter diagram seems to indicate that the higher the price of a new car, the higher its corresponding second hand price. The scales used on the axis will determine how marked this difference appears to be. This is one reason why further analysis is essential.

5

(i)

(ii) The largest proportion of people live in semi-detached houses (about 30%). Over 25% live in terraced houses and about 20% live in detached houses. About 15% live in flats or maisonettes and about 10% live in some other type of accommodation.

6

(i) Detached houses.

(ii) Terraced houses.

(iii) About 1/3 or 30% of employers/managers live in detached houses but only about 10% of skilled workers live in this type of accommodation. More skilled workers than managers live in terraced houses. About the same proportion live in flats or maisonettes and 'other' accommodation. Marginally more skilled workers lived in semi-detached houses although the proportions in this category were very similar.

(iv) From the first pie chart labelled 'all households', most people live in semi-detached houses although almost as many live in terraced houses.

7

(i)

The graph indicates that the cost is roughly linearly proportional to the weight.

(ii) The percentage of households with only 1 person remains very low until the graph reaches the age of 40. It then increases until at the age of 80, 40% of households are one person households. The graph for 4 people has a peak at about 40 when about 70% of households consist of 4 people. The graph then falls off quickly, probably because a lot of people are married with two children and the children begin to leave home.

The graph for three people increases up to 30, decreases until 40 then increases and decreases again. This supports the notion that families tend to consist of 2 children and the children begin to leave home when the parents are about 40.

The graph for 2 people decreases up till 40 then increases up till 60 before falling off gradually. This also supports the above theory. After the age of 60 there are more deaths which explains why the graph decreases.

(iii)

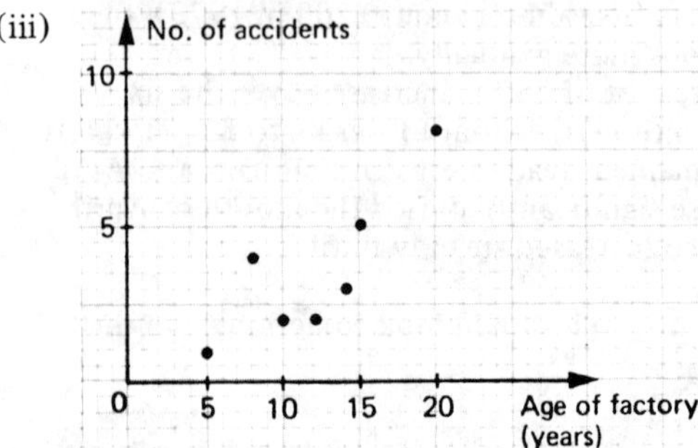

The trend seems to indicate that the number of accidents per 1000 employees increases with the age of the factory—although we would need more information to be more precise.

(iv) In 1951 solid fuel accounted for the largest proportion (about 80%). This dropped to about 30% in 1977, a large decrease. Gas increased from about 10% to nearly 50%, probably due to the introduction of North Sea gas. Electricity also increased substantially from about 5% to nearly 20%. Oil increased a little from about 3% to about 5%.

Section 4.3 Solutions

1

(i) (a)

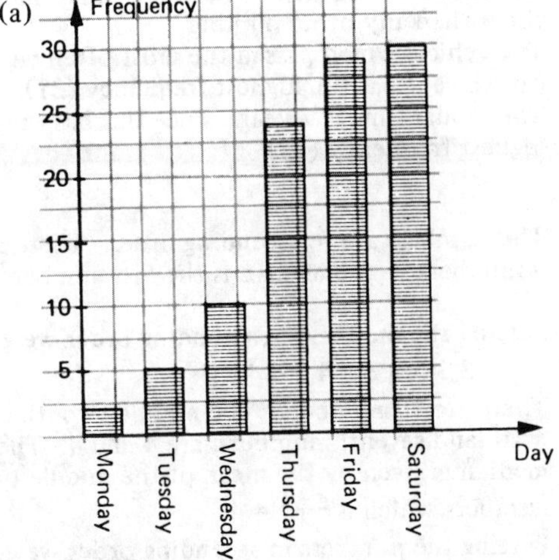

(b) Saturday.
(c) Monday.

(ii) (a) About 12,000.
(b) About 21,000.
(c) The figures are generally increasing although there was a slight fall in 1974.

2

(i) (a)

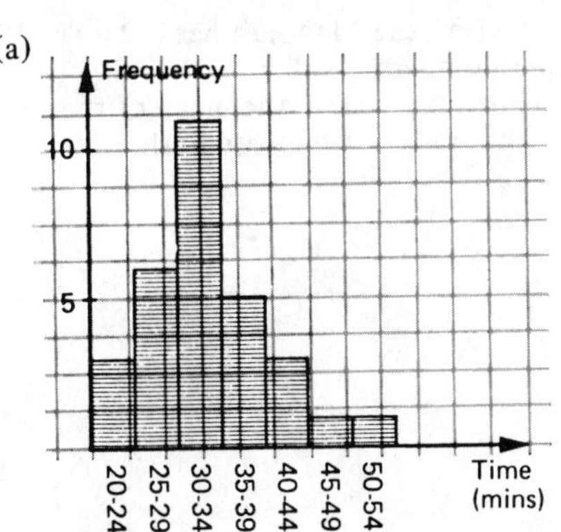

(b) Each interval covers 5 minutes. For example, 20, 21, 22, 23 and 24 minutes make one interval. Notice that the intervals do not overlap.
(c) The interval is 30-34 minutes.

(ii) (a) 10 years. For example, 0, 1, 2, 3, 4, 5, 6, 7, 8, and 9 years make one interval.
(b) 10-19 years.
(c) The number of people in their 70s is about 4000.
The number of people in their 30s is about 7000.
So the ratio is 4000:7000 or 4:7.
(d) We need to find the total number of people in the sample. This is given by adding the frequencies for each column. It's about
8000 + 9000 + 8000 + 7000 + 6500 + 7000 + 6000 + 4000 + 1500 = 57,000
So the approximate sample size is 57,000.

3

(i) About 700.
(ii) About 2000.
(iii) In each age group, non-smokers contribute the largest numbers.
There are more heavy smokers in the middle age brackets than aged 16-24.

4

(i) About 10%.
(ii) Just over 50%.
(iii) About 10% of under 25s live alone. The proportion then drops before beginning to rise again after the age of 45.

5

(i)

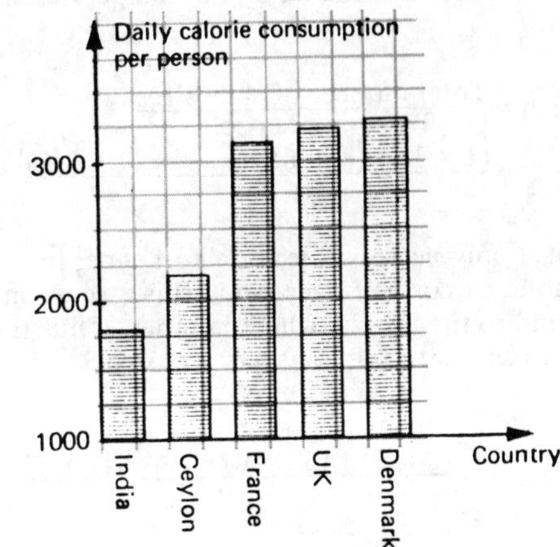

(ii) (a)

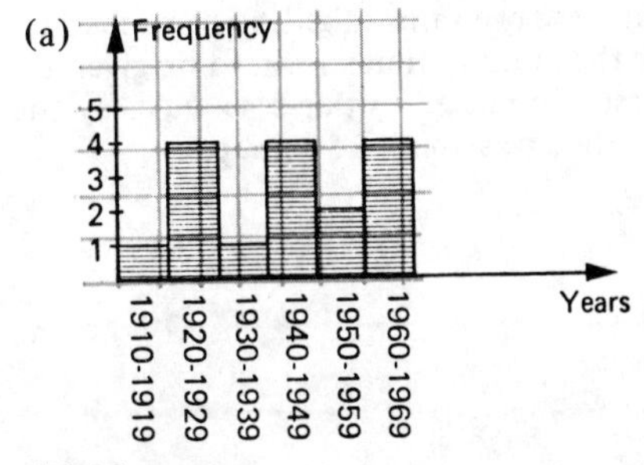

(b) 10 years.

(iii) (a) The numbers increased steadily to a maximum in 1951.
(b) The numbers increased to 1921 then fell in 1951.
(c) In 1841 and 1891 more people were involved in mining than white collar jobs. The relationship was reversed in 1921 and 1951. See parts (a) and (b) for further information.

(iv) Very few young people owned their house outright, but this proportion increased steadily until about 40% of over 70 year olds owned their property outright. The proportions with a mortgage reflect this position since this percentage increased a little before decreasing to almost nothing. The proportion renting from the local authority remained steady—between 30 and 40%. More young people rent accommodation privately—furnished or unfurnished. The proportion in this category decreases then increases again after the age of 60.

Section 4.4 Solutions

1

(i) $$\text{Mean} = \frac{52 + 49 + 64 + 56 + 63 + 66}{6} = 58{\cdot}3 \simeq 58$$
(Rounded to the nearest whole number.)

(ii) $$\text{Mean} = \frac{\text{Total of temperatures}}{31} = 19{\cdot}8^\circ\text{C}$$
(Use a calculator to add up the temperatures.)

2

(i) $$\text{Mean} = \frac{\text{Total number of shoe sizes}}{\text{Total number sold}}$$
$$= \frac{(1 \times 3) + (3 \times 3{\cdot}5) + \ldots + (1 \times 7{\cdot}5)}{57}$$
$$= 5{\cdot}25$$
But it only makes sense to talk of shoes in whole sizes or half sizes—so in this case we must round to the nearest whole number. Thus the mean shoe size sold that day was size 5.

3

(ii) $$\text{Mean} = \frac{(3 \times 5) + (5 \times 10) + \ldots + (1 \times 30)}{30}$$
$$= 16{\cdot}33 \text{ minutes}$$
which is 16 minutes to the nearest minute. In this case even this might be too accurate since the journey times seem to be given to the nearest 5 minutes. In this case we'd say the mean time was about 15 minutes.

3

(i) Putting the information into a frequency table we get

Number	Frequency
1	9
2	7
3	5
4	8
5	8
6	5
Total	42

The mode is 1 because it occurs 9 times; more times than any other number.

(ii) The vehicle which passed the most often was the car, since it has the highest frequency (21).

(iii) The modal time is 7 years, since this has the highest frequency (15).

4

(i) The numbers are in ascending order. There are 9 numbers so the median is the 5th number: 11.

(ii) Putting the numbers in ascending order we get

3 3 3 3 3   4 4 4 4 5 5

There are 12 numbers. The middle two, the sixth and seventh numbers, are 4 and 4. The median is given by the mean of the middle two numbers which is $\frac{4+4}{2} = 4$.

(iii) Putting the numbers in ascending order we get

3 4 7 8   10 10 12 14

There are 8 numbers. The middle two, the 4th and 5th numbers, are 8 and 10. The median is given by the mean of the middle two numbers, which is $\frac{8+10}{2} = 9$.

5

By inspection the 11th number is 2. The 12th to 22nd numbers are all 3.
The median is given by the mean of the 15th and 16th numbers which are both 3. So the median is 3.

6

(i) From the bar chart the mode is 13 years, since it has the highest column.

(ii) Rewriting the information in a frequency table we get

Age	Frequency
12	2
13	9
14	7
15	2
16	7
17	3
Total	30

$$\text{Mean} = \frac{(2 \times 12) + (9 \times 13) + \ldots + (3 \times 17)}{30}$$

= 14·4 years or 14 years to the nearest year

(iii) By inspection both the 15th and 16th ages are 14. So the median is 14 years.

7

(i) (a) Mean = 4·86
(b) Mode = 4
(c) Median = 4

(ii) (a) Mean = 16·57
(b) Mode = 17
(c) Median = 17

(iii) Mean = 8·6 hours
Mode = 5 hours
Median = 8 hours

(iv) We can get an approximate value for the mean as follows:

$$\text{Mean} = \frac{(11 \times 1) + (4 \times 2) + (6 \times 3) + (4 \times 4) + (5 \times 5)}{30}$$

$\simeq$ £2·60

The modal amount spent was under £1.
The median amount spent was about £2.
(The 15th number is between £1·00 and £1·99 the 16th number is between £2·00 and £2·99. £2 is therefore a reasonable approximation.)

Section 4.5 Solutions

1

(i) (a) $(57)^2 = 3249$
(b) $(73)^3 = 389{,}017$
(c) $\frac{1}{\sqrt{67}} = 0{\cdot}1222$
(d) $\frac{1}{\sqrt{590}} = \frac{1}{\sqrt{10 \times 59}} = 0{\cdot}0412$

(ii) (a) 0·1344
(b) 0·3671
(c) $\sqrt{p(1-p)} = 0{\cdot}4386$
$\sqrt{1-p^2} = 0{\cdot}6726$

So when $p = 0{\cdot}74$, $\sqrt{1-p^2}$ is bigger than $\sqrt{p(1-p)}$.

2

(i) $\sqrt{59} = \sqrt{59{\cdot}0} = 7{\cdot}681$

(ii) $\sqrt{59{\cdot}9} = 7{\cdot}740$

(iii) $\sqrt{75{\cdot}6} = 8{\cdot}695$

3

(i) (a) The dollar equivalent for £3·50 must be halfway between \$6·9 and \$9·2. So

$$£3{\cdot}50 \simeq \frac{\$6{\cdot}9 + \$9{\cdot}2}{2} \simeq \$8{\cdot}05 \simeq \$8$$

(This is an alternative way of finding the equivalent of a *halfway point*. It is just the mean of the dollar equivalent of the neighbouring whole numbers.)

(b) $£8{\cdot}50 \simeq \frac{\$18{\cdot}4 + \$20{\cdot}7}{2} \simeq \$19{\cdot}55$

(ii) (a) $\sqrt{4{\cdot}5} \simeq \frac{2 + 2{\cdot}24}{2} \simeq 2{\cdot}12$.
(b) $\sqrt{6{\cdot}5} \simeq \frac{2{\cdot}45 + 2{\cdot}65}{2} = 2{\cdot}55$

(In fact $\sqrt{4{\cdot}5} = 2{\cdot}121$ and $\sqrt{6{\cdot}5} = 2{\cdot}550$ to three decimal places.)

4

(i) $R_{12} = 84$

(ii) The number of TV sets sold in March = T_3 = 130.

(iii) $T_2 = 167$ and $R_2 = 125$. Therefore $T_2 + R_2 = 167 + 125 = 292$.

(iv) T_i stands for the number of television sets sold in the ith month.

5

(i) (a) $\sum_{i=1}^{i=4} N_i = N_1 + N_2 + N_3 + N_4$
$= 100 + 170 + 200 + 150 = 620$

(b) $\sum_{i=5}^{i=8} S_i = S_5 + S_6 + S_7 + S_8$
$= 150 + 100 + 200 + 180 = 630$

(c) The mean number of second hand cars sold each year was

$$\frac{\sum_{i=1}^{i=8} S_i}{8} = \frac{1260}{8} = 157{\cdot}5 \simeq 158 \text{ cars}$$

(ii) (a) $\sum_{i=4}^{i=10} T_i = T_4 + T_5 + T_6 + T_7 + T_8 + T_9 + T_{10}$

$= 798$

(b) $\sum_{i=2}^{i=5} R_i = R_2 + R_3 + R_4 + R_5 = 563$

(c) $\sum_{i=7}^{i=12} R_i = R_7 + R_8 + R_9 + R_{10} + R_{11} + R_{12}$

$= 598$

(d) The mean number of TV sets sold each month was

$$\frac{\sum_{i=1}^{i=12} T_i}{12} = 119{\cdot}8 \doteq 120 \text{ sets.}$$

6

(i) (a) 4624 (b) 8·3666 (c) 7·7136 (d) 0·0391

(ii) (a) 7·810 (b) 8·093 (c) 8·096

(iii) (a) D_i = number of deciduous trees planted in year 197*i*. C_i = number of conifers planted in year 197*i*. (You may have used different letters; it doesn't matter as long as you make it clear what the letters stand for.)

(b)

Year i	D_i	C_i
0	D_0 (10)	C_0 (3)
1	D_1 (12)	C_1 (5)
2	D_2 (19)	C_2 (11)
3	D_3 (31)	C_3 (13)
4	D_4 (43)	C_4 (8)
5	D_5 (41)	C_5 (8)
6	D_6 (36)	C_6 (6)
7	D_7 (30)	C_7 (4)

(iv) (a) 118 (b) 95 (c) 137 (d) 504 (e) 791
(f) 119 sets